T0182084

The Information Retrieval Series   Volume 39

More information about this series at http://www.springer.com/series/6128

Krisztian Balog

Entity-Oriented Search

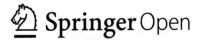

Krisztian Balog
University of Stavanger
Stavanger, Norway

ISSN 1387-5264
The Information Retrieval Series
ISBN 978-3-030-06749-6   ISBN 978-3-319-93935-3   (eBook)
https://doi.org/10.1007/978-3-319-93935-3

Printed on acid-free paper

This Springer imprint is published by the registered company Springer Nature Switzerland AG
The registered company address is: Gewerbestrasse 11, 6330 Cham, Switzerland

Szüleimnek

Preface

I have not yet reached my goal... But I forget what is behind, and I struggle for what is ahead. I run toward the goal, so I can win the prize of being called to heaven. This is the prize God offers because of what Christ Jesus has done.

(Philippians 3:12–14, CEV)

The idea of writing this book stemmed from a series of tutorials that I gave with colleagues on "entity linking and retrieval for semantic search." There was no single text on this topic that would cover all the material that I wished to introduce to someone who is new to this field. With this book, I set out to fill that gap. I hope that by making the book open access, many will be able to use it and benefit from it.

For me, writing this book, in many ways, was like running a marathon. No one forced me to do it, yet I thought that—for some reason—it'd be a good idea to challenge myself to do it. Then, along the way, there comes inevitably a point where one asks: Why am I doing this to myself? But then, in the end, crossing the finish line certainly feels like an accomplishment. In time, this experience might even be remembered as if it was a walk in the park.[1] In any case, it was a good run.

I wish to express my gratitude to a number of people who played a role in making this book happen. First of all, I would like to thank Ralf Gerstner, executive editor for Computer Science at Springer, for seeing me through to the successful completion of this book and for always being a gentleman when it came to my deadline extension requests. I also want to thank the Information Retrieval Series editors Maarten de Rijke and ChengXiang Zhai for the comments on my book proposal.

A very special thanks to Jamie Callan and to anonymous Reviewer #2 for reviewing the book and for making numerous valuable suggestions for improvements.

The following colleagues provided feedback on drafts of specific chapters at various stages of completion, and I would like to thank them for their insightful comments: Marek Ciglan, Arjen de Vries, Kalervo Järvelin, Miguel Martinez, Edgar

---

[1] Note to self: No, it wasn't.

Meij, Kjetil Nørvåg, Doug Oard, Heri Ramampiaro, Ralf Schenkel, Alberto Tonon, and Chenyan Xiong.

I want to thank Edgar Meij and Daan Odijk for the collaboration on the entity linking and retrieval tutorials, which planted the idea of this book. Working with you was always easy, enjoyable, and fun. My gratitude goes to all my co-authors for the joint work that contributed to the material that is presented in this book.

I am especially grateful to the Department of Electrical Engineering and Computer Science at the University of Stavanger for providing a pleasant work environment, where I could devote a substantial amount of time to writing this book.

I would like to thank my PhD students for giving me their honest opinion and offering constructive criticism on drafts of the book. They are, in gender-first-then-alphabetical order: Faegheh Hasibi, Jan Benetka, Heng Ding, Darío Garigliotti, Trond Linjordet, and Shuo Zhang. Special thanks, in addition, to Faegheh for the thorough checking of technical details and for suggestions on the organization of the material; to Darío for tidying up my references; to Jan for prettifying the figures and illustrations; to Trond for injecting entropy and for the careful proofreading and numerous suggestions for language improvements; to Shuo and Heng for the oriental perspective and for telling me that I use too many words.

Last but not least, I want to thank my friends and family for their outstanding support throughout the years. You know who you are.

Stavanger, Norway   Krisztian Balog
April 2018

Website

http://eos-book.org

This book is accompanied by the above website. The website provides a variety of supplementary material, corrections of mistakes, and related resources.

Contents

Acronyms

EF	Entity frequency
EL	Entity linking
ELQ	Entity linking in query
ER	Entity retrieval
IEF	Inverse entity frequency
INEX	Initiative for the Evaluation of XML Retrieval
IR	Information retrieval
KB	Knowledge base
KG	Knowledge graph
KR	Knowledge repository
LM	Language models
LTR	Learning-to-rank
NLP	Natural language processing
SDM	Sequential dependence model
SERP	Search engine result page
SPO	Subject-predicate-object (triple)
TREC	Text Retrieval Conference

Notation

Throughout this book, unless stated otherwise, the notation used is as follows:

Symbol	Meaning
$c(x)$	Total count of x
$c(x; y)$	Count of x in the context of y
$c(x, y; z)$	Number of times x and y co-occur in the context of z
d	Document ($d \in \mathcal{D}$)
$\mathcal{D}$	Document collection
$\mathcal{D}_q(k)$	Top-k ranked documents for query q
e	Entity ($e \in \mathcal{E}$)
$\mathcal{E}$	Entity catalog (set of all entities)
$\mathcal{E}_q(k)$	Top-k ranked entities for query q
$\mathcal{K}$	Knowledge base (set of SPO triples)
$\mathcal{L}_e$	Set of links of an entity e
l_x	Representation length of x ($l_x = \sum_{t \in \mathcal{V}} c(t; x)$)
q	Query
t	Term (string token, $t \in \mathcal{V}$)
$\mathcal{T}_e$	Types of entity e ($\mathcal{T}_e \subset \mathcal{T}$)
$\mathcal{T}$	Type taxonomy
$\mathcal{V}$	Vocabulary of terms
$|X|$	Cardinality of set X
Z	Normalization factor
$\mathbb{1}(x)$	Binary indicator function (returns 1 if x is true, otherwise 0)

Chapter 1
Introduction

Search engines have become part of our daily lives. We use Google (Bing, Yandex, Baidu, etc.) as the main gateway to find information on the Web. With a certain type of content in mind, we may search directly on a particular site or service, e.g., on Facebook or LinkedIn for people, organizations, and events; on Amazon or eBay for products; or on YouTube or Spotify for music. Even on our smartphones, we are increasingly reliant on search functionality to find contacts, email, notes, calendar entries, apps, etc. We have grown accustomed to expect a search box somewhere near the top of the screen, and we have also increased our expectations of the quality and speed of the responses to our searches.

On the highest level of abstraction, the field of information retrieval (IR) is concerned with developing technology for matching *information needs* with *information objects*. What we put in the search box, i.e., the *query*, is an expression of our information need. It may range from a few simple keywords (e.g., *"Bond girls"*) to a proper natural language question (e.g., *"What are good digital cameras under $300?"*). The search engine then responds with a ranked list of items, i.e., information objects. Traditionally, these items were documents. In fact, IR has been seen as synonymous with document retrieval by many. The past decade, however, has seen an enormous development in search technology. As regular users, we have witnessed first-hand the transitioning of search engines into "answering engines." Today's contemporary web search engines return rich search result pages, which include direct displays of entities, facts, and other structured results instead of merely a list of documents ("ten blue links"), as illustrated in Fig. 1.1. A primary enabling component behind these advanced search services is the availability of large-scale structured knowledge repositories (called *knowledge bases*), which organize information around specific things or objects (which we will be referring to as *entities*). The objective of this book is to give a detailed account of the developments of a decade of IR research that have enabled us to search for "things, not strings."

© The Author(s) 2018
K. Balog, *Entity-Oriented Search*, The Information Retrieval Series 39,
https://doi.org/10.1007/978-3-319-93935-3_1

Fig. 1.1 An example of a rich search result page from the Google search engine. The panel on the right-hand side of the page is an example of an *entity card*

1.1   What Is an Entity?

Informally, an entity is a "thing" or "object" that can be referred to. Common types of entities include, e.g., people, organizations, products, locations, and events. Producing a precise definition, as we shall see, turns out to be quite challenging. A commonly accepted definition of an entity is as follows:

> An entity is an object or concept in the real world that can be distinctly identified.

However, this definition is not without complications. Let us take the entity "Superman" as an example. Does it refer to the fictional comic book superhero, to the comic book itself, or to the actor who is playing the character in the movie adaptation? Entity identity is a hard question to tackle. Part of the issue is related to defining "the" (real) world. Any attempt to resolve this is likely to lead to a long philosophical debate about "existence." Therefore, we will resort to a more pragmatic and data-oriented approach. For that, we go all the way back to database management systems of the 1970s, where the importance of entities, as meaningful units for organizing information, has been recognized. The entity-relationship (ER) model proposed by Chen [11] in 1976 is a high-level conceptual data model that "incorporates some of the important semantic information about the real world" [11]. The ER model revolves around real-world entities and the associations among them. Both entities and relationships are described by means of their properties (attribute-value pairs). Further, an entity is an instance of a given entity type (i.e., a semantic class). We capture these key facets of entities in the following definition:

Definition 1.1 An *entity* is a uniquely identifiable object or thing, characterized by its name(s), type(s), attributes, and relationships to other entities.

We circumvent the "existential" questions by restricting our universe to some particular registry of entities, which we will refer to as the *entity catalog*. Thus, we consider that an entity "exists" if an only if it is an entry in the given entity catalog.

Definition 1.2 An *entity catalog* is collection of entries, where each entry is identified by a unique ID and contains the name(s) of the corresponding entity.

The entity catalog defines the universe of entities by providing entities with unique identifiers. While this alone can turn out to be surprisingly useful, we typically have more knowledge about entities (regarding their types, attributes, and relationships). We will shortly come back to the question of how to represent this knowledge, in Sect. 1.1.3.

1.1.1 Named Entities vs. Concepts

Entities are most commonly thought of as real-world objects represented by a proper noun. There are, in fact, two main classes of entities that may be distinguished:

- *Named entities* are real-world objects that can be denoted by a proper noun. Examples include specific persons, locations, organizations, products, events, etc.
- *Concepts* are abstract objects, including, but not limited to, mathematical and philosophical concepts (e.g., "distance," "axiom," "quantity"), physical concepts and natural phenomena (e.g., "gravity," "force," "wind"), psychological concepts (e.g., "emotion," "thought," "identity"), and social concepts (e.g., "authority," "human rights," "peace").

These two classes generally correspond to the dichotomy between concrete and abstract objects in philosophy. It is worth noting that the distinction between concrete/abstract objects has a curious status in contemporary philosophy, with many plausible ways of drawing the line between the two [34].

As far as our work is concerned, this distinction is mostly of a philosophical nature. From a technical perspective, the exact same methods may be used for named entities and/or concepts. Thus, unless stated otherwise, whenever we write *entity* in this book, we mean both of them. Nevertheless, the focus of practical application scenarios is, more commonly than not, restricted to named entities.

1.1.2 Properties of Entities

We shall collectively refer to all information associated with an entity (e.g., the unique identifier, names, types, attributes, and relationships) as *entity properties*. Let us now explore each of these properties in a bit more detail.

Unique identifier: Entities need to be uniquely identifiable. There must be a one-to-one correspondence between each entity identifier (ID) and the (real-world or fictional) object it represents (i.e., within a given entity catalog; the same entity may exist under different identifiers in other catalogs). Examples of entity identifiers from past IR benchmarking campaigns include email addresses for people (within an organization), Wikipedia page IDs (within Wikipedia), and unique resource identifiers (URIs, within Linked Data repositories).

Name(s): Entities are known and referred to by their name—usually, a proper noun. Unlike IDs, names do not uniquely identify entities; multiple entities may share the same name (e.g., "Michael Jordan"). Also, the same entity may be known by more than a single name (e.g., "Barack Obama," "President Obama," "Barack Hussein Obama II"). These alternative names are called *surface forms* or *aliases*. Humans can easily resolve the ambiguity of entity references from the context of the mention most of the time. For machines, automatically disambiguating entity references presents many challenges.

Type(s): Entities may be categorized into multiple entity types (or *types* for short). Types can also be thought of as containers (semantic categories) that group together entities with similar properties. An analogy can be made to object-oriented programming, whereby an entity of a type is like an instance of a class. The set of possible entity types are often organized in a hierarchical structure, i.e., a *type taxonomy*. For example, the entity Albert Einstein is an instance of the type "scientist," which is a subtype of "person."

Attributes: The characteristics or features of an entity are described by a set of attributes. Different types of entities are typically characterized by different sets of attributes. For example, the attributes of a person include the date and place of birth, weight, height, parents, spouses, etc. The a Attributes of a populated place include latitude, longitude, population, postal code(s), country, continent, etc.

Notice that some of the items in these lists are entities themselves, e.g., locations or persons. We do not treat those as attributes but consider them separately, as relationships. Attributes always have literal values; optionally, they may also be accompanied by data type information (such as number, date, geographic coordinate, etc.).

Relationships: In the words of Booch [9]: "an object by itself is intensely uninteresting." Relationships describe how two entities are associated to each other. From a linguistic perspective, entities may be thought of as proper nouns and relationships between them as verbs. For example, "Homer *wrote* the Odyssey" or "The General Theory of Relativity *was discovered* by Albert Einstein." Relationships may also be seen as "typed links" between entities.

1.1.3 Representing Properties of Entities

Information about entities can be represented and stored in semi-structured or in structured form.

> **Definition 1.3** A *knowledge repository* (KR) is a catalog of entities that contains entity type information, and (optionally) descriptions or properties of entities, in a semi-structured or structured format.

Wikipedia is a classic example of a knowledge repository. Each article in Wikipedia is an entry that describes a particular entity. Articles are also assigned to categories (which can be seen as entity types) and contain hyperlinks to other articles (thereby indicating the presence of a relationship between two entities, albeit not the type of the relationship). Wikipedia articles also contain information about attributes and relationships of entities, but not in a structured form.

To organize and store information about entities in a structured form, one needs a knowledge representation model. The *Resource Description Framework* (RDF), which we will discuss in detail in Sect. 2.3.1.2, is the prevalent standard for describing entities (and, more generally, resources). An entity can be represented as a set of RDF statements. These statements may be seen as *facts* or *assertions* about that entity. A *knowledge base* is a structured knowledge repository for storing and organizing statements about entities.

> **Definition 1.4** A *knowledge base* (KB) is a structured knowledge repository that contains a set of facts (assertions) about entities.

According to our definition, all knowledge bases are also knowledge repositories, but the reverse is not true.

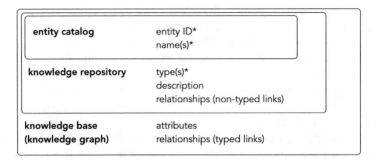

Fig. 1.2 Illustration of the relationship between entity catalog, knowledge repository, and knowledge base, each complementing and extending the previous concept. The entity properties marked with * are mandatory

```
<dbr:Kimi_Raikkonen>

      <foaf:name>          "Kimi Räikkönen"
      <dbo:birthPlace>     <dbr:Espoo>
      <dbo:nationality>    <dbr:Finland>
      <dct:description>    "Finnish race driver"
      <dbo:birthDate>      "1979-10-17"
      <rdf:type>           <dbo:RacingDriver>
      <dct:subject>        <dbc:Finnish_racing_drivers>
      <dct:subject>        <dbc:Ferrari_Formula_One_drivers>
      <rdfs:comment>       "Kimi-Matias Räikkönen [...] nicknamed "The Ice Man",
                           is a Finnish racing driver currently driving for
                           Ferrari in Formula One. [...]"
```

Listing 1.1 Excerpt from the DBpedia knowledge base entry of KIMI RÄIKKÖNEN

Conceptually, entities in a knowledge base may be seen as nodes of a graph, with the relationships between them as (labeled) edges. Thus, especially when this graph nature is emphasized, a knowledge base may also be referred to as a *knowledge graph* (KG). Figure 1.2 shows the relationship between these concepts.

To give an idea of what a knowledge base entry of an entity looks like, we refer to Listing 1.1. This particular example is from DBpedia knowledge base, showing an excerpt from the entry of the entity KIMI RÄIKKÖNEN who is displayed on the entity card in Fig. 1.1. We are going to cover knowledge bases and the RDF model in greater detail in Chap. 2.

1.2   A Brief Historical Outlook

Before delving into the topic of entity-oriented search, it is important to put things in historical context. Therefore, in this section, we present a broad perspective on developments within multiple fields of computer science, in particular information retrieval (IR), databases (DB), natural language processing (NLP), and the Semantic

Web (SW). Even though they have developed largely independently of each other, concentrated on separate problems, and operated on different types of data, they seem to converge on a common theme: *entities* as units for capturing, storing, organizing, and accessing information.

1.2.1 Information Retrieval

According to an early definition by Salton [35] from 1968, "Information retrieval is a field concerned with the structure, analysis, organization, storage, searching, and retrieval of information." From its inception, IR has always kept a strong focus on evaluating the *effectiveness* of systems: "determining the *relevance* of items, retrieved by a search engine, relative to a user's information need" [36]. The launch of the Text REtrieval Conference (TREC) series in 1992, co-sponsored by the US National Institute of Standards and Technology (NIST) and the US Department of Defense, has had a profound impact on the field, by standardizing retrieval evaluation through the creation of large *test collections*. TREC was followed by Asian and European sister events, the NII Test Collection for IR Systems (NTCIR) in 1999, and the Conference and Labs of the Evaluation Forum (CLEF, formerly Cross-Language Evaluation Forum) in 2000. These benchmarking campaigns follow an annual cycle. Each edition features a number of specific tasks, which are thematically organized into different "tracks." By looking at the development of these tracks, one can get a good overview of how the focus of research in IR has shifted over the years.

Up to the mid-1990s, the field has primarily focused on documents as the unit of retrieval. Driven by the motto "users want answers, not documents," a new front of IR research has emerged with the arrival of the TREC Question Answering track in 1999. *Question answering* systems respond with a short, focused answer to a question formulated in natural language, e.g., "Who invented the paper clip?" or "How many calories are there in a Big Mac?" The *expert finding* task at TREC Enterprise track (2005–2008) concentrated on answering a more specific type of question: "Who are the experts on topic X?" Here, the input is a keyword query, specifying the area of expertise (e.g., "XML schema"), and the system answers this by returning a ranked list of people. The INEX Entity Ranking (2007–2009) and the TREC Entity (2009–2011) tracks broadened the scope of answers (from people) to arbitrary entity types, laying the groundwork for the area of *entity retrieval*. With the transitioning from documents to entities as the units of retrieval also came an increased reliance on structured data sources, known as *knowledge bases*. The TREC Knowledge Base Acceleration track (2012–2014) aimed at developing technology that can aid humans in maintaining and expanding information stored about entities in knowledge bases.

In addition to research developments in academia, the search industry (and especially major web search engines, like Google) has also played an prominent role in shaping the field. Search has become a commodity, and users have grown accustomed to expressing their information needs using short keyword queries, and

Table 1.1 Comparison of database systems and information retrieval, based on [40]

	Database systems	Information retrieval
Data type	Numbers, short strings	Text
Foundation	Algebraic/logic based	Probabilistic/statistics based
Search paradigm	Boolean retrieval	Ranked retrieval
Queries	Structured query languages	Free text queries
Evaluation criteria	Efficiency	Effectiveness (user satisfaction)
User	Programmer	Nontechnical person

getting—most of the time—relevant results almost instantly. At the same time, the massive volumes of usage data collected from users allows for improved methods, by harnessing the "wisdom of the crowds." As Liu [24] explains, "given the amount of potential training data available, it has become possible to leverage machine learning technologies to build effective ranking models." Such models exploit a large number of features by means of discriminative learning, known as "learning-to-rank" [24].

1.2.2 Databases

"A database management system is a software system that enables the creation, maintenance, and use of large amounts of data" [1]. This definition suggests that database systems and information retrieval have a lot in common. This is indeed the case, yet DB and IR emphasize very different aspects of information management. Databases contain highly structured data, which is queried by expert users (i.e., programmers) using formal query languages, like SQL. The focus is on precise query processing and efficiency. IR systems, on the other hand, "understand queries as approximate, best-effort formulations of the user's information needs" [40]. Search is an interactive process, which often involves multiple query reformulations upon the inspection of results. Table 1.1 summarizes the traditional differences between DB and IR systems. Given the complementary foci and techniques in DB and IR, the two fields can benefit from each other's developments. For instance, IR can profit from efficient indexing structures, whereas DB can make use of natural language search interfaces and probabilistic ranking mechanisms from IR. While the traditional boundaries between these two fields still exist, they are getting blurred. Entity retrieval is a cross-over application area between IR and DB that requires flexible ranking on text, categorical, and numerical attributes. Additionally, the search also needs to be able to cope with "no answers" and "too many answers." Searching online product catalogs is a good illustrative example, where users issue keyword queries but also use various filters (e.g., via faceting) to narrow down the scope of results. Many of these queries could be answered more or less exactly, but many others will require probabilistic scoring and ranking.

As we have already discussed in Sect. 1.1, it has been realized very early on in the database field that entities offer a disciplined way of handling data. The *entity-relationship* approach of Chen [11] was originally proposed as a semantic data model, to provide a better representation of real-world entities. Entity-relationship diagrams, which are built up of entities, relationships, and attributes, are now normally used as a conceptual modeling technique [7]. The field of databases recognized the need for an entity-centric view of web content about the same time as IR did [13, 40]. The recent focus in databases—within our interest area— has primarily been on developing indexing schemes that facilitate efficient query processing [10, 12], and on interpreting queries with the help of structured data, i.e., translating keyword queries to structured queries [18, 31, 38, 41].

Additionally, the field of databases also deals with a range of data integration and data quality problems, such as record linkage (a.k.a. entity resolution) [14, 16] or schema mapping [33]. We consider these being outside the scope of this book.

1.2.3 Natural Language Processing

Most research in *natural language processing* (or *computational linguistics*) aims to capture the meaning of text. One might divide NLP problems into (1) low-level parsing and segmentation tasks, (2) linguistic annotations, and (3) end-user applications. Common text parsing and segmentation tasks include sentence breaking, word segmentation, stemming, and lemmatization. Linguistic annotation tasks include part-of-speech tagging, word sense disambiguation, named entity recognition and disambiguation, coreference resolution, temporal tagging, semantic role labeling, and dependency parsing. These annotations are meant to yield deeper representations that are closer to meaning and may be exploited in real-world applications. End-user applications include, among others, information extraction, machine translation, text summarization, sentiment analysis, and question-answering. For us, the most relevant of these is *information extraction* (IE), which "refers to the automatic extraction of structured information such as entities, relationships between entities, and attributes describing entities from unstructured sources" [37].

There are two main modes in which an IE system may be deployed: one is to annotate text with the identified mentions of structured information, another is to populate a knowledge base with the extracted information. Information extraction is narrower in scope than full text understanding—which is still beyond our capabilities today. Nevertheless, identifying entities and relationships makes it possible to capture, to a large extent, what a given piece of text is about. Furthermore, entities can serve as a pivot for connecting unstructured text and structured knowledge bases. While rooted in NLP, the problem area of extracting structured information from unstructured sources now engages the IR, DB, machine learning, and Web communities as well. Over time, the scope of IE systems was expanded to include the extraction of not only atomic elements (entities and relations) but of higher-order structures as well, such as tables and lists [15, 25, 29].

Up until the late 1980s, most NLP systems employed rule-based approaches, which relied heavily on linguistic theory. Then came the "statistical revolution," introducing machine learning algorithms for language processing that could learn from manually annotated corpora [22]. The current state of the art "draws far more heavily on statistics and machine learning than it does on linguistic theory" [22]. Today, a broad range of robust, efficient, and scalable techniques for shallow NLP processing (as opposed to deep linguistic analysis) are available [30].

1.2.4 Semantic Web

The *Semantic Web* is a relatively young field, especially compared to the other three (IR, DB, NLP). The term was coined by Tim Berners-Lee, referring to an envisioned extension of the original Web. While the original Web is a medium of documents for people (i.e., the Web of Documents), the Semantic Web is meant to be a Web of "actionable information," i.e., an environment that enables intelligent agents to carry out sophisticated tasks for users. The Semantic Web is "a Web of relations between resources denoting real world objects, i.e., objects such as people, places and events" [19]. The challenge of the Semantic Web, as explained in the 2001 Scientific American by Berners-Lee et al. [6], is "to provide a language that expresses both data and rules for reasoning about the data." Thus, from the late 1990s and throughout the 2000s, a great deal of effort was expended toward establishing standards for *knowledge representation*. Several important technologies were introduced:

- The Universal Resource Identifier (URI), to be able to uniquely identify "things" (i.e., entities, which are called *resources*);
- The eXtensible Markup Language (XML), to add structure to web pages;
- The Resource Description Framework (RDF), to encode meaning in a form of (sets of) triples;
- Various serializations for storing and transmitting RDF data, e.g., Notation-3, Turtle, N-Triples, RDFa, and RDF/JSON;
- The SPARQL query language, to retrieve and manipulate RDF data;
- A large palette of techniques to describe and define vocabularies, including the RDF Schema (RDFS), the Simple Knowledge Organization System (SKOS), and the Web Ontology Language (OWL).

These technologies together form a layered architecture, referred to as the *Semantic Web Stack*.

In terms of large-scale, agent-based mediation with heterogeneous data, the Semantic Web is a dream that has not (yet) come true. The Semantic Web movement, nevertheless, has resulted in structured data on a previously unprecedented scale. As a terminological distinction, *Semantic Web* is often used to refer to the various standards and technologies, while the data that is being published using Semantic Web standards is called *Linked Data* or the *Web of Data*. Linked data may be exposed as semantic mark-up embedded within HTML pages or as entire datasets (i.e., knowledge bases) published as RDF (e.g., DBpedia or Wikidata). A key idea is that resources that refer to the same real-world entity may be interlinked across different sources.

Ontologies, for automated inference or for integrating heterogeneous data, have seen little adoption in the search industry. Recent efforts are geared toward speaking the same language using a shared vocabulary. Schema.org is a collaborative activity by major search providers (including Google, Microsoft, Yahoo, and Yandex) in order to define a standard for semantic markup. At the time of writing, over 10 million sites use Schema.org to mark up their web pages and email messages.

Regarding information access, it was realized that formal, structured query languages, like SPARQL, are unsuitable for ordinary users, who would prefer simple keyword search. Thus, the Semantic Web community has adopted IR-style ranking models for retrieving specific entities [8, 17, 27].

1.3 Entity-Oriented Search

We use the term *entity-oriented search* to refer to a broad range of information access tasks where entities are used as information objects, instead of or in addition to documents.

> **Definition 1.5** *Entity-oriented search* is the search paradigm of organizing and accessing information centered around entities, and their attributes and relationships.

The significance of this information access paradigm is twofold:

- From a user perspective, entities are natural units for organizing information. We care about and mostly think in terms of real-world things and their connections. Allowing users to interact with specific entities offers a richer and more effective user experience than what is provided by conventional document-based retrieval systems.
- From a machine perspective, entities allow for a better understanding of search queries, of document content, and even of users (e.g., their context and preferences). Entities enable search engines to be more intelligent.

1.3.1 A Bird's-Eye View

Figure 1.3 shows a high-level overview of an entity-oriented search system. At first glance, one might say that this looks a lot like any conventional (i.e., document-

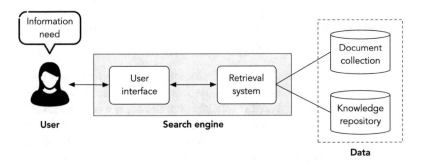

Fig. 1.3 Architecture of an entity-oriented search system

oriented) retrieval system. While that observation is indeed valid from this distance, there is a single, yet important difference on the data end. The document collection is complemented with a knowledge repository. The knowledge repository contains, at the bare minimum, an entity catalog: a dictionary of entity names and unique identifiers. Typically, the knowledge repository also contains the descriptions and properties of entities in semi-structured (e.g., Wikipedia) or structured format (e.g., Wikidata, DBpedia). Commonly, the knowledge repository also contains ontological resources (e.g., a type taxonomy).

Next, we briefly look at the three main components depicted on Fig. 1.3, moving from left to right.

1.3.1.1 Users and Information Needs

Users may articulate their information needs in many different ways. These are sometimes referred to as *search paradigms* [4]. Traditionally, keyword, structured, and natural language queries are distinguished [4]. We complement this list with two additional categories.

Keyword queries   Thanks to major web search engines, keyword queries have become the "dominating *lingua franca* of information access" [2]. Keyword queries are also known as *free text queries*: "a query in which the terms of the query are typed freeform into the search interface, without any connecting search operators (such as Boolean operators)" [26]. Keyword queries are easy to formulate, but—by their very nature—are imprecise.

Structured queries   Structured data sources (databases and knowledge bases) are traditionally queried using formal query languages (such as SQL or SPARQL). These queries are very precise. However, formulating them requires a "knowledge of the underlying schema as well as that of the query language" [3]. Structured queries are primarily intended for expert users and well-defined, precise information needs.

Keyword++ queries   We use the term keyword++ query (coined in [3]) to refer to keyword queries that are complemented with additional structural elements. For example, when users supply target categories or various filters via faceted search interfaces, those extra pieces of input constitute the ++ part. With well-designed user interfaces, supplying these does not induce a cognitive load on the user. Keyword++ queries may be seen as "fielded" keyword queries.

Natural language queries   Information needs can be formulated using natural language, the same way as one human would express it to another in an everyday conversation. Often, natural language queries take a question form. Also, such queries are increasingly more spoken aloud with voice search, instead of being typed [28].

Zero-query   The traditional way of information access is reactive: the search system responds to a user-initiated query. Proactive systems, on the other hand, "anticipate and address the user's information need, without requiring the user to issue (type or speak) a query" [5]. The *zero-query search* paradigm can be expressed with the slogan "the query is the user." In practice, the context of the user is used to infer information needs.

Sawant and Chakrabarti [39] refer to queries typically sent to search engines as "telegraphic queries." These are not well-formed grammatical sentences or questions. Keywords could also be described as "shallow" natural language queries. For example, most users would simply issue *"birth date neil armstrong."* With voice search being increasingly more prevalent, especially on mobile devices, alternatively, the user could ask the question: *"When was Neil Armstrong born?"* Bast et al. [4] point out that "keyword search and natural language search are less clearly delineated than it may seem." The distinction often depends on the processing technique used rather than the query text itself. In this book, we will concentrate on keyword (and keyword++) queries. We note that the same techniques may be applied for natural language queries as well (but will likely yield suboptimal results).

1.3.1.2 Search Engine

At this high-level perspective, the search engine consists of two main parts: the user interface and the retrieval system. The former takes care of the interaction with the user, from the formulation of the information need to the presentation of search results. The "single search box" paradigm became extremely popular thanks to major web search engines. Recently, natural language interfaces have also been receiving increased attention. These allow users to pose a (possibly complex) question in natural language (instead of merely a list of keywords). The retrieval system interprets the search request and compiles a response. Modern web search engine result pages are composed of a ranked list of documents (web pages), entity cards, direct answers, and other knowledge panels, along with further entity recommendations and suggestions for query reformulations. In vertical search, the result list comprises a ranked list of entities, possibly grouped by entity type. Our main focus in this book will be on how to generate entity-oriented responses.

1.3.1.3 Data

We distinguish between three main types of data.

Unstructured data   can be found in vast quantities in a variety of forms: web
   pages, spreadsheets, emails, blogs, tweets, medical records, etc. Without making
   any assumptions about the format, all these may be treated as textual documents,
   i.e., a sequence of words.
Semi-structured data   is characterized by the lack of rigid, formal structure. Typically, it contains tags or other types of markup to separate textual content from
   semantic elements. Semi-structured data is "self-describing," i.e., "the schema is
   contained within the data and is evolving together with the content" [3].
Structured data   adheres to a predefined (fixed) schema and is typically organized in a tabular format—think of relational databases. The schema serves as
   a blueprint of how the data is organized, describes how real-world entities are
   modeled, and imposes constraints to ensure the consistency of the data.

In Fig. 1.3, the document collection is an unstructured or semi-structured data
source. The knowledge repository may be either in semi-structured (e.g., RDF) or
in structured format (e.g., a relational database). One of the challenges in entity-
oriented search is that information about a given entity has to be collected and
aggregated across noisy, heterogeneous, and potentially conflicting data sources,
both unstructured and structured.

1.3.2   Tasks and Challenges

Next, we identify a number of specific tasks, and related challenges, that we will
be concerned with in this book. These can be organized around three main thematic
areas. In fact, these themes largely correspond to the three parts of the book.

1.3.2.1   Entities as the Unit of Retrieval

According to various studies, 40–70% of queries in web search mention or target
specific entities [20, 23, 32]. These queries commonly seek a particular entity,
albeit often an ambiguous one (e.g., "*harry potter*") or a list of entities (e.g.,
"*doctors in barcelona*"). Such queries are better answered by returning a ranked
list of entities, as opposed to a list of documents. We refer to this as the task
of *entity retrieval*. There are three main challenges involved here: (1) how to
represent information needs, (2) how to represent entities (using both unstructured
and structured datasets), and (3) how to match those representations. One of the
most exciting opportunities in entity retrieval is how to leverage the additional
structure associated with entities in the knowledge repository—attributes, types, and
relationships—to improve retrieval effectiveness.

1.3.2.2 Entities for Knowledge Representation

Entities help to bridge the gap between the worlds of unstructured and structured data: they can be used to semantically enrich unstructured text, while textual sources may be utilized to populate structured knowledge bases.

Recognizing mentions of entities in text and associating these mentions with the corresponding entries in a knowledge base is known as the task of *entity linking*. Entities allow for a better understanding of the meaning of text, both for humans and for machines. While humans can relatively easily resolve the ambiguity of entities, based on the context in which they are mentioned, for machines this presents many difficulties and challenges.

The knowledge base entry of an entity summarizes what we know about that entity. As the world is constantly changing, so are new facts surfacing. Keeping up with these changes requires a continuous effort from editors and content managers. This is a demanding task at scale. By analyzing the contents of documents in which entities are mentioned, this process—of finding new facts or facts that need updating—may be supported, or even fully automated. We refer to this as the problem of *knowledge base population*.

1.3.2.3 Entities for an Enhanced User Experience

Besides being meaningful retrieval and information organization units, entities can improve the user experience throughout the entire search process. This starts with query assistance services that can aid users in articulating their information needs. Next, entities may be utilized for improved content understanding, by connecting entities and facts to queries and documents. For example, they make it possible to automatically direct requests to specific services or verticals (sites dedicated to a specific segment of online content). When presenting retrieval results, knowledge about entities may be used to complement the traditional document-oriented search results (i.e., the "ten blue links") with various information boxes and knowledge panels (like it is shown in Fig. 1.1). Finally, entities may be harnessed for providing contextual recommendations. See, e.g., the "People also search for" section on Fig. 1.1.

1.3.3 Entity-Oriented vs. Semantic Search

Entity-oriented and semantic search are often mentioned in the same context, and even treated as casual synonyms by many. The question inevitably arises: What is the difference between the two (if any)?

There is no agreed definition of *semantic search*, in fact, the term itself is highly contested. One of the first published references to the term appeared in a 2003 paper by Guha et al. [19]: "Semantic Search attempts to augment and improve traditional search results (based on Information Retrieval technology) by using data from the Semantic Web." Since the Semantic Web is primarily organized around real-world objects and their relationships, according to this definition, entity-oriented search could indeed be seen as synonymous with semantic search. According to a more recent definition attributed to John [21], "Semantic Search is defined as search for information based on the intent of the searcher and contextual meaning of the search terms, instead of depending on the dictionary meaning of the individual words in the search query."

We prefer to take a broader view on semantic search, which is as follows.

Definition 1.6 *Semantic search* encompasses a variety of methods and approaches aimed at aiding users in their information access and consumption activities, by understanding their context and intent.

This definition emphasizes the overall high-level objective, an improved user experience, without restricting the techniques to explicit semantics. This definition includes, among others, implicit semantics, such as term dependencies, topic models, or latent space models. Furthermore, we do not limit semantic search to the traditional keyword-based search paradigm. As such, proactive recommendations also fall under the umbrella of semantic search. Simply put, semantic search is *broader* than entity-oriented search. Entities, nonetheless, play a leading role in it.

Throughout this book, our notion of semantics will be the following: references to meaningful, i.e., machine understandable (ontological or linguistic) structures.

1.3.4   Application Areas

Where can entity-oriented search technology be applied? Obviously, web search is the most prominent application area, but it is certainly not the only one. Entities play a major role in a wide range of information access scenarios, including enterprise search, domain-specific and vertical search (e.g., e-commerce, automotive industry, medical search, legal information, scholarly literature, job search, and travel), social networking, and intelligence services. Unlike web search, most of these focus on a single or at most a handful of entity types in a given domain. Furthermore, entities have an important function in question answering systems and in personal digital assistants.

1.4 About the Book

The book aims to cover all facets of entity-oriented search—where "search" can be interpreted in the broadest sense of information access—from a unified point of view, and provide a coherent and comprehensive overview of the state of the art. This work is the first synthesis of research in this broad and rapidly developing area. Selected topics are discussed in depth, with the intention of establishing foundational techniques and methods for future research and development. A range of other topics are treated at a survey level, with numerous pointers to relevant literature for those interested. We also identify open issues and challenges along the way, and conclude with a roadmap for future research.

1.4.1 Focus

The book is firmly rooted in information retrieval, and it thus bears the characteristics of the field. Developments are motivated and driven by specific use-cases, with theory, evaluation, and application all being interconnected. A strong focus on data is maintained throughout the book—after all, it is the data that dictates to a large extent what can be done.

We deliberately refrain from reporting evaluation results from specific studies; the absolute values of those evaluation scores may be largely influenced by, among others, the various data (pre-)processing techniques, choice of tools, and parameter settings. A direct comparison of results from different studies (performed by different groups/individuals) may thus be misleading. Nevertheless, we indicate overall performance ranges on standard benchmark suites. A great deal of attention is given to evaluation methodology and to available resources, such as datasets, software tools, and frameworks.

To remain focused, we shall follow a language agnostic approach and use English as our working language (as, indeed, most test collections are in English). Languages with markedly different syntax, morphology, or compositional semantics may need additional processing techniques. The discussion of those is outside the scope of this book.

1.4.2 Audience and Prerequisites

The primary target audience of this book are researchers and graduate students. It is our hope that readers with a theoretical inclination will find it as useful as will those with a practical orientation.

An understanding of basic probability and statistics concepts is required for most models and algorithms that are discussed in the book. A general background in information retrieval (i.e., familiarity with the main components of a search engine and traditional document retrieval models, such as BM25 and language models, and with basics of retrieval evaluation) is sufficient to follow the material. Further, a basic understanding of machine learning concepts and algorithms for supervised learning is assumed. It was our intention to make the book as self-contained as possible. Therefore, standard retrieval models, learning-to-rank methods, and IR evaluation measures will be briefly explained when we come across them for the first time, in Chap. 3.

1.4.3 Organization

The book is divided into three main parts, sandwiched by introductory and concluding chapters.

- The first two chapters, *Introduction* and *Meet the Data*, introduce the basic concepts, provide an overview of entity-oriented search tasks, and present the various types and sources of data that will be used throughout the book.
- **Part I** deals with the core task of *entity ranking*: given a textual query, possibly enriched with additional elements or structural hints, return a ranked list of entities. This core task is examined in a number of different flavors, using both structured and unstructured data collections, and various query formulations. In all these cases, the output is a ranked list of entities. The main questions guiding this part are:

  - How to represent entities and information needs, and how to match those representations?
  - How to exploit unique properties of entities, namely, types and relationships, to improve retrieval performance?

  Specifically, Chap. 3 introduces models purely for the text-based ranking of entities. Chapter 4 presents advanced models capable of leveraging structured information associated with entities, such as entity types and relationships. As these two chapters build on each other, the reader is advised to read them sequentially.
- **Part II** is devoted to the role of entities in *bridging unstructured and structured data*. The following two questions are addressed:

  - How to recognize and disambiguate entity mentions in text and link them to structured knowledge repositories?
  - How to leverage massive volumes of unstructured (and semi-structured) data to populate knowledge bases with new information about entities?

  Chapters 5 and 6 may be read largely independent of each other and of other chapters of the book.

- **Part III** explores how entities can enable search engines to understand the concepts, meaning, and intent behind the query that the user enters into the search box, and provide rich and focused responses (as opposed to merely a list of documents)—a process known as *semantic search*. As we have discussed earlier, semantic search is not a single method or approach, but rather a collection of techniques. We present those techniques by dividing them into three broad categories: understanding information needs (Chap. 7), leveraging entities in document retrieval (Chap. 8), and utilizing entities for an enhanced search experience (Chap. 9). Chapters 7–9 are relatively autonomous and can be read independently of each other, but they build on concepts and tools from Parts I and II.
- The final chapter, *Conclusions and Future Directions*, concludes the book by discussing limitations of current approaches and suggests directions for future research.

1.4.4 Terminology and Notation

This section provides a detailed description of the terminological and notational conventions that will be used throughout the book.

Terminology Great care has been taken to use the following "reserved keywords" only in their explicitly defined senses.

- *Entity description*: Textual (term-based) entity representation created with the purpose of retrieval.
- *Entity mention*: Text span that is referring to a specific entity.
- *Knowledge repository*: A semi-structured or structured data collection that contains a catalog of entities with unique identifiers, along with other information about entities (such as entity descriptions, entity types, and links between entities). Examples include Wikipedia, DBpedia, Freebase, etc.
- *Knowledge base*: A structured knowledge repository that contains facts (assertions) about entities (including specific attributes and relationships). In this book, these facts are represented as a set of subject-predicate-object (SPO) triples, according to the RDF data model. For example, DBpedia is a knowledge base, but Wikipedia is not.
- *Knowledge graph*: When viewed as a graph, we refer to a knowledge base as a knowledge graph. This name is reserved for the contexts where the graph nature of the data is utilized.
- *Term*: Atomic unit of text tokenization and indexing (i.e., a "word").

Typography We adhere to certain typographical conventions.

- Whenever referring to a particular entity, the name of that entity is typeset in small capitals, e.g., JOHN SMITH.

- We typeset queries in italics, e.g., *"example search query."* We include these queries in verbatim, as they appear in the given dataset, i.e., without correcting grammar or capitalization.
- When quoting data from a knowledge repository, it is typeset in `typewriter` font.

Selected definitions, key concepts, and ideas are highlighted in gray boxes throughout the book.

Mathematical Notation   We adopt the following notational conventions.

- Sequences of elements of the same type (such as vectors, lists, etc.) are denoted as $\langle x_1, \ldots, x_n \rangle$.
- Tuples, i.e., ordered collections of elements of different types, are denoted as $(x_1, \ldots, x_n)$.
- Set-like variables are denoted by capital calligraphic letters, e.g., $\mathcal{D}$ for documents, $\mathcal{E}$ for entities, $\mathcal{T}$ for the taxonomy of types, $\mathcal{V}$ for the vocabulary of terms, etc. Graphs represent an exception with vertices and edges denoted as V and E, respectively (as the calligraphic versions of those letters are already taken).
- Matrices are denoted by bold capital roman letters (e.g., **A**) and vectors are denoted by bold small roman letters (e.g., **w**).
- We occasionally use the semicolon to group the input variables of a function, to show which are specific to the given *target* (before semicolon) and which are more contextual (after semicolon). For example, $c(t, e; d)$ denotes the number of times the term t and entity e co-occur in a particular document d. The semicolon is not more than a reading aid, and there is no mathematical difference between the comma and the semicolon.
- Some functions, like weight ($w()$), score ($score()$), or similarity ($sim()$), are formulated differently in the various works that this book draws upon. However, these functions are named similarly (though their arguments may vary) because they play similar roles in their respective contexts.
- Performance measures are typeset in roman font, e.g., F1 or NDCG.
- The symbol $\times$ denotes multiplication, while $\cdot$ is reserved for the dot product.

References

1. Abiteboul, S., Hull, R., Vianu, V. (eds.): Foundations of Databases: The Logical Level. 1st edn. Addison-Wesley Publishing Co. (1995)
2. Agarwal, G., Kabra, G., Chang, K.C.C.: Towards rich query interpretation: walking back and forth for mining query templates. In: Proceedings of the 19th international conference on World wide web, WWW '10, pp. 1–10. ACM (2010). doi: 10.1145/1772690.1772692

3. Balog, K.: Semistructured data search. In: Ferro, N. (ed.) Bridging Between Information Retrieval and Databases, *Lecture Notes in Computer Science*, vol. 8173, pp. 74–96. Springer (2014). doi: 10.1007/978-3-642-54798-0_4

4. Bast, H., Buchhold, B., Haussmann, E.: Semantic search on text and knowledge bases. Found. Trends Inf. Retr. **10**(2-3), 119–271 (2016). doi: 10.1561/1500000032

5. Benetka, J.R., Balog, K., Nørvåg, K.: Anticipating information needs based on check-in activity. In: Proceedings of the 10th ACM International Conference on Web Search and Data Mining, WSDM '17, pp. 41–50. ACM (2017). doi: 10.1145/3018661.3018679

6. Berners-Lee, T., Hendler, J., Lassila, O.: The semantic web. Scientific American **284**(5), 34–43 (2001)

7. Beynon-Davies, P.: Database Systems. 3rd edn. Palgrave, Basingstoke, UK (2004)

8. Blanco, R., Mika, P., Vigna, S.: Effective and efficient entity search in RDF data. In: Proceedings of the 10th International Conference on The Semantic Web, ISWC '11, pp. 83–97. Springer (2011). doi: 10.1007/978-3-642-25073-6_6

9. Booch, G.: Object Oriented Design with Applications. Benjamin-Cummings Publishing Co., Inc. (1991)

10. Chakrabarti, S., Kasturi, S., Balakrishnan, B., Ramakrishnan, G., Saraf, R.: Compressed data structures for annotated web search. In: Proceedings of the 21st International Conference on World Wide Web, WWW '12, pp. 121–130. ACM (2012). doi: 10.1145/2187836.2187854

11. Chen, P.P.S.: The entity-relationship model–toward a unified view of data. ACM Trans. Database Syst. **1**(1), 9–36 (1976). doi: 10.1145/320434.320440

12. Cheng, T., Chang, K.C.C.: Beyond pages: Supporting efficient, scalable entity search with dual-inversion index. In: Proceedings of the 13th International Conference on Extending Database Technology, EDBT '10, pp. 15–26. ACM (2010). doi: 10.1145/1739041.1739047

13. Cheng, T., Yan, X., Chang, K.C.C.: EntityRank: Searching entities directly and holistically. In: Proceedings of the 33rd International Conference on Very Large Data Bases, VLDB '07, pp. 387–398 (2007)

14. Christen, P.: A survey of indexing techniques for scalable record linkage and deduplication. IEEE Trans. on Knowl. and Data Eng. **24**(9), 1537–1555 (2012). doi: 10.1109/TKDE.2011.127

15. Cohen, W.W., Hurst, M., Jensen, L.S.: A flexible learning system for wrapping tables and lists in HTML documents. In: Proceedings of the 11th International Conference on World Wide Web, WWW '02, pp. 232–241. ACM (2002). doi: 10.1145/511446.511477

16. Elmagarmid, A.K., Ipeirotis, P.G., Verykios, V.S.: Duplicate record detection: A survey. IEEE Trans. on Knowl. and Data Eng. **19**(1), 1–16 (2007). doi: 10.1109/TKDE.2007.9

17. Fetahu, B., Gadiraju, U., Dietze, S.: Improving entity retrieval on structured data. In: In Proceedings of the 14th International Semantic Web Conference. Springer (2015). doi: 10.1007/978-3-319-25007-6_28

18. Ganti, V., He, Y., Xin, D.: Keyword++: A framework to improve keyword search over entity databases. Proc. VLDB Endow. **3**(1-2), 711–722 (2010). doi: 10.14778/1920841.1920932

19. Guha, R., McCool, R., Miller, E.: Semantic search. In: Proceedings of the 12th International Conference on World Wide Web, WWW '03, pp. 700–709. ACM (2003). doi: 10.1145/775152.775250

20. Guo, J., Xu, G., Cheng, X., Li, H.: Named entity recognition in query. In: Proceedings of the 32nd international ACM SIGIR conference on Research and development in information retrieval, SIGIR '09, pp. 267–274. ACM (2009)

21. John, T.: What is semantic search and how it works with Google search (2012)

22. Johnson, M.: How the statistical revolution changes (computational) linguistics. In: Proceedings of the EACL 2009 Workshop on the Interaction Between Linguistics and Computational Linguistics: Virtuous, Vicious or Vacuous?, ILCL '09, pp. 3–11. Association for Computational Linguistics (2009)

23. Lin, T., Pantel, P., Gamon, M., Kannan, A., Fuxman, A.: Active objects. In: Proceedings of the 21st international conference on World Wide Web, WWW '12, pp. 589–598. ACM (2012). doi: 10.1145/2187836.2187916

24. Liu, T.Y.: Learning to Rank for Information Retrieval. Springer (2011)

25. Liu, Y., Bai, K., Mitra, P., Giles, C.L.: TableSeer: Automatic table metadata extraction and searching in digital libraries. In: Proceedings of the 7th ACM/IEEE-CS Joint Conference on Digital Libraries, JCDL '07, pp. 91–100. ACM (2007). doi: 10.1145/1255175.1255193

26. Manning, C.D., Raghavan, P., Schütze, H.: Introduction to Information Retrieval. Cambridge University Press (2008)

27. Pérez-Agüera, J.R., Arroyo, J., Greenberg, J., Iglesias, J.P., Fresno, V.: Using BM25F for semantic search. In: Proceedings of the 3rd International Semantic Search Workshop, SEMSEARCH '10. ACM (2010). doi: y10.1145/1863879.1863881

28. Pichai, S.: Google I/O 2016 keynote (2016)

29. Pinto, D., McCallum, A., Wei, X., Croft, W.B.: Table extraction using conditional random fields. In: Proceedings of the 26th Annual International ACM SIGIR Conference on Research and Development in Information Retrieval, SIGIR '03, pp. 235–242. ACM (2003). doi: 10.1145/860435.860479

30. Piskorski, J., Yangarber, R.: Information extraction: Past, present and future. In: Multi-source, Multilingual Information Extraction and Summarization, pp. 23–49. Springer (2013). doi: 10.1007/978-3-642-28569-1_2

31. Pound, J., Hudek, A.K., Ilyas, I.F., Weddell, G.: Interpreting keyword queries over web knowledge bases. In: Proceedings of the 21st ACM International Conference on Information and Knowledge Management, CIKM '12, pp. 305–314. ACM (2012). doi: 10.1145/2396761.2396803

32. Pound, J., Mika, P., Zaragoza, H.: Ad-hoc object retrieval in the web of data. In: Proceedings of the 19th international conference on World wide web, WWW '10, pp. 771–780. ACM (2010). doi: 10.1145/1772690.1772769

33. Qian, L., Cafarella, M.J., Jagadish, H.V.: Sample-driven schema mapping. In: Proceedings of the 2012 ACM SIGMOD International Conference on Management of Data, SIGMOD '12, pp. 73–84. ACM (2012). doi: 10.1145/2213836.2213846

34. Rosen, G.: Abstract objects. In: Zalta, E.N. (ed.) The Stanford Encyclopedia of Philosophy (Spring 2017 Edition) (2017)

35. Salton, G.: Automatic Information Organization and Retrieval. McGraw Hill Text (1968)

36. Sanderson, M.: Test collection based evaluation of information retrieval systems. Found. Trends Inf. Retr. **4**(4), 247–375 (2010). doi: 10.1561/1500000009

37. Sarawagi, S.: Information extraction. Found. Trends databases **1**(3), 261–377 (2008). doi: 10.1561/1900000003

38. Sarkas, N., Paparizos, S., Tsaparas, P.: Structured annotations of web queries. In: Proceedings of the 2010 ACM SIGMOD International Conference on Management of Data, SIGMOD '10, pp. 771–782. ACM (2010). doi: 10.1145/1807167.1807251

39. Sawant, U., Chakrabarti, S.: Learning joint query interpretation and response ranking. In: Proceedings of the 22nd International Conference on World Wide Web, WWW '13, pp. 1099–1109. ACM (2013). doi: 10.1145/2488388.2488484

40. Weikum, G.: DB & IR: both sides now. In: Proceedings of the 2007 ACM SIGMOD International Conference on Management of Data, SIGMOD '07, pp. 25–30. ACM (2007). doi: 10.1145/1247480.1247484
41. Yu, J.X., Qin, L., Chang, L.: Keyword search in relational databases: A survey. IEEE Data Eng. Bull. **33**(1), 67–78 (2010)

Chapter 2
Meet the Data

This chapter introduces the basic types of data sources, as well as specific datasets and resources, that we will be working with in later chapters of the book. These may be placed on a spectrum of varying degrees of structure, from unstructured to structured data, as shown in Fig. 2.1.

unstructured   semi-structured   structured

Fig. 2.1 The data spectrum

On the *unstructured* end of the spectrum we have plain text. Typically, these are documents written in natural language.[1] As a matter of fact, almost any type of data can be converted into plain text, including web pages, emails, spreadsheets, and database records. Of course, such a conversion would result in an undesired loss of internal document structure and semantics. It is nevertheless always an option to treat data as unstructured, by not making any assumptions about the particular data format. Search in unstructured text is often referred to as *full-text search*.

On the opposite end of the spectrum there is *structured* data, which is typically stored in relational databases; it is highly organized, tabular, and governed by a strict schema. Search in this type of data is performed using formal query languages, like SQL. These languages allow for a very precise formulation of information needs, but require expert knowledge of the query language and of the underlying database schema. This generally renders them unsuitable for ordinary users.

The data we will mostly be dealing with is neither of two extremes and falls somewhere "in the middle." Therefore, it is termed *semi-structured*. It is

---

[1]Written in natural language does not imply that the text has to be grammatical (or even sensible).

© The Author(s) 2018

K. Balog, *Entity-Oriented Search*, The Information Retrieval Series 39,
https://doi.org/10.1007/978-3-319-93935-3_2

Table 2.1 Comparison of unstructured, semi-structured, and structured data search

	Unstructured	Semi-structured	Structured
Unit of retrieval	Documents	Objects	Tuples
Schema	No	Self-describing	Fixed
Queries	Keyword	Keyword++	Formal languages

characterized by the lack of a fixed, rigid schema. Also, there is no clear separation between the data and the schema; instead, it uses a self-describing structure (tags or other markers). Semi-structured data can be most easily viewed as a combination of unstructured and structured elements. Let us point out that text is rarely completely without structure. Even simple documents typically have a title (or a filename, that is often meaningful). In HTML documents, markup tags specify elements such as headings, paragraphs, and tables. Emails have sender, recipient, subject, and body fields. What is important to notice here is that these document elements or fields may or may not be present. This differs from structured data, where every field specified by the schema ahead of time must be given some permitted value. Therefore, documents with optional, self-describing elements naturally belong to the category of semi-structured data. Furthermore, relational database records may also be viewed as semi-structured data, by converting them to a set of hierarchically nested elements. Performing such conversions can in fact simplify data processing for entity-oriented applications. Using a semi-structured entity representation, all data related to a given entity is available in a single entry for that entity. Therefore, no aggregation via foreign-key relationships is needed. Table 2.1 summarizes data search over the unstructured-structured spectrum.

The remainder of this chapter is organized according to the main types of data sources we will be working with: the Web (Sect. 2.1), Wikipedia (Sect. 2.2), and knowledge bases (Sect. 2.3).

2.1   The Web

The World Wide Web (WWW), commonly known simply as "the Web," is probably the most widely used information resource and service today. The idea of the Web (what today would be considered Web 1.0) was introduced by Tim Berners-Lee in 1989. Beginning in 2002, a new version, dubbed "Web 2.0" started to gain traction, facilitating a more active participation by users, such that they changed from mere consumers to become also creators and publishers of content. The early years of the Web 2.0 era were landmarked by the launch of some of today's biggest social media sites, including Facebook (2004), YouTube (2005), and Twitter (2006). Finally, the Semantic Web (or Web 3.0) was proposed as an extension of the current Web [3, 20]. It represents the next major evolution of the Web that enables data to be understood by computers, which could then perform tasks intelligently on behalf of

Table 2.2 Publicly available web crawls

Name	Time period	Size	#Documents
ClueWeb09 full[a]	Jan 2009–Feb 2009	5 TB	1B
ClueWeb09 (Category B)		230 GB	50M
ClueWeb12[b]	Feb 2012–May 2012	5.6 TB	733M
ClueWeb12 (Category B)		400 GB	52M
Common Crawl[c]	May 2017	58 TB	2.96B
KBA stream corpus 2014[d]	Oct 2011–Apr 2013	10.9 TB	1.2B

Size refers to compressed data
[a]https://lemurproject.org/clueweb09/
[b]https://lemurproject.org/clueweb12/
[c]http://commoncrawl.org/2017/06/may-2017-crawl-archive-now-available/
[d]http://trec-kba.org/kba-stream-corpus-2014.shtml

users. The term *Semantic Web* refers both to this as-of-yet-unrealized future vision and to a collection of standards and technologies for knowledge representation (cf. Sect. 1.2.4).

Web pages are more than just plain text; one of their distinctive characteristics is their hypertext structure, defined by the HTML markup. HTML tags describe the internal document structure, such as headings, paragraphs, lists, tables, and so on. Additionally, HTML documents contain hyperlinks (or simply "links") to other pages (or resources) on the Web. Links are utilized in at least two major ways. First, the networked nature of the Web may be leveraged to identify important or authoritative pages or sites. Second, many of the links also have a textual label, referred to as *anchor text*. Anchor text is "incredibly useful for search engines because it provides some extra description of the page being pointed to" [23].

2.1.1 Datasets and Resources

We introduce a number of publicly available web crawls that have been used in the context of entity-oriented search. Table 2.2 presents a summary.

ClueWeb09/12 The ClueWeb09 dataset consists of about one billion web pages in 10 languages,[2] collected in January and February 2009. The crawl aims to be a representative sample of what is out there on the Web (which includes SPAM and pornography). ClueWeb09 was used by several tracks of the TREC conference. The data is distributed in gzipped files that are in WARC format. About half of the collection is in English; this is referred to as the "Category A" subset. Further, the first segment of Category A, comprising about 50 million pages, is referred to

---

[2]English, Chinese, Spanish, Japanese, French, German, Portuguese, Arabic, Italian, and Korean.

as the "Category B" subset.[3] The Category B subset also includes the full English Wikipedia. These two subsets may be obtained separately if one does not need the full collection.

ClueWeb12 is successor to the ClueWeb09 web dataset, collected between February and May 2012. The crawl was initially seeded with URLs from ClueWeb09 (with the highest PageRank values, and then removing likely SPAM pages) and with some of the most popular sites in English-speaking countries (as reported by Alexa[4]). Additionally, domains of tweeted URLs were also injected into the crawl on a regular basis. A blacklist was used to avoid sites that promote pornography, malware, and the like. The full dataset contains about 733 million pages. Similarly to ClueWeb09, a "Category B" subset of about 50 million English pages is also made available.

Common Crawl Common Crawl[5] is a nonprofit organization that regularly crawls the Web and makes the data publicly available. The datasets are hosted on Amazon S3 as part of the Amazon Public Datasets program.[6] As of May 2017, the crawl contains 2.96 billion web pages and over 250 TB of uncompressed content (in WARC format). The Web Data Commons project[7] extracts structured data from the Common Crawl and makes those publicly available (e.g., the Hyperlink Graph Dataset and the Web Table Corpus).

KBA Stream Corpus The KBA Stream Corpus 2014 is a focused crawl, which concentrates on news and social media (blogs and tweets). The 2014 version contains 1.2 billion documents over a period of 19 months (and subsumes the 2012 and 2013 KBA Stream Corpora). See Sect. 6.2.5.1 for a more detailed description.

2.2   Wikipedia

Wikipedia is one of the most popular web sites in the world and a trusted source of information for many people. Wikipedia defines itself as "a multilingual, web-based, free-content encyclopedia project supported by the Wikimedia Foundation and based on a model of openly editable content."[8] Content is created through the collaborative effort of a community of users, facilitated by a *wiki* platform. There are various mechanisms in place to maintain high-quality content, including the verifiability policy (i.e., readers should be able to check that the information comes

---

[3]The Category B subset was mainly intended for research groups that were not yet ready at that time to scale up to one billion documents, but it is still widely used.

[4]http://www.alexa.com/.

[5]http://commoncrawl.org/.

[6]https://aws.amazon.com/public-datasets/.

[7]http://webdatacommons.org/.

[8]https://en.wikipedia.org/wiki/Wikipedia:About.

from a reliable source) and a clear set of editorial guidelines. The collaborative editing model makes it possible to distribute the effort required to create and maintain up-to-date content across a multitude of users. At the time of writing, Wikipedia is available in nearly 300 languages, although English is by far the most popular, with over five million articles. As stated by Mesgari et al. [15], "Wikipedia may be the best-developed attempt thus far to gather all human knowledge in one place."

What makes Wikipedia highly relevant for entity-oriented search is that most of its entries can be considered as (semi-structured) representations of entities. At its core, Wikipedia is a collection of pages (or articles, i.e., encyclopedic entries) that are well interconnected by hyperlinks. On top of that, Wikipedia offers several (complementary) ways to group articles, including categories, lists, and navigation templates. In the remainder of this section, we first look at the anatomy of a regular Wikipedia article and then review (some of the) other, special-purpose page types. We note that it is not our aim to provide a comprehensive treatment of all the types of pages in Wikipedia. For instance, in addition to the encyclopedic content, there are also pages devoted to the administration of Wikipedia (discussion and user pages, policy pages and guidelines, etc.); although hugely important, these are outside our present scope of interest.

2.2.1   The Anatomy of a Wikipedia Article

A typical Wikipedia article focuses on a particular entity (e.g., a well-known person, as shown in Fig. 2.2) or concept (e.g., "democracy").[9] Such articles typically contain, among others, the following elements (the letters in parentheses refer to Fig. 2.2):

- Title (I.)
- Lead section (II.)

  - Disambiguation links (II.a)
  - Infobox (II.b)
  - Introductory text (II.c)

- Table of contents (III.)
- Body content (IV.)
- Appendices and bottom matter (V.)

  - References and notes (V.a)
  - External links (V.b)
  - Categories (V.c)

---

[9]We refer back to Sect. 1.1.1 for a discussion on the difference between entities and concepts.

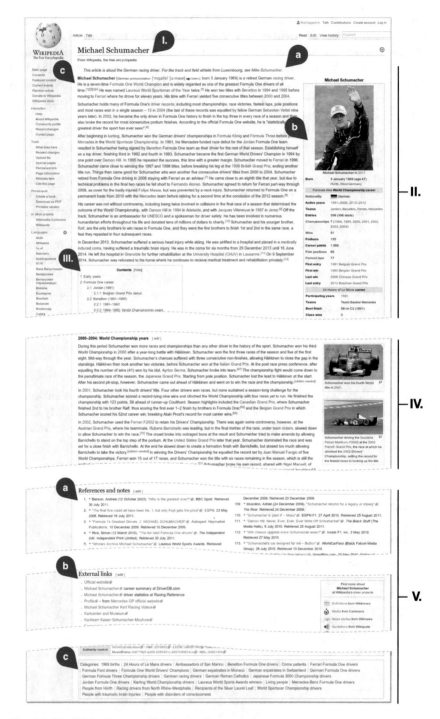

Fig. 2.2 The Wikipedia page of MICHAEL SCHUMACHER

The lead section of a Wikipedia article is the part between the title heading and the table of contents. It serves as an introduction to the article and provides a summary of its contents. The lead section may contain several (optional) elements, including disambiguation links, maintenance tags, infobox, image, navigational boxes, and introductory text. We will further elaborate on the title, infobox, and introductory text elements below.

The main body of the article may be divided into sections, each with a section heading. The sections may be nested in a hierarchy. When there are at least four sections, a navigable table of contents gets automatically generated and displayed between the lead section and the first heading.

The body of the article may be followed by optional appendix and footer sections, including internal links to related Wikipedia articles ("see also"), references and notes (that cite sources), further reading (links to relevant publications that have not been used as sources), internal links organized into navigational boxes, and categories.

2.2.1.1 Title

Each Wikipedia article is uniquely identified by its *page title*. The title of the page is typically the most common name for the entity (or concept) described in the article. When the name is ambiguous, the pages of the other namesakes are disambiguated by adding further qualifiers to their title within parentheses. For instance, MICHAEL JORDAN refers to the American (former) professional basketball player, and the page about the English footballer with the same name has the title MICHAEL JORDAN (FOOTBALLER). Note that the page title is case-sensitive (except the first character). For special pages, the page title may be prefixed with a namespace, separated with a colon, e.g., "Category:German racing drivers." We will look at some of the Wikipedia namespaces later in this section.

2.2.1.2 Infobox

The infobox is a panel that summarizes information related to the subject of the article. In desktop view, it appears at the top right of the page, next to the lead section; in mobile view it is displayed at the very top of the page. In the case of entity pages, the infobox summarizes key facts about the entity in the form of property-value pairs. Therefore, infoboxes represent an important source for extracting structured information about entities (cf. Sect. 2.3.2). A large number of infobox templates exist, which are created and maintained collaboratively, with the aim to standardize information across articles that belong to the same category. Infoboxes, however, are "free form," meaning that what ultimately gets included in the infobox of a given article is determined through discussion and consensus among the editors.

```
Schumacher holds many of Formula One's [[List of Formula One driver records|
driver records]], including most championships, race victories, fastest laps,
pole positions and most races won in a single season - 13 in [[2004 Formula
One season|2004]] (the last of these records was equalled by fellow German
[[Sebastian Vettel]] 9 years later). In [[2002 Formula One season|2002]],
he became the only driver in Formula One history to finish in the top three
in every race of a season and then also broke the record for most consecutive
podium finishes. According to the official Formula One website, he is
"statistically the greatest driver the sport has ever seen".
```

Listing 2.1 Wikitext markup showing internal links to other Wikipedia pages

2.2.1.3 Introductory Text

Most Wikipedia articles include an introductory text, the "lead," which is a brief summary of the article—normally, no more than four paragraphs long. This should be written in a way that it creates interest in the article. The first sentence and the opening paragraph bear special importance. The first sentence "can be thought of as the definition of the entity described in the article" [11]. The first paragraph offers a more elaborate definition, but still without being too detailed. DBpedia, e.g., treats the first paragraph as the "short abstract" and the full introductory text as the "long abstract" of the entity (cf. Sect. 2.3.2).

2.2.2 Links

Internal links are an important feature of Wikipedia as they allow "readers to deepen their understanding of a topic by conveniently accessing other articles."[10] Listing 2.1 shows the original wiki markup for the second paragraph of the introductory text from Schumacher's Wikipedia page from Fig. 2.2. Links are created by enclosing the title of a target page in double square brackets ([[...]]). Optionally, an alternative label, i.e., *anchor text*, may be provided after the vertical bar (|). Linking is governed by a detailed set of guidelines. A key rule given to editors is to link only the first occurrence of an entity or concept in the text of the article.

The value of links extends beyond navigational purposes; they capture semantic relationships between articles. In addition, anchor texts are a rich source of entity name variants. Wikipedia links may be used, among others, to help identify and disambiguate entity mentions in text (cf. Chap. 5).

---

[10]https://en.wikipedia.org/wiki/Wikipedia:Manual_of_Style/Linking.

2.2.3 Special-Purpose Pages

Not all Wikipedia articles are entity pages. In this subsection and the next, we discuss two specific kinds of special-purpose pages.

2.2.3.1 Redirect Pages

Each entity in Wikipedia has a dedicated article and is uniquely identified by the page title of that article. The page title is the most common (canonical) name of the entity. Entities, however, may be referred to by multiple names (aliases). The purpose of *redirect pages* is to allow entities to be referred to by their name variants. Additionally, redirect pages are also often created for common misspellings of entity names. Redirect pages have no content themselves, they merely act as pointers from alternative surface forms to the canonical entity page. Whenever the user visits a redirect page, she will automatically be taken to the "main" article representing the entity. For example, pages redirecting to UNITED STATES include acronyms (U.S.A., U.S., USA, US), foreign translations (ESTADOS UNIDOS), misspellings (UNTIED STATES), and synonyms (YANKEE LAND). Mind that Wikipedia page titles are unique, thus each redirect may refer to a single entity, i.e., the most popular entity with that name (e.g., OBAMA redirects to BARACK OBAMA).

2.2.3.2 Disambiguation Pages

Disambiguation pages serve a reverse role: They are created for ambiguous names and list all entities that share that name. That is, they enumerate all possible meanings of a name. Disambiguation pages are always suffixed by "(disambiguation)" in the title. For example, the BENJAMIN FRANKLIN (DISAMBIGUATION) page lists eight different people, four ships, seven art and literature productions, along with a number of other possible entities, including the $100 bill and various buildings.

2.2.4 Categories, Lists, and Navigation Templates

Wikipedia offers three complementary ways to group related articles together: categories, lists, and navigation templates. These are independent of each other, and each method follows a particular set of guidelines and standards. It is not uncommon that a topic is simultaneously covered by a category, a list, and a navigation template. For example, the "Formula One constructors" article is grouped in all three ways: as a category, a list, and a template.

2.2.4.1 Categories

Categories mainly serve navigational purposes: they provide navigational links for the reader to browse sets of related pages; see V.c on Fig. 2.2. Each Wikipedia article should be assigned to at least one category; most articles are members of several categories. Each article designates to what categories it belongs, and the category page is automatically populated based on what articles declare membership to that category.

Category pages can be distinguished from regular articles by the "Category:" prefix in the page title. That is, a category is a page itself in the *Category* namespace (all other content we have discussed so far is in the *main* namespace). Each category page contains an introductory text (that can be edited like an article), and two automatically generated lists: one with subcategories and another with articles that belong to the category. Different kinds of categories may be distinguished, with the first two being of primary importance:

- *Topic categories* are named after a topic (usually corresponding to a Wikipedia article with the same name on that topic), e.g., "Formula One."
- *Set categories* are named after a particular class (usually in plural), e.g., "German racing drivers."
- *Set-and-topic categories* are a combination of the above two types, e.g., "Formula One drivers of Brawn."
- *Container categories* only contain other categories.
- *Universal categories* provide a comprehensive list of articles that are otherwise divided into subcategories, e.g., "1969 births."
- *Administration categories* are mainly used by editors for management purposes, e.g., "Clean up categories."

Categories are also organized in a hierarchy; each category should be a subcategory of some other category (except a single root category, called "Contents"). This categorization, however, is not a well-defined "is-a" hierarchy, but a (directed) graph; a category may have multiple parent categories and there might be cycles along the path to ancestors. There are various alternative ways to turn this graph into a tree, depending on where we start, i.e., what are selected as top-level categories. Below are two possible starting points:

- Fundamental categories[11] define four fundamental ontological categories: physical entities ("physical universe"), biological entities ("life"), social entities ("society"), and intellectual entities ("concepts").
- Wikipedia's portal for Categories[12] provides another starting point with a set of 27 main categories covering most of the knowledge domains.

---

[11] https://en.wikipedia.org/wiki/Category:Fundamental_categories.

[12] https://en.wikipedia.org/wiki/Portal:Contents/Categories.

According to Wikipedia's guidelines, the general rule to categorization, apart from certain exceptions, is that an article (1) should be categorized as low down in the category hierarchy as possible and (2) should usually not be in both a category and its subcategory.

2.2.4.2 Lists

Lists, as contrasted with categories, provide a means for manual categorization of articles. Lists have a number of advantages over categories. They can be maintained from a centralized location (at the list page itself), and there is more control over the presentation of the content (order of items, formatting, etc.). Importantly, they can also include "missing" articles, that is, items that do not have a Wikipedia page yet. Unfortunately, lists are more difficult to process automatically. Also, for some topics (e.g., people from a particular country), lists would be infeasible to maintain, due to the large number of entries.

2.2.4.3 Navigation Templates

Navigation templates are manual compilations of links that may be included in multiple articles and edited in a central place, i.e., the template page. They provide a navigation system with consistent look and organization for related articles. Navigation templates are meant to be compact and should offer a useful grouping of the linked articles (e.g., by topic, by era, etc.); for that, they may use custom formatting, beyond standard lists or tables. Template inclusion is bidirectional: every article that includes a given navigation template should also be contained as a link in that template. Like categories and lists, templates can also be utilized, e.g., for the task of completing a set of entities with other semantically related entities.

2.2.5 Resources

Wikipedia is based on the MediaWiki software,[13] which is a free open source wiki package. MediaWiki uses an extensible lightweight wiki markup language. Wikipedia may be downloaded in various formats, including XML and SQL dumps or static HTML files.[14] Page view statistics are also made publicly available for download.[15] Wikipedia's content is also accessible via the MediaWiki API in

---

[13] https://www.mediawiki.org.

[14] https://dumps.wikimedia.org/.

[15] https://dumps.wikimedia.org/other/analytics/.

various formats (including JSON and XML).[16] There exists a broad selection of tools for browsing, editing, analyzing, and visualizing Wikipedia.[17] In addition to the official MediaWiki parser, a number of alternative and special-purpose parsers have been created.[18]

2.3 Knowledge Bases

It is important to realize that Wikipedia has been created for human consumption. For machines, the content in this form is hardly accessible. Specifically, it is not a machine-readable structured knowledge model. In our terminology, Wikipedia is a knowledge repository.

The first usage of the term *knowledge base* is connected to expert systems, dating back to the 1970s. Expert systems, one of the earliest forms of (successful) AI software, are designed to solve (or aid humans in solving) complex problems, such as medical diagnosis, by reasoning about knowledge [6]. These systems have rather different data needs than what is supported by relational databases ("tables with strings and numbers"[19]). For such systems, knowledge needs to be represented explicitly. A knowledge base is comprised of a large set of assertions about the world. To reflect how humans organize information, these assertions describe (specific) entities and their relationships. An AI system can then solve complex tasks, such as participating in a natural language conversation, by exploiting the KB.[20]

One of the earliest attempts at such an AI system was the Cyc project that started in 1984 with the objective of building a comprehensive KB to represent everyday commonsense knowledge. The development has been ongoing for over 30 years now and is still far from finished. The main limitation of Cyc is that it relies on human knowledge engineers. In a system with ever-growing complexity, it becomes increasingly difficult to add new objects. While the Cyc project is still alive, it appears that manually codifying knowledge using formal logic and an extensive ontology is not the way forward with respect to the ultimate goal of natural language understanding.

Knowledge bases are bound to be incomplete; there is always additional information to be added or updated. Modern information access approaches embrace this inherent incompleteness. A KB is often regarded as a "semantic backbone" and used in combination with unstructured resources. Instead of relying on large

---

[16]https://www.mediawiki.org/wiki/API:Main_page.

[17]https://en.wikipedia.org/wiki/Wikipedia:Tools.

[18]https://www.mediawiki.org/wiki/Alternative_parsers.

[19]We admit that this is a gross oversimplification.

[20]To make this possible, the AI system also requires an *ontology* (a finite set of rules governing object relationships) and a (logical) inference engine.

ontologies, generally a rather lightweight approach is taken by using some form of subsumption ("is-a") ontology. When the emphasis is on the relationships between entities, a knowledge base is often referred to as a *knowledge graph*.

There exist general purpose as well as domain-specific knowledge bases. DBpedia and YAGO are academic projects that each derive a KB automatically from Wikipedia. Freebase is a community-curated KB that was the open core of Google's Knowledge Graph. It was, however, closed down in 2015 and the data is currently being transferred to Wikidata. All major search providers have their own proprietary knowledge base. Examples include Google's Knowledge Graph,[21] Microsoft's Satori,[22] and Facebook's Entity Graph.[23] Unfortunately, there is very little information available about these beyond popular science introductions.

Before discussing a number of specific knowledge bases, we first explain some fundamentals.

2.3.1 A Knowledge Base Primer

Knowledge bases will be instrumental to most tasks and approaches that will be discussed in this book. Thus, in this section, we explain the core underlying concepts, as well as the RDF data model for representing knowledge in a structured format.

A knowledge base may be divided into two layers:

- On the *schema level* lies a knowledge model, which defines semantic classes (i.e., entity types), the relationships between classes and instances, properties that (instances of) classes can have, and possibly additional constraints and restrictions on them (e.g., range of allowed values for a certain property). Classes are typically organized in a subsumption hierarchy (i.e., a *type taxonomy*).
- The *instance level* comprises a set of assertions about specific entities, describing their names, types, attributes, and relationships with each other.

The predominant language on the Web for describing instances is RDF, which we shall introduce in greater detail in Sect. 2.3.1.2. The KB schema may be encoded using a declarative language, such as RDFS (for expressing taxonomical relationships)[24] or OWL (for full-fledged ontological modeling).[25]

---

[21]https://googleblog.blogspot.no/2012/05/introducing-knowledge-graph-things-not.html.

[22]https://blogs.bing.com/search/2014/03/31/150-million-more-reasons-to-love-bing-everyday/.

[23]http://www.technologyreview.com/news/511591/facebook-nudges-users-to-catalog-the-real-world/.

[24]https://www.w3.org/TR/rdf-schema/.

[25]https://www.w3.org/OWL/.

2.3.1.1   Knowledge Bases vs. Ontologies

In information retrieval and natural language processing, knowledge bases have become central to machine understanding of natural language over the past decade. More recently, and especially in the search industry, often the term *knowledge graph* is used. In the fields of artificial intelligence and the Semantic Web, people have been using *ontologies* for similar, or even more ambitious, goals since the 1990s. The question naturally arises: What is the difference between a knowledge base and an ontology? Is it only a matter of choice of terminology or is there more to it? The simple answer is that a knowledge base can be represented as an ontology. For example, the YAGO knowledge base refers to itself as an ontology [21]. But the two are not exactly the same.

To understand the difference, we should first clarify what an ontology is. The word carries several connotations, depending on the particular discipline and research field. Perhaps the most widely cited definition is by Gruber [9], which states that "an ontology is an explicit specification of a conceptualization." According to Navigli [16], "an ontology is a set of definitions in a formal language for concepts that describe the world of interest, including the relationships that connect these concepts." Put simply, an ontology is a means to formalizing knowledge. Building blocks of an ontology include (1) *classes* (or *concepts*), (2) *objects* (or *instances*), (3) *relations*, connecting classes and objects to one another, (4) *attributes* (or *properties*), representing relations intrinsic to specific objects, (5) *restrictions* on relations, and (6) *rules* and *axioms*, which are assertions in a logical form [16]. The conceptual design of an ontology revolves around the possible concepts and relations that are to be encoded. The instance level (i.e., individual objects) may not even be involved in the process or included in the ontology. Knowledge bases, on the other hand, place the main emphasis on individual objects and their properties. The formal constraints imposed by the ontology on those objects are only of interest when the knowledge base is being populated with new facts.

In summary, both knowledge bases and ontologies attempt to capture a useful representation of a (physical or virtual) world, with the overall objective to solve (complex) problems. Ontologies are schema-oriented ("top-down" design) and focus on describing the concepts and relationships within a given domain with the highest possible expressiveness. Conversely, knowledge bases are fact-oriented ("bottom-up" design), with an emphasis on describing specific entities.

2.3.1.2   RDF

The Resource Description Framework (RDF) is a language designed to describe "things," which are referred to as *resources*. A resource denotes either an entity (object), an entity type (class), or a relation. Each resource is assigned a Uniform Resource Identifier (URI), making it uniquely and globally identifiable. Each RDF statement is a triple, consisting of *subject*, *predicate*, and *object* components.

- The *subject* is always a URI, denoting a resource.
- The *predicate* is also a URI, corresponding to a relationship or property of the subject resource.
- The *object* is either a URI (referring to another resource) or an (optionally typed) literal.

Consider the following piece of information:

Michael Schumacher (born 3 January 1969) is a retired German racing driver, who raced in Formula One for Ferrari.

It may be represented as the following set of RDF statements (often referred to as *SPO triples*, for short). Each triple represents an atomic factual statement in the knowledge base. URIs are enclosed in angle brackets and are shortened for improved readability (see Table 2.4 for the full URIs of the namespaces dbr, foaf, etc.); literal values are in quotes.

`<dbr:Michael_Schumacher>`	`<foaf:name>`	`"Schumacher, Michael"`
`<dbr:Michael_Schumacher>`	`<dbo:birthPlace>`	`<dbr:West_Germany>`
`<dbr:Michael_Schumacher>`	`<dbo:birthDate>`	`"1969-01-03"`
`<dbr:Michael_Schumacher>`	`<rdf:type>`	`<dbo:RacingDriver>`
`<dbr:Michael_Schumacher>`	`<dct:subject>`	`<dbc:Ferrari_Formula_One_drivers>`

Note that the expressivity of the RDF representation depends very much on the vocabulary of predicates. This particular example is taken from DBpedia (which we shall introduce in detail in the next section), where the object `<dbc:Ferrari_Formula_One_drivers>` identifies a certain category of entities and the predicate `<dct:subject>` assigns the subject entity to that category. Using the DBpedia Ontology, there is (currently) no way to express this relationship more precisely, emphasizing that the person *was driving* for that team but not anymore.

Mind that a relationship between two entities is a directed link (labeled with a predicate). For instance, the fact that SCHUMACHER won the 1996 SPANISH GRAND PRIX may be described as either of the following two statements:

`<dbr:1996_Spanish_Grand_Prix>`	`<dbp:firstDriver>`	`<dbr:Michael_Schumacher>`
`<dbr:Michael_Schumacher>`	`<dbp:firstDriverOf>`	`<dbr:1996_Spanish_Grand_Prix>`

In reality, only the first one is an actual triple; the second is a made up example, as there is no `<dbp:firstDriverOf>` predicate in DBpedia. Even if there was one, this second triple would only introduce redundant information. What is important here is that information pertinent to a given entity is contained in triples with that entity standing as either subject or object. When the entity stands as object, the triple may be reversed by taking the inverse of the predicate, like:

`<dbr:Michael_Schumacher>` is **`<dbp:firstDriver>`** of `<dbr:1996_Spanish_Grand_Prix>`

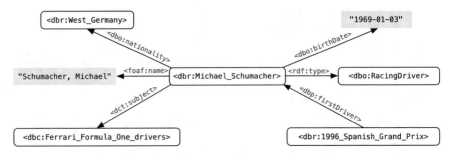

Fig. 2.3 Excerpt of an RDF graph (taken from DBpedia). URIs (i.e., entities) are represented by rounded rectangles, literals (i.e., attribute values) are denoted by shaded rectangles

Conceptually, the set of RDF triples forms a large, directed, labelled graph (referred to as the *RDF graph*). Each RDF triple corresponds to a pair of nodes in the graph (subject and object), connected by an edge (predicate). Figure 2.3 displays the graph corresponding to the triples from our running example.

Note that RDF describes the instance level in the knowledge base. To cope with the knowledge base schema (concepts and relations), an ontology representation language is needed, such as RDFS or OWL. Essentially, what RDFS and OWL provide are vocabularies for ontological modeling. RDFS (RDF Schema) provides a vocabulary for encoding taxonomies and is often preferred when lightweight modeling is sufficient. OWL (Web Ontology Language) builds upon RDFS and comes with the full expressive power of description logics. For the serialization, storage, and retrieval of RDF data we refer to Sect. 2.3.7.

2.3.2   DBpedia

In layman's terms, DBpedia[26] is a "database version of Wikipedia." More precisely, DBpedia is a knowledge base that is derived by extracting structured data from Wikipedia [12]. One powerful aspect of DBpedia is that it is not a result of a one-off process but rather of a continuous community effort, with numerous releases since its inception in 2007. Over the years, DBpedia has developed into an interlinking hub in the Web of Data (which will be discussed in Sect. 2.3.6). Another distinguishing feature of DBpedia is that it is available in multiple languages. We base our discussion below on the latest release that is available at the time of writing, DBpedia 2016-10, and especially on the English version. For an overview of DBpedia's evolution over time and for details on the localized versions, we refer to [12]. Due to DBpedia's importance as a resource, we shall provide an in-depth treatment of its main components.

---

[26]http://dbpedia.org/.

2.3.2.1 Ontology

The DBpedia Ontology is a cross-domain ontology that has been manually created based on the most frequently used Wikipedia infoboxes (cf. II.b on Fig. 2.2). The current version contains 685 classes, which are organized in a six-level deep subsumption hierarchy. The ontology is intentionally kept this shallow so that it can be easily visualized and navigated. Each class within the ontology is described by a number of properties; for each property, the range of possible values is also defined. Figure 2.4 shows a small excerpt from the DBpedia Ontology.

The maintenance of the ontology is a community effort that is facilitated by the DBpedia Mappings Wiki. Using this wiki, users can collaboratively create and edit mappings from different infobox templates to classes and properties in the DBpedia Ontology. These mappings are instrumental in extracting high-quality data as they alleviate problems arising from the heterogeneity of Wikipedia's infoboxes. We elaborate more on this in the following subsection. Crowdsourcing turned out to be a powerful tool for extending and refining the ontology. The number of properties has grown from 720 in 2009, to 1650 in 2014, and to 2795 in 2016. The number of classes has increased at a similar pace, from 170 in 2009, to 320 in 2014, and to 685 in 2016. The current version of DBpedia describes 4.58 million entities, out of which 4.22 million are classified in the ontology. The largest ontology classes

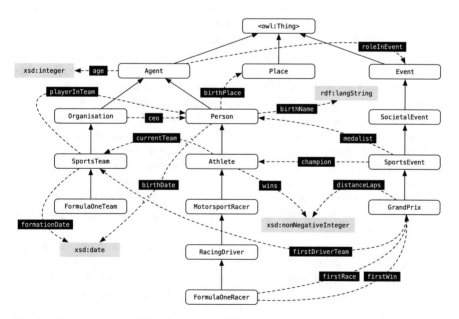

Fig. 2.4 Excerpt from the DBpedia Ontology. Classes are represented by rounded rectangles where arrows with solid lines indicate subclass relationships (from subclass to superclass). Properties are denoted by the dashed arrows with black labels. Value types are shown in gray boxes. Unless specifically indicated, classes/values are in the dbo namespace (cf. Table 2.4)

include persons (1.4M), places (735k), creative works like music albums and films (411k), and organizations (241k).

2.3.2.2   Extraction

The DBpedia extraction framework follows a pipeline architecture, where the input is a Wikipedia article and the output is a set of RDF statements extracted from that article. The framework encompasses a number of different purpose-built extractors; some of these are designed to grab a single property (e.g., the abstract or a link to an image depicting the entity), while others deal with specific parts of Wikipedia pages (e.g., infoboxes). The DBpedia extractors can be categorized into four main types:

Raw infobox extraction   The most important source of structured information are the infoboxes. These list the main facts about a given entity as property-value pairs. The raw infobox extractor directly translates all Wikipedia infobox properties to DBpedia predicates. There is no normalization performed either on properties (i.e., RDF predicates) or values (i.e., RDF objects). This generic extraction method provides complete coverage of all infobox data. The predicates with the dbp prefix in the example triples in Sect. 2.3.1.2 are the results of this raw infobox extraction.

Mapping-based infobox extraction   One major issue with infoboxes is inconsistency. A wide range of infobox templates are used in Wikipedia, which evolve over time. As a result, the same type of entity may be described by different templates; these templates may use different names for the same property (e.g., `birthplace` vs. `placeofbirth`). Further, attribute values may be expressed using a range of alternative formats and units of measurement. In order to enforce consistency, which is an important dimension of data quality, a homogenization of properties and values is necessary. This normalization (or homogenization) is done against the DBpedia Ontology and is made possible by the community-provided mappings specified in the DBpedia Mappings Wiki. Not only are predicates normalized but literal object values are also canonicalized to basic units according to the assigned datatypes. The mapping-based approach significantly increases the quality compared to the raw infobox data. The RDF statements generated by this extractor can be distinguished by the dbo prefix of the predicates.

Feature extraction   A number of specialized extractors are developed for the extraction of a single feature from an article. These include, among others, abstract, categories, disambiguations, external links, geo-coordinates, homepage, image, label, page links, and redirects; we refer to Table 2.3 for the descriptions of these.

Statistical extraction   The extractors in this last group are not part of the DBpedia "core." They were created with the intent to provide resources that can support computational linguistics tasks [14]. Unlike the core extractors, which are essentially rule-based, these employ statistical estimation techniques. Some of

Table 2.3   A selection of specific feature extractors in DBpedia

Name	Predicate	Description
Abstract	`dbo:abstract`	The first lines of the Wikipedia article
Categories	`dc:subject`	Wikipedia categories assigned to the article
Disambiguation	`dbo:wikiPageDisambiguates`	Disambiguation links
External links	`dbo:wikiPageExternalLink`	Links to external web pages
Geo-coordinates	`georss:point`	Geographical coordinates
Homepage	`foaf:homepage`	Link to the official homepage of an instance
Image	`foaf:depiction`	Link to the first image on the Wikipedia page
Label	`rdfs:label`	The page title of the Wikipedia article
Page links	`dbo:wikiPageWikiLink`	Links to other Wikipedia articles
Redirect	`dbo:wikiPageRedirects`	Wikipedia page to redirect to

See Table 2.4 for the URI prefixes

them deviate further from the regular extractors in that they aggregate data from all Wikipedia pages as opposed to operating on a single article. The resulting datasets include grammatical gender (for entities of type person), lexicalizations (alternative names for entities and concepts), topic signatures (strongest related terms), and thematic concepts (the main subject entities/concepts for Wikipedia categories).

2.3.2.3   Datasets and Resources

The output of each DBpedia extractor, for each language, is made available as a separate dataset. All datasets are provided in two serializations: as Turtle (N-triples) and as Turtle quads (N-Quads, which include context). The datasets can be divided into the following categories:

- *DBpedia Ontology*: The latest version of the ontology that was used while extracting all datasets.
- *Core datasets*: All infobox-based and specific feature extractors (including the ones listed in Table 2.3) belong here.
- *Links to other datasets*: DBpedia is interlinked with a large number of knowledge bases. The datasets in this group provide links to external resources both on the instance level (`owl:sameAs`), e.g., to Freebase and YAGO, and on the schema level (`owl:equivalentClass` and `owl:equivalentProperty`), most notably to schema.org.
- *NLP datasets*: This last group corresponds to the output of the statistical extractors.

Namespaces and Internationalization   The generic DBpedia URI namespaces are listed in the upper block of Table 2.4. As part of the internationalization efforts, some datasets are available both in *localized* and in *canonicalized* version.

Table 2.4 Main URI namespaces used in DBpedia

Prefix	URL	Description
DBpedia namespaces		
dbr	http://dbpedia.org/resource/	One-to-one mapping between Wikipedia articles and DBpedia resources
dbp	http://dbpedia.org/property/	Properties from raw infobox extraction
dbo	http://dbpedia.org/ontology/	DBpedia Ontology
External namespaces[a]		
dc	http://purl.org/dc/elements/1.1/	Dublin core
foaf	http://xmlns.com/foaf/0.1/	Friend of a friend (FOAF)
georss	http://www.georss.org/georss/	Geographically encoded objects for RSS
owl	http://www.w3.org/2002/07/owl#	W3C web ontology language
rdf	http://www.w3.org/1999/02/22-rdf-syntax-ns#	Standard W3C RDF vocabulary
rdfs	http://www.w3.org/2000/01/rdf-schema#	Extension of the basic RDF vocabulary

[a]The list is not intended to be exhaustive

The localized datasets include everything from the given language's Wikipedia and use language-specific URIs (http://<lang>.dbpedia.org/resource/ and http://<lang>.dbpedia.org/property/). The canonicalized datasets, on the other hand, only contain resources that exist in the English edition of Wikipedia as well; here, the generic (language-agnostic) URI namespace is used.

SPARQL Endpoint The various datasets are not only available for download but can also be accessed and queried via a public SPARQL endpoint.[27] The endpoint is hosted using the Virtuoso Universal Server.

DBpedia Live Content on Wikipedia is changing. For example, in April 2016, there were 3.4M edits on the English Wikipedia.[28] New DBpedia versions are released fairly regularly, at least once per year. However, for certain applications or usage scenarios, this rate of updating might be too slow. Instead of the infrequent batch updates, a live synchronization mechanism would be preferable. DBpedia Live[29] is a module that does exactly this: It processes Wikipedia updates real-time and keeps DBpedia up-to-date. DBpedia Live offers a separate SPARQL endpoint.[30] Additionally, the "changesets" (added and removed triples) are made available and can be applied on top of a local copy of DBpedia with the help of a sync tool.[31]

---

[27]http://dbpedia.org/sparql.

[28]https://stats.wikimedia.org/EN/SummaryEN.htm.

[29]http://live.dbpedia.org/.

[30]http://live.dbpedia.org/sparql.

[31]https://github.com/dbpedia/dbpedia-live-mirror.

2.3.3 YAGO

YAGO[32] (which stands for Yet Another Great Ontology) is a similar effort to DBpedia in that it extracts structured information from Wikipedia, such that each Wikipedia article becomes an entity. Although they share similar aims, the underlying systems and philosophy are quite different. While DBpedia stays close to Wikipedia and aims to simply provide an RDF version of it, YAGO focuses on achieving high precision and consistent knowledge. Instead of relying on mappings collaboratively created by a community, YAGO's extraction is facilitated by expert-designed declarative rules. Each fact in YAGO is annotated with a confidence value. According to an empirical evaluation, the accuracy of the contained facts is about 95% [21].

2.3.3.1 Taxonomy

Another key difference between DBpedia and YAGO lies in the typing of entities. While DBpedia employs a small, manually curated ontology, YAGO constructs a deep subsumption hierarchy of entity types by connecting Wikipedia categories with WordNet concepts. WordNet[33] is a large lexical resource that groups words into sets of cognitive synonyms (synsets). Each synset expresses a distinct concept; ambiguous words (i.e., words with multiple meanings) belong to multiple synsets.

Wikipedia categories are organized in a directed graph, but this is not a strict hierarchy (cf. Sect. 2.2.4.1). Moreover, the relations between categories merely reflect the thematic structure. "Thus, the hierarchy is of little use from an ontological point of view" [21]. Hence, YAGO establishes a hierarchy of classes, where the upper levels are based on WordNet synsets and the leaves come from (a subset of the) Wikipedia leaf categories. This results in over 568K entity types, hierarchically organized in 19 levels.

2.3.3.2 Extensions

There have been two major extensions to the original YAGO knowledge base. YAGO2 [10] anchors knowledge in time and space, i.e., places entities and facts on their spatial and temporal dimension. Specifically, timestamps are defined for four main entity types: people, groups, artifacts, events. It is argued that "these four types cover almost all of the cases where entities have a meaningful existence time" [10]. Location is extracted for entities that have a "permanent spatial extent on Earth" [10], such as countries, cities, mountains, and rivers. A new super-

---

[32]http://www.mpi-inf.mpg.de/departments/databases-and-information-systems/research/yago-naga/yago/.

[33]http://wordnet.princeton.edu/.

class (yagoGeoEntity) is introduced to the YAGO taxonomy, which groups together all geo-entities. Geo-entities are harvested from two sources: Wikipedia (based on associated geographical coordinates) and GeoNames[34] (a freely available geographical database). In summary, YAGO2 associates over 30 million facts with their occurrence time, and over 17 million facts with the location of occurrence. The time of existence is known for 47% and the location is known for 30% of all entities. According to a sample-based manual assessment, YAGO2 has a precision of 95%.

YAGO3 [13] extends YAGO to multiple languages, by running the YAGO extraction system on different language editions of Wikipedia. This brings in one million new entities and seven million facts over the original (English-only) YAGO.

2.3.3.3   Resources

YAGO is publicly available for download in TSV or Turtle formats, either in its entirety or just specific portions.[35] The YAGO type hierarchy is also offered in DBpedia as an alternative to the DBpedia Ontology. Mappings (i.e., "same-as" links) between YAGO and DBpedia instances are provided in both directions.

2.3.4   Freebase

Freebase[36] is an open and large collaborative knowledge base [5]. It was launched in 2007 by the software company Metaweb, which was acquired by Google in 2010. Freebase "was used as the open core of the Google Knowledge Graph" [19]. In December 2014, Google announced that it would shut down Freebase and help with the transfer of content from Freebase to Wikidata.[37] The migration process is briefly elaborated on in the next subsection.

The content of Freebase has been partially imported from open data sources, such as Wikipedia, MusicBrainz,[38] and the Notable Names Database (NNDB).[39] Another part of the data comes from user-submitted wiki contributions. Freebase encouraged users to create entries for less popular entities (which would have not made it to Wikipedia). Instead of using controlled ontologies, Freebase adopted a folksonomy approach, in which users could use types much like tags. Each type has a number of properties (i.e., predicates) associated with it.

---

[34]http://www.geonames.org.

[35]http://yago-knowledge.org.

[36]https://developers.google.com/freebase/.

[37]https://plus.google.com/109936836907132434202/posts/bu3z2wVqcQc.

[38]https://musicbrainz.org/.

[39]http://www.nndb.com/.

Google has made some important data releases using Freebase, specifically, entity annotations for the ClueWeb09 and ClueWeb12 corpora (cf. Sect. 5.9.2) and for the KBA Stream Corpus 2014 (cf. Sect. 6.2.5.1). The latest Freebase dump, from 31 March 2015, is still available for download. It contains 1.9 billion triples and about 39 million entities (referred to as *topics* in Freebase).

2.3.5   Wikidata

Wikidata[40] is a free collaborative knowledge base operated by the Wikimedia Foundation [22]. Its goal is to provide the same information as Wikipedia, but in a structured format. Launched in October 2012, Wikidata "has quickly become one of the most active Wikimedia projects" [22]. As of 2017, it has over 7K active monthly contributors (those making at least 5 edits per month). Unlike the previous KBs we have discussed, Wikidata does not consider statements as facts, but rather as *claims*, each having a list of references to sources supporting that claim. Claims can contradict each other and coexist, thereby allowing opposing views to be expressed (e.g., different political positions). Essentially, claims are property-value pairs for a given *item*, which is Wikidata lingo for an entity. There is support for two special values: "unknown" (e.g., a person's exact day of death) and "no value" (e.g., Australia has no bordering countries); these cases are different from data being incomplete. Claims can also have additional subordinate property-value pairs, called *qualifiers*. Qualifiers can store contextual information (e.g., the validity time for an assertion, such as the population of a city in a certain year, according to a particular source). Importantly, Wikidata is multilingual by design and uses language-independent entity IDs.

Wikidata relies on crowdsourced manual curation to ensure data quality. With the retirement of Freebase, Google decided to offer the content of Freebase to Wikidata. This migration, which is still underway at the time of writing, is not without challenges. One of the main difficulties is rooted in the "cultural" differences between the two involved communities; they "have a very different background, subtly different goals and understandings of their tasks, and different requirements regarding their data" [19]. One specific challenge is that Wikidata is eager to have references for their statements, which are not present in Freebase. Such references are obtained from the Google Knowledge Vault [8], then checked and curated manually by Wikidata contributors using a purpose-built tool, called the *Primary Sources Tool*; we refer to [19] for details. As of June 2017, Wikidata contains over 158 million statements about 26.9 million entities. Wikidata offers copies of its

---

[40]https://wikidata.org/.

Easy Chicken Satay Recipe - Allrecipes.com

https://www.allrecipes.com/recipe/132929/easy-chicken-satay/ ▾

★★★★★ Rating: 4.7 - 305 reviews - 2 hr 45 min - 418 cal

Toss marinade with the **chicken**, cover, and marinate for at least 2 hours. Bring 1 cup coconut milk, 1 tablespoon curry powder, peanut butter, **chicken** stock, and 1/4 cup brown sugar to a simmer in a saucepan over medium-high heat. Simmer for 5 minutes, stirring constantly, until smooth and thickened.

Fig. 2.5 Result snippet from a Google search result page

```
<section class="ar_recipe_index full-page" itemscope
        itemtype="http://schema.org/Recipe">
    <link href="http://allrecipes.com/recipe/132929/easy-chicken-satay/"
        itemprop="url" />
    <meta itemprop="mainEntityOfPage" content="True" />
```

Listing 2.2 Excerpt from a recipe's HTML page annotated with Microdata meta tags. Source: http://allrecipes.com/recipe/132929/easy-chicken-satay/

content for download in JSON, RDF, and XML formats,[41] and also provides access via a search API.[42]

2.3.6 The Web of Data

The amount of structured data available on the Web is growing steadily. A large part of it is contained in various knowledge bases, like DBpedia and Freebase. In addition, increasingly more data is being exposed in the form of semantic annotations added to traditional web pages using metadata standards, such as Microdata, RDFa, and JSON-LD. There is a strong incentive for websites for marking up their content with semantic metadata: It allows search engines to enhance the presentation of the site's content in search results. An important standardization development was the introduction of `schema.org`, a common vocabulary used by major search providers (including Google, Microsoft, and Yandex) for describing commonly used entity types (including people, organizations, events, products, books, movies, recipes, etc.). Figure 2.5 displays a snippet from a Google search result page; Listing 2.2 shows an excerpt from the HTML source of the corresponding HTML page with Microdata annotations.

Historically, data made available in RDF format was referred to as *Semantic Web* data. One of the founding principles behind the Semantic Web is that data should be interlinked. These principles were summarized by Berners-Lee [2] in the following four simple rules:

---

[41] http://www.wikidata.org/wiki/Wikidata:Database_download.

[42] https://query.wikidata.org/.

1. Use URIs as names for things.
2. Use HTTP URIs so that people can look up those names.
3. When someone looks up a URI, provide useful information.
4. Include links to other URIs, so that people can discover more things.

The term *Linked Data* (LD) refers to a set of best practices for publishing structured data on the Web. The key point about Linked Data is that it enables to connect entities (or, generally speaking, resources) across multiple knowledge bases. This is facilitated by a special "same-as" predicate, `<owl:sameAs>`, basically saying that the subject and object resources connected by that predicate are the same. For example, the following two statements connect the representations of the entity MICHAEL SCHUMACHER across DBpedia, Freebase, and Wikidata:

`<dbr:Michael_Schumacher>`	`<owl:sameAs>`	`<fb:m.053w4>`
`<dbr:Michael_Schumacher>`	`<owl:sameAs>`	`<wikidata:Q9671>`

These "same-as" links connect all Linked Data into a single global data graph. Mind that the term Linked Data should be used to describe the publishing practice, not the data itself. A knowledge base published using LD principles should be called *Linked Dataset*. To avoid the terminological confusion, we shall refer to the collection of structured data exposed on the Web in machine understandable format as the *Web of Data* (emphasizing the difference in nature to the traditional *Web of Documents*). *Linked Open Data* (LOD) may also be used as a casual synonym, emphasizing the fact that Linked Data is released under an open license. Figure 2.6 shows the Linked Open Data cloud, where edges indicate the presence of "same-as" links between two datasets (knowledge bases). Notice that DBpedia is a central hub here.

2.3.6.1   Datasets and Resources

In this book, we focus on two particular data collections, BTC-2009 and Sindice-2011, that have been used in the information retrieval community for entity-oriented research. For a more extensive list of datasets, tools, and Linked Data resources, see http://linkeddata.org/.

BTC-2009   The Billion Triples Challenge 2009 dataset (BTC-2009)[43] was created for the Semantic Web Challenge in 2009 [4]. The dataset was constructed by combining the crawls of multiple semantic search engines during February–March 2009. It comprises 1.1 billion RDF triples, describing 866 million distinct resources, and amounts to 17 GB in size (compressed).

---

[43]https://km.aifb.kit.edu/projects/btc-2009/.

Fig. 2.6 Excerpt from the Linking Open Data cloud diagram 2014, by Max Schmachtenberg, Christian Bizer, Anja Jentzsch, and Richard Cyganiak. Source: http://lod-cloud.net/

Sindice-2011 The Sindice-2011 dataset [7] was created in 2011 for the TREC Entity track [1] with the aim to provide a more accurate reflection of the at-the-time current Web of Data. The data has been collected by the Sindice semantic search engine [18] between 2009 and 2011. Sindice-2011 contains 11 billion RDF statements, describing 1.7 billion entities. The dataset is 1.3 TB in size (uncompressed).

2.3.7 Standards and Resources

RDF, RDFS, and OWL are all standards of the World Wide Web Consortium (W3C),[44] which is the main international standards organization for the World Wide Web. There exist numerous serializations for RDF data, e.g., Notation-3, Turtle, N-Triples, RDFa, and RDF/JSON. The choice of serialization depends on the context and usage scenario. For example, Turtle is probably the easiest serialization to use for human consumption and manipulation. If large volumes of data need to be interchanged between systems, then producing data dumps in N-Triples format is a common choice. If only HTML documents are produced, then RDFa is preferred. SPARQL[45] is a structured query language for retrieving and manipulating RDF data, and is also a W3C standard. Triplestores are special-purpose databases designed for storing and querying RDF data. Examples of triplestores include Apache Jena,[46] Virtuoso,[47] and RDF-3X [17].

2.4 Summary

This chapter has introduced the different kinds of data, from unstructured to structured, that we will be using in the coming chapters. The order in which we have discussed them—first the Web, then Wikipedia, and finally knowledge bases—reflects how the research focus in entity-oriented search is shifting toward relying increasingly more on structured data sources, and specifically on knowledge bases. Knowledge bases are rich in structure, but light on text; they are of high quality, but are also inherently incomplete. This stands in stark contrast with the Web, which is a virtually infinite source of heterogeneous, noisy, and text-heavy content that comes with limited structure. Their complementary nature makes the combination of knowledge bases and the Web a particularly attractive and fertile ground for entity-oriented (re)search.

---

[44]https://www.w3.org/.

[45]https://www.w3.org/TR/sparql11-query/.

[46]https://jena.apache.org/.

[47]https://virtuoso.openlinksw.com/.

References

1. Balog, K., Serdyukov, P., de Vries, A.P.: Overview of the TREC 2011 Entity track. In: The Twentieth Text REtrieval Conference Proceedings, TREC '11. NIST (2012)
2. Berners-Lee, T.: Linked data (2009)
3. Berners-Lee, T., Hendler, J., Lassila, O.: The semantic web. Scientific American **284**(5), 34–43 (2001)
4. Bizer, C., Mika, P.: Editorial: The semantic web challenge, 2009. Web Semantics: Science, Services and Agents on the World Wide Web **8**(4) (2010)
5. Bollacker, K., Evans, C., Paritosh, P., Sturge, T., Taylor, J.: Freebase: A collaboratively created graph database for structuring human knowledge. In: Proceedings of the 2008 ACM SIGMOD International Conference on Management of Data, SIGMOD '08, pp. 1247–1250. ACM (2008). doi: 10.1145/1376616.1376746
6. Buchanan, B.G., Shortliffe, E.H.: Rule Based Expert Systems: The Mycin Experiments of the Stanford Heuristic Programming Project (The Addison-Wesley Series in Artificial Intelligence). Addison-Wesley Publishing Co. (1984)
7. Campinas, S., Ceccarelli, D., Perry, T.E., Delbru, R., Balog, K., Tummarello, G.: The Sindice-2011 dataset for entity-oriented search in the web of data. In: 1st International Workshop on Entity-Oriented Search, EOS '11 (2011)
8. Dong, X., Gabrilovich, E., Heitz, G., Horn, W., Lao, N., Murphy, K., Strohmann, T., Sun, S., Zhang, W.: Knowledge Vault: A web-scale approach to probabilistic knowledge fusion. In: Proceedings of the 20th ACM SIGKDD International Conference on Knowledge Discovery and Data Mining, KDD '14, pp. 601–610. ACM (2014). doi: 10.1145/2623330.2623623
9. Gruber, T.R.: A translation approach to portable ontology specifications. Knowl. Acquis. **5**(2), 199–220 (1993). doi: 10.1006/knac.1993.1008
10. Hoffart, J., Suchanek, F.M., Berberich, K., Weikum, G.: YAGO2: A spatially and temporally enhanced knowledge base from Wikipedia. Artificial Intelligence **194**, 28–61 (2013). doi: 10.1016/j.artint.2012.06.001
11. Kazama, J., Torisawa, K.: Exploiting Wikipedia as external knowledge for named entity recognition. In: Proceedings of the 2007 Joint Conference on Empirical Methods in Natural Language Processing and Computational Natural Language Learning, EMNLP-CoNLL '07, pp. 698–707. Association for Computational Linguistics (2007)
12. Lehmann, J., Isele, R., Jakob, M., Jentzsch, A., Kontokostas, D., Mendes, P., Hellmann, S., Morsey, M., van Kleef, P., Auer, S., Bizer, C.: DBpedia - A large-scale, multilingual knowledge base extracted from Wikipedia. Semantic Web Journal (2012)
13. Mahdisoltani, F., Biega, J., Suchanek, F.M.: YAGO3: A knowledge base from multilingual Wikipedias. In: Seventh Biennial Conference on Innovative Data Systems Research, CIDR '15 (2015)
14. Mendes, P.N., Jakob, M., Bizer, C.: DBpedia for NLP: A multilingual cross-domain knowledge base. In: Proceedings of the Eight International Conference on Language Resources and Evaluation, LREC '12. ELRA (2012)
15. Mesgari, M., Okoli, C., Mehdi, M., Nielsen, F.Å., Lanamäki, A.: "The sum of all human knowledge": A systematic review of scholarly research on the content of Wikipedia. Journal of the Association for Information Science and Technology **66**(2), 219–245 (2015). doi: 10.1002/asi.23172
16. Navigli, R.: Ontologies. In: Mitkov, R. (ed.) Ontologies. Oxford University Press (2017)
17. Neumann, T., Weikum, G.: RDF-3X: a risc-style engine for RDF. Proc. VLDB Endow. **1**(1), 647–659 (2008). doi: 10.14778/1453856.1453927
18. Oren, E., Delbru, R., Catasta, M., Cyganiak, R., Stenzhorn, H., Tummarello, G.: Sindice.com: a document-oriented lookup index for open linked data. Int. J. Metadata Semant. Ontologies **3**(1), 37–52 (2008). doi: 10.1504/IJMSO.2008.021204

19. Pellissier Tanon, T., Vrandečić, D., Schaffert, S., Steiner, T., Pintscher, L.: From Freebase to Wikidata: The great migration. In: Proceedings of the 25th International Conference on World Wide Web, WWW '16, pp. 1419–1428. International World Wide Web Conferences Steering Committee (2016). doi: 10.1145/2872427.2874809
20. Shadbolt, N., Berners-Lee, T., Hall, W.: The semantic web revisited. IEEE Intelligent Systems **21**(3), 96–101 (2006). doi: 10.1109/MIS.2006.62
21. Suchanek, F.M., Kasneci, G., Weikum, G.: YAGO: A core of semantic knowledge. In: Proceedings of the 16th International Conference on World Wide Web, WWW '07, pp. 697–706. ACM (2007). doi: 10.1145/1242572.1242667
22. Vrandečić, D., Krötzsch, M.: Wikidata: A free collaborative knowledge base. Commun. ACM **57**(10), 78–85 (2014). doi: 10.1145/2629489
23. Zhai, C., Massung, S.: Text Data Management and Analysis: A Practical Introduction to Information Retrieval and Text Mining. ACM and Morgan & Claypool (2016)

Part I
Entity Ranking

Part I is dedicated to the problem of *entity ranking*: Given an input query, return a ranked list of entities. Entity ranking is a multifaceted problem involving a variety of interrelated factors, such as the task at hand (ad-hoc entity ranking, list completion, related entity finding, etc.), query formulation (from keyword-only to queries with additional components, such as target types or example entities), and data source (unstructured, semi-structured, structured, as well as their combinations). Chapter 3 considers keyword queries and focuses on obtaining term-based representations for entities, referred to as *entity descriptions*. Once created, these entity descriptions can be ranked using traditional document-based retrieval models. Chapter 4 presents semantically informed retrieval models that utilize specific characteristics of entities (attributes, types, and relationships) for retrieval. Some of these methods assume a semantically enriched *keyword++ query*. The entity ranking methods discussed in this part lay the foundations for all the various approaches discussed in the rest of the book. As we will see, entity ranking turns out to be an indispensable tool to address many sub-tasks in the systems that we will subsequently discuss.

Chapter 3
Term-Based Models for Entity Ranking

We have established in Chap. 1 that entities are natural and meaningful units of retrieval. To recap, according to our working definition, an entity is a uniquely identifiable "thing," a typed object, with name(s), attributes, and relationships to other entities. Examples of some of the most frequent types of entities include people, locations, organizations, products, and events. Returning specific entities, instead of a mere list of documents, can provide better answers to a broad range of information needs. The way these information needs are expressed may vary from short keyword queries to full-fledged natural language questions (cf. Sect. 1.3.1.1). In this chapter, we adhere to the "single search box" paradigm, which accepts "free text" search queries, and simply treat queries as sequences of words, referred to hereinafter as *terms*. The task we are going to focus on is *ad hoc entity retrieval*: answering a one-off free text query, representing the user's underlying information need, with a ranked list of entities. The fundamental question we are concerned with then is: How do we perform relevance matching of queries against entities?

In Chap. 2, we have introduced large-scale knowledge repositories, like Wikipedia and DBpedia, that are devoted to organizing information around entities in a (semi-)structured format. Having information accumulated about entities in a knowledge repository is indeed helpful, yet it is not a prerequisite. As we shall see later in this chapter, it is possible to rank entities without (pre-existing) direct representations, as long as they can be recognized and identified uniquely in documents. The main idea of this chapter can be summarized as follows: If textual representations can be constructed for entities, then the ranking of these representations ("entity descriptions") becomes straightforward by building on traditional document retrieval techniques (such as language models or BM25). Accordingly, the bulk of our efforts in this chapter, described in Sect. 3.2, will revolve around assembling term-based entity representations from various sources, ranging from unstructured to structured. In the simplest case, these entity representations are based on the bag-of-words model, which considers the frequency of words but disregards their order. We will also introduce an extension to multiple document fields, as a mechanism for preserving some of the structure associated with entities.

© The Author(s) 2018   57
K. Balog, *Entity-Oriented Search*, The Information Retrieval Series 39,
https://doi.org/10.1007/978-3-319-93935-3_3

	Query
Table 3.1 Example entity search queries taken from various benchmarking campaigns [4]	martin luther king
	disney orlando
	Apollo astronauts who walked on the Moon
	Winners of the ACM Athena award
	EU countries
	Hybrid cars sold in Europe
	birds cannot fly
	Who developed Skype?
	Which films starring Clint Eastwood did he direct himself?

Finally, we will detail how to preserve term position information, which allows for employing more sophisticated proximity-aware ranking models.

Section 3.3 presents baseline retrieval models for ranking term-based representations. Readers familiar with traditional document retrieval techniques may choose to skip this section. In Sect. 3.4, we take a small detour and consider a different retrieval approach in which entities are not modeled directly. Evaluation measures and test collections are discussed in Sect. 3.5.

3.1   The Ad Hoc Entity Retrieval Task

Entity retrieval is the task of answering queries with a ranked list of entities. Ad hoc refers to the standard form of retrieval in which the user, motivated by an ad hoc information need, initiates the search process by formulating and issuing a query.[1] The search query might come in a variety of flavors, ranging from a few keywords to proper natural language questions; see Table 3.1 for examples. Adhering to the "single search box" paradigm, we do not make any assumptions about the specific query format and accept any written or spoken text that is entered in a search box. This free text input is then treated as a sequence of *keywords* (without any query language operators, such as Boolean operators). We further assume that a *catalog* of entities is available, which contains unique identifiers of entities. For matching the query against entities in the catalog, both unstructured and structured information sources may be utilized. The formal definition of the task, then, is as follows.

> **Definition 3.1** Given a keyword query q and an entity catalog $\mathcal{E}$, *ad hoc entity retrieval* is the task of returning a ranked list of entities $\langle e_1, \ldots, e_k \rangle, e_i \in \mathcal{E}$ with respect to each entity's relevance to q. The relevance of entities is inferred based on a collection of unstructured and/or (semi-) structured data.

---

[1] More precisely, ad hoc refers to the nature of the information need, i.e., "a temporary information need that might disappear once the need is satisfied" [78].

How to go about solving this task? We can build on the extensive body of work on document retrieval that has established models and algorithms for matching queries against documents. One way to frame the problem of entity ranking is to imagine that an entity description or "profile" document is to be compiled for each entity in the catalog, which contains all there is to know about the given entity, based on the available data. Once those descriptions are created, entities can be ranked using existing document retrieval algorithms. We detail these two steps, constructing and ranking term-based entity representations, in Sects. 3.2 and 3.3, respectively. Alternatively, entity retrieval may also be approached without requiring explicit representations to be built for each entity. This possibility is discussed in Sect. 3.4.

Before we proceed further, a note of clarification on entity identifiers vs. representations. Strictly speaking, we return a ranked list of entity identifiers and establish relevance based on each entity's representation. Since the identifier of an entity is uniquely associated with its representation, for convenience, we disregard the distinction in our notation and simply write e for an entity *or* its identifier.

3.2 Constructing Term-Based Entity Representations

The first step in addressing the problem of ad hoc entity retrieval is to create a "profile" document for each entity in our catalog, which includes all that we know about that entity. This will serve as the textual representation of the given entity, referred to as *entity description* for short. One can assemble entity descriptions by considering the textual contexts, from a large document collection, in which the entities occur. For popular entities, such descriptions may already be readily available (think of the Wikipedia page of an entity). Also, there is a large amount of information about entities organized and stored in knowledge bases in RDF format. Reflecting the historical development of the field, we will proceed and consider the underlying data sources in exactly this order: unstructured document corpora (Sect. 3.2.1), semi-structured documents (Sect. 3.2.2), and structured knowledge bases (Sect. 3.2.3).

For all types of data sources, our main focus will be on estimating *term count*, denoted as $c(t; e)$, which is the number of times term t appears in the description constructed for entity e. With that at hand, we can define standard components of (bag-of-words) retrieval models analogously to how they are defined for documents:

- *Entity length* (l_e) is the total number of terms in the entity description:

$$l_e = \sum_{t \in \mathcal{V}} c(t; e) .$$

Table 3.2 Notation used in this chapter

Symbol	Meaning
$c(x; y)$	Count (raw frequency) of x in y
$c(x, y; z)$	Number of times x and y co-occur in z
$c_o(x, y; z)$	Number of times x and y co-occur in z in that exact order
$c_w(x, y; z)$	Number of times x and y co-occur in z in an unordered window of size w
d	Document ($d \in \mathcal{D}$)
$\mathcal{D}$	Document collection
e	Entity ($e \in \mathcal{E}$)
$\mathcal{E}$	Entity catalog (set of all entities)
f	Field ($f \in \mathcal{F}$)
$\mathcal{F}$	Set of fields
f_e	Field f of entity e
$\mathcal{K}$	Knowledge base
l_x	Length of x ($l_x = \sum_{t \in \mathcal{V}} c(t; x)$)
$\bar{l}_x$	Mean representation length of x
q	Query (bag of terms $t \in q \implies c(t; q) > 0$ or sequence of terms $q = \langle q_1, \ldots, q_n \rangle$)
t	Term ($t \in \mathcal{V}$)
$\mathcal{V}$	Vocabulary of terms
$w(e, d)$	Weight of association between entity e and document d
x_t	Sequence of terms in x ($x_t = \langle t_1, \ldots, t_{l_x} \rangle$)

- *Term frequency* (TF) is computed as the normalized term count in the entity description:

$$TF(t, e) = \frac{c(t; e)}{l_e}, \tag{3.1}$$

  but other TF weighting variants (e.g., log normalization) may also be used.
- *Entity frequency* (EF) is the number of entities in which a term occurs:

$$EF(t) = |\{e \in \mathcal{E} : c(t; e) > 0\}| .$$

- *Inverse entity frequency* (IEF) is defined as:

$$IEF(t) = \log \frac{|\mathcal{E}|}{EF(t)}, \tag{3.2}$$

  where $|\mathcal{E}|$ is the total number of entities in the catalog.

We shall also discuss how the sequential ordering of terms may be preserved in entity descriptions, which is needed for proximity-based retrieval models.

Table 3.2 summarizes the notation that will be used throughout the chapter.

```
<head>
  <title>Formula One Surprise: Schumacher Is Second</title>
  ...
  <classifier class="indexing_service" type="descriptor">
             Automobile Racing</classifier>
  <classifier class="indexing_service" type="descriptor">
             Belgian Grand Prix</classifier>
  <classifier class="online_producer" type="taxonomic_classifier">
             Top/News/Sports</classifier>
  ...
  <person class="indexing_service">Raikkonen, Kimi</person>
  <person class="indexing_service">Schumacher, Michael</person>
  ...
</head>
<body>
  ...
  <p>One of the most predictable Formula One seasons yielded one of the most
unpredictable of races yesterday in the Belgium Grand Prix at Spa-Francorchamps,
  where, for the first time this year, Michael Schumacher completed a race in a
position other than first.</p>
  <p>His second-place finish to Kimi Raikkonen, though, was all he needed to
win his seventh driver's title.</p>
  <p>"Of course, I would have preferred to have taken the title with a win,
but it was not possible," he said. "The better man won today, but I am quite
happy with what I have achieved."</p>
  ...
</body>
```

Listing 3.1 Excerpt from an article (ID: 1607874) from The New York Times Annotated Corpus [67]

3.2.1   Representations from Unstructured Document Corpora

Let us consider a setting where ready-made entity descriptions are unavailable. This is typical of a scenario when one wants to find entities in an arbitrary document collection, like a web crawl or an enterprise's intranet. How can entity descriptions be constructed in this case? This question has been extensively studied in the context of *expert finding* [3],[2] and the approach we shall present below is based on the concept of *candidate models* [2].

The main idea is to have documents annotated with the entities that are referenced in them, and then use documents as a proxy to connect terms and entities. Annotations may be performed on the document level, like tags (see Listing 3.1), or on the level of individual entity occurrences, referred to as entity *mentions* (see Fig. 3.1). We discuss each of those cases in turn. For now, we take it for granted that entity annotations, either on the document or on the mention level, are available to us. These may be provided by human editors or by a fully automated (entity linking) process.[3]

---

[2]In expert finding (a.k.a. *expert search*), the task is to automatically identify, based on a document collection, people who are experts on a given query topic; see Sect. 3.5.2.1.

[3]We will come back to this issue in Chap. 5, which is devoted entirely to the problem of annotating text with entities.

The 2004 Belgian Grand Prix (formally the LXI Belgian Grand Prix) was a **Formula_One** motor race held on 29 August 2004, at the **Circuit_de_Spa-Francorchamps**, near the town of **Spa, Belgium**. It was the 14th race of the **2004_Formula_One_Season**. The race was contested over 44 laps and was won by **Kimi_Raikkonen**, taking his and **McLaren**'s only race win of the season from tenth place on the grid. Second place for **Michael_Schumacher** won him his seventh world championship, after beating third-placed **Rubens_Barrichello**.

Fig. 3.1 Illustration of mention-level entity annotations; terms representing entities are in bold-face. The text is from Wikipedia: https://en.wikipedia.org/wiki/2004_Belgian_Grand_Prix

Using documents as a bridge between terms and entities, term counts can be computed using the following general formula:

$$\tilde{c}(t;e) = \sum_{d \in \mathcal{D}} c(t,e;d)\, w(e,d)\,. \tag{3.3}$$

Here, we use tilde to emphasize that $\tilde{c}(t;e)$ values are pseudo-counts; in later parts of the chapter we will not necessarily make this distinction. The method underlying Eq. (3.3) bases the term pseudo-count $\tilde{c}(t;e)$ on the number of co-occurrences between a term and an entity in a particular document, $c(t,e;d)$, weighted by the strength of the *association* between the entity and the document, $w(e,d)$. These two components depend on the granularity of entity annotations (i.e., whether document-level or mention-level), as we shall detail in the next two subsections.

3.2.1.1 Document-Level Annotations

Listing 3.1 shows an excerpt from an article from The New York Times Annotated Corpus. As part of The New York Times' indexing procedures, the article has been manually tagged by a staff of library scientists with persons, places, organizations, titles, and topics, using a controlled vocabulary [67]. This particular example serves as an illustration of the case where we know which entities are mentioned in the document, but we do not have information about the positions or frequencies of those mentions.

> The process of creating an entity's description may be viewed as simply concatenating the contents of all documents that are tagged with the given entity.

Formally, let $e \in d$ denote the fact that e is mentioned in d. For convenience, we introduce the shorthand notation $\mathcal{D}_e$ for the set of documents that are annotated with entity e: $\mathcal{D}_e = \{d \in \mathcal{D} : e \in d\}$. The components of Eq. (3.3) are then estimated as follows.

Term-document-entity co-occurrences are equal to the term count in the document if it mentions the entity, and are zero otherwise:

$$c(t, e; d) = \begin{cases} c(t; d), & e \in d \\ 0, & e \notin d . \end{cases}$$

Document-entity weights are taken to be binary:

$$w(e, d) = \begin{cases} 1, & e \in d \\ 0, & e \notin d . \end{cases} \tag{3.4}$$

This naïve approach (also known as the *Boolean model of associations* [5]) is based on rather strong simplifying assumptions. It does not matter (1) where and how many times the entity e is mentioned in the document, and (2) what other entities occur (and how many times) in the same document. However, given the limited information we have (and especially that we do not have the frequencies of entity mentions), this choice is reasonable. Experimental results have shown that this simple method is robust and surprisingly effective [2].

Putting our choices together, term pseudo-counts can be computed according to:

$$\tilde{c}(t; e) = \sum_{d \in \mathcal{D}_e} c(t; d) .$$

While pseudo-counts suffice for bag-of-words retrieval models, for proximity-aware retrieval models, term position information needs to be preserved. An easy way of implementing this is by constructing entity descriptions indeed as a concatenation of all documents that mention a given entity. Although the ordering of documents is unimportant, it is vital to keep track of document boundaries (e.g., by inserting NIL-terms between each pair of subsequent documents).

3.2.1.2 Mention-Level Annotations

It is not unreasonable to assume that all entity mentions within documents are annotated (with unique identifiers from the entity catalog). Imagine that all entity mentions are replaced with the corresponding entity ID tokens; see Fig. 3.1 for an example. It is important to emphasize that here we assume that *all* entity occurrences are annotated in documents. With manually annotated documents, that is often not the case; e.g., Wikipedia guidelines dictate that only the first appearance of each entity is to be linked in an article. How can we estimate the term pseudo-counts in Eq. (3.3) then?

A general principle for creating semantic representations of words was summarized by the British linguist J.R. Firth in his often-quoted statement: "You shall know a word by the company it keeps" [31]. The same principle can also be applied to entities: We shall represent an entity by the terms in its company, i.e., terms

---

... for his ninth victory of 2001. **Michael_Schumacher**, who had clinched his fourth ...
... On lap 14, Takuma Sato hit **Michael_Schumacher**'s car from behind, causing both to ...
                              Michael_Schumacher won the season-opening Formula ...
... which paved the way for the **Michael_Schumacher** era of Ferrari dominance ...
... started 1-2 ahead of Ferrari's **Michael_Schumacher**.

Fig. 3.2 Example contexts in which a given entity is mentioned; mentions of the entity are replaced with the entity ID and are typeset in boldface. These contexts collectively represent the entity

within a window of text around the entity's mention. Figure 3.2 illustrates this idea by showing a set of snippets mentioning a certain entity; the combination of all such contexts across the document collection produces an accurate representation of the entity. This idea dates back to Conrad and Utt [24], who extracted and combined all paragraphs in a corpus, mentioning a given entity, to represent that entity. Ten years later, Raghavan et al. [62] provided a more formal framework to construct bag-of-words representations of entities by considering fixed sized word windows surrounding entity mentions.

To formalize this idea we introduce the following notation. Let $d_t = \langle t_1, \ldots, t_i, \ldots, t_{l_d} \rangle$ be a document, where 1, i, and l_d are absolute positions of terms within d, and l_d is the length of the document. Let $\mathbb{1}_d(i,t)$ be an indicator function that returns the value 1 if the term at position i in d is t, and 0 otherwise. We further assume that entity occurrences have been replaced with unique identifiers that behave as regular terms (i.e., are not broken up into multiple tokens). Accordingly, $\mathbb{1}_d(i,e)$ will yield the value 1 if entity e appears at position i in d, and 0 otherwise.

Term-Document-Entity Co-occurrences We can use positional information, i.e., the distance between terms and entity mentions, to weigh the strength of co-occurrence between t and e. There are a number of different ways to accomplish this; we will detail two specific approaches.

Fixed-Sized Windows For any particular window size w, we set $c(t,e;d)$ to the number of times term t and entity e co-occur at a distance of at most w in document d:

$$c(t,e;d) = \sum_{i=1}^{l_d} \mathbb{1}_d(i,t) \sum_{\substack{j=1 \\ |i-j| \le w}}^{l_d} \mathbb{1}_d(j,e) . \tag{3.5}$$

Prior work has used window sizes ranging from 20 to 250 [2].

Proximity Kernels The previous approach can be taken one step further by capturing the intuition that terms closer to the mention of an entity should be given more importance than terms appearing farther away. Petkova and Croft [60] formalize this intuition using proximity kernels. A proximity kernel is a function that is used for distributing the "count" of a term that appears at position i across other positions.

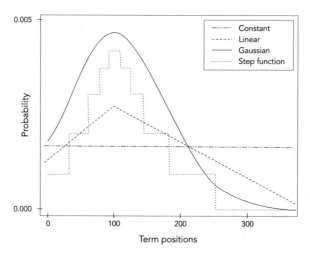

Fig. 3.3 Illustration of proximity kernels. Image is based on [60]

As Lv and Zhai [46] explain, "we would like to pretend that the same word has also occurred at all other positions with a discounted count such that if the position is closer to i, the propagated count for word w at that position would be larger than the propagated count at a position farther away, even though both propagated counts would be less than one." Figure 3.3 shows various proximity kernels with $i = 100$. The proximity-aware term-entity co-occurrence in a given document is defined as:

$$c(t,e;d) = \frac{1}{Z} \sum_{i=1}^{l_d} \mathbb{1}_d(i,t) \sum_{j=1}^{l_d} \mathbb{1}_d(j,e)\, k(i,j)\,,$$

where $k(i,j)$ is a proximity-kernel function and Z serves as a normalizing constant, which may be omitted. Note that these counts will no longer be integers, but real numbers.

The simplest kernel is the *constant function*, which ignores all position information and thus corresponds to the bag-of-words representation: $k(i,j) = 1/l_d$. The *passage kernel* corresponds to the fixed-sized window approach in Eq. (3.5):

$$k(i,j) = \begin{cases} 1, & |i - j| \leq w \\ 0, & \text{otherwise}\,. \end{cases}$$

Any non-uniform, non-increasing function can be considered as a proximity-based kernel. Following [46, 60], we present two such kernel functions:

- *Gaussian kernel*:
$$k(i,j) = \exp\left(\frac{-(i - j)^2}{2\sigma^2}\right)\,.$$

- *Triangle kernel*:

$$k(i,j) = \begin{cases} 1 - \frac{|i-j|}{\sigma}, & |i - j| \leq \sigma \\ 0, & \text{otherwise} . \end{cases}$$

Both these kernels involve a single parameter σ, which controls the shape of the kernel curves, i.e., how quickly the propagation of terms tails off. Petkova and Croft [60] compared three non-uniform kernels (Gaussian kernel, triangle kernel, and step function) and found them to deliver very similar empirical performance, with no significant differences among them.

Document-Entity Weights The default option for estimating the strength of the association between a document and an entity is to repeat what we have done before for document-level annotations, in Eq. (3.4), by setting $w(e,d)$ to either 0 or 1. This time, however, we have more information to base our estimate on. We want $w(e,d)$ to reflect not merely the presence of an association between an entity and a document, but also the strength of that association. This specific task has been studied in [5], in the context of expert finding, where the authors proposed to use well-established term weighting schemes from IR, like TF-IDF. The rationale behind TF-IDF is that a term that is mentioned frequently in a document is important to that document (TF), whereas it becomes less important when it occurs in many documents (IDF). This sounds a lot like what we would need, only that we need it for entities, not for terms.

Imagine, for a moment, that all regular (non-entity) word tokens have been filtered out from all documents in our collection; documents now are comprised of entities only. Those entities could then be weighted as if they were terms. This is exactly how we will go about implementing "entity weighting." Since TF-IDF is based on a bag-of-words representation, we do not need to actually remove regular terms. The only information we need is the frequency of mentions of an entity in a document, $c(e; d)$. Using the notation from earlier, this entity frequency can be computed as:

$$c(e; d) = \sum_{i=1}^{l_d} \mathbb{1}_d(i, e) .$$

We can then define the document-entity weight as follows:

$$w(e,d) = \frac{c(e; d)}{\sum_{e' \in d} c(e'; d)} \log \frac{|\mathcal{D}|}{|\mathcal{D}_e|} , \tag{3.6}$$

where $\mathcal{D}_e$ is the set of documents that are annotated with entity e, and $|\mathcal{D}|$ denotes the total number of documents in the collection.

Fig. 3.4 Web page of the movie THE MATRIX from IMDb (http://www.imdb.com/title/tt0133093/)

The same objective may also be accomplished using language models, where smoothing with the background model plays the role of IDF. The interested reader is referred to Balog and de Rijke [5] for details.

3.2.2 Representations from Semi-structured Documents

A significant portion of web content is already organized around entities. There exist many sites with a large collection of web pages dedicated to describing or profiling entities—think of the Wikipedia page of MICHAEL SCHUMACHER from Chap. 2, for instance (cf. Fig. 2.2). Another example is shown in Fig. 3.4, this time a web page of a movie from the Internet Movie Database (IMDb). Documents that represent entities are rarely just flat text. Wikipedia pages are made up of elements such as title, introductory text, infobox, table of contents, body content, etc. (cf. Fig. 2.2). The IMDb page in Fig. 3.4 has dedicated slots for movie name, synopsis, cast, ratings, etc. In fact, all web pages are annotated with structural markup using HTML.

The standard way of incorporating the internal document structure into the retrieval model is through the usage of *document fields*. Fields correspond to specific parts or segments of the document, such as title, introductory text, infobox, etc. in the case of Wikipedia, or movie name, synopsis, directors, etc., in the case of IMDb.

Table 3.3 Fielded entity description created for THE MATRIX, corresponding to Fig. 3.4

Field	Content
Name	The Matrix
Genre	Action, Sci-Fi
Synopsis	A computer hacker learns from mysterious rebels about the true nature of his reality and his role in the war against its controllers
Directors	Lana Wachowski (as The Wachowski Brothers), Lilly Wachowski (as The Wachowski Brothers)
Writers	Lilly Wachowski (as The Wachowski Brothers), Lana Wachowski (as The Wachowski Brothers)
Stars	Keanu Reeves, Laurence Fishburne, Carrie-Anne Moss
Catch-all	The Matrix Action, Sci-Fi A computer hacker learns from mysterious rebels about the true nature of his reality and his role in the war against its controllers. Lana Wachowski (as The Wachowski Brothers), Lilly Wachowski (as The Wachowski Brothers) Lilly Wachowski (as The Wachowski Brothers), Lana Wachowski (as The Wachowski Brothers) Keanu Reeves, Laurence Fishburne, Carrie-Anne Moss

Fields can be defined rather flexibly; certain parts of the document may be considered for multiple fields or not assigned to any field at all. Furthermore, not all fields can necessarily be found in all documents (e.g., a relatively small portion of movie pages have an "awards" section; for most movies this would be empty, and therefore, is not even shown). This is why we consider these representations semi-structured, and not structured; we do not require them to follow a rigid, formal schema.

Let $\mathcal{F}$ denote the set of possible fields and $f \in \mathcal{F}$ denote a specific field. We write f_e to refer to field f in the description of entity e. The contents of these fields are typically extracted using *wrappers* (also called *template-based extractors*), which are designed specifically to deal with a particular type of document (e.g., Wikipedia or IMDb pages) [41]. (To reduce the effort and maintenance cost associated with developing wrappers manually, automated wrapper induction has been an active area of research, see, e.g., [28].) Assuming that the contents of fields have been obtained, we can define term statistics as before, but this time all computations take place on the field-level, as opposed to the document-level. Specifically, we let $c(t; f_e)$ be the number of times term t appears in field f of the description of e, and l_{f_e} is the total number of terms in that field.

Table 3.3 shows the fielded entity description corresponding to Fig. 3.4. Notice that in addition to the regular fields, a special "catch-all" field has also been introduced, which amasses the contents of all fields (or sometimes of the entire document).[4] The rationale behind having a catch-all field is twofold. First, it can be used to narrow the search space by filtering entities that are potentially good candidates for a given query, and consider only those for scoring (thereby improves

---

[4]Fields provide a non-exclusive and non-complete partitioning of the document's content, therefore the union of all fields does not necessarily equal the whole document.

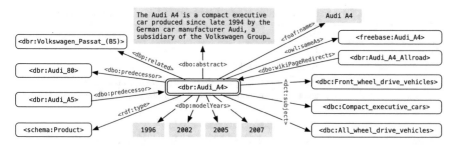

Fig. 3.5 Excerpt from an RDF graph. Rounded rectangles represent entities (URIs) and shaded rectangles denote attributes (literal values)

efficiency). Second, fielded entity descriptions are often sparse. When computing the relevance between an entity and a query, a more robust estimate can be obtained by combining the field-level matching scores with an entity-level matching score, enabled by the catch-all field (hence it improves effectiveness).

While our focus in this section has been on a setting where documents have a one-to-one correspondence with entities, fielded representations may also be realized when entity descriptions are created by aggregating information from multiple documents (cf. Sect. 3.2.1). The only difference is that the term pseudo-counts need to be computed on the field-level as opposed to the document-level. (That is, a given field in the description of an entity, f_e, is constructed by concatenating the contents of the respective field, f_d, from all documents in the collection that mention e.)

3.2.3   Representations from Structured Knowledge Bases

Over the past few years, an increasing amount of structured data has been published (and interlinked) on the Web. We have introduced some large general-purpose knowledge bases, including DBpedia and Freebase, in Sect. 2.3. These organize information around entities using the RDF data model (cf. Sect. 2.3.1.2). To recap, each entity is uniquely identified by its URI (Uniform Resource Identifier) and its properties are described in the form of subject-predicate-object (SPO) triples. Conceptually, a set of SPO triples forms a directed graph, where each edge is labelled with a URI (the predicate) and each node is either a URI or a literal. Our focus, here, is on the immediate vicinity of the entity node: nodes that are directly connected to it (irrespective of the direction of the edge), as shown in Fig. 3.5. Or, in terms of SPO triples, we consider all those triples where the entity appears either as subject or object. How can we turn this into a term-based entity representation?

Let us note, before we continue with answering the above question, that while our focus will be on RDF data, from the perspective of constructing term-based entity representations, this is conceptually no different from having data stored in relational databases, where the same information is available through fields and foreign-key relationships.

Table 3.4 Excerpt from the fielded entity description of AUDI A4, corresponding to Fig. 3.5

Field	Content
Name	Audi A4
Name variants	Audi A4 ... Audi A4 Allroad
Attributes	The Audi A4 is a compact executive car produced since late 1994 by the German car manufacturer Audi, a subsidiary of the Volkswagen Group [...] ... 1996 ... 2002 ... 2005 ... 2007
Types	Product ... Front wheel drive vehicles ... Compact executive cars ... All wheel drive vehicles
Outgoing relations	Volkswagen Passat (B5) ... Audi 80
Incoming relations	Audi A5
`<foaf:name>`	Audi A4
`<dbo:abstract>`	The Audi A4 is a compact executive car produced since late 1994 by the German car manufacturer Audi, a subsidiary of the Volkswagen Group [...]
Catch-all	Audi A4 ... Audi A4 ... Audi A4 Allroad ... The Audi A4 is a compact executive car produced since late 1994 by the German car manufacturer Audi, a subsidiary of the Volkswagen Group [...] ... 1996 ... 2002 ... 2005 ... 2007 ... Product ... Front wheel drive vehicles ... Compact executive cars ... All wheel drive vehicles ... Volkswagen Passat (B5) ... Audi 80 ... Audi A5

Values coming from different SPO triples are separated by "...". For the sake of illustration, we consider `<foaf:name>` and `<dbo:abstract>` as "top predicates," which are also included on their own. The other fields are the result of predicate folding

The goal is to obtain entity descriptions by assembling text from all SPO triples that are about a given entity. In the previous section, we have discussed how to represent entities using a fielded structure. Can the same technique be used here? It would seem logical, as predicates naturally correspond to fields. There is one important difference though. Earlier we had only a handful of fields; even in the case of fairly detailed entity pages, the number of entity properties (i.e., fields) would be in the 10s. On the other hand, the number of distinct predicates in a knowledge base is typically in the 1000s. Yet, most entities have only a small number of predicates associated with them. The huge number of possible fields, coupled with the sparsity of the data, makes the estimation of field weights computationally prohibitive.

A commonly used workaround is *predicate folding*: grouping predicates together into a small set of predefined categories. This way, we are back to having a handful of fields in the entity descriptions, and the estimation of field weights may be carried out without problems. Predicates may be grouped based, among other alternatives, on their type [54, 59] or on their (manually determined) importance [12]. Table 3.4 shows the fielded entity description created for our example entity. In what follows, we shall discuss the main steps involved with obtaining such representations from RDF data.

3.2.3.1 Predicate Folding

Deciding which predicates to fold is a hard and still open problem. Two main predicate mapping schemes may be distinguished in past work: based on *predicate importance* or based on *predicate type*. Blanco et al. [12] employ the first scheme and differentiate between three types of fields according to weight levels: important, neutral, and unimportant. Important and unimportant predicates are selected manually, with the remainder being neutral. Grouping based on predicate type has been the more dominant approach. The following fields have been identified (with interpretations slightly varying across different studies). Unless stated otherwise, the entity appears as the subject of the SPO triple and the object is the field value (with URIs resolved, which is explained in Sect. 3.2.3.2). In our examples, we will use the namespaces from DBpedia (cf. Table 2.4).

- *Name* contains the name(s) of the entity. The two main predicates mapped to this field are `<foaf:name>` and `<rdfs:label>`. One might follow a simple heuristic and additionally consider all predicates ending with "name," "label," or "title" [54].
- *Name variants* (aliases) may be aggregated in a separate field. In DBpedia, such variants may be collected via Wikipedia redirects (via `<dbo:wikiPageRedi-rects>`) and disambiguations (using `<dbo:wikiPageDisambiguates>`). Note that in both these cases some simple inference is involved.[5] This field may also contain anchor texts of links to the entity (via `<dbo:wikiPageWikiLinkText>` in DBpedia) or names of the entity in other knowledge bases (collected via `<owl:sameAs>` relations).
- *Attributes* includes all objects with literal values, except the ones already included in the *name* field. In some cases, the name of the predicate may also be included along with the value, e.g., "founding date 1964" [81] (vs. just the value part, "1964").
- *Types* holds all types (categories, classes, etc.) to which the entity is assigned; commonly, `<rdf:type>` is used for types. In DBpedia, `<dct:subject>` is used for assigning Wikipedia categories, which may also be considered as entity types. In Freebase, the "profession" attribute for people or the "industry" attribute for companies have similar semantics.
- *Outgoing relations* contains all URI objects, i.e., names of entities (or resources in general) that the subject entity links to; if the *types* or *name variants* fields are used, then those predicates are excluded. As with attributes, values might be prefixed with the predicate name, e.g., "spouse Michelle Obama" [81].
- *Incoming relations* is made up of subject URIs from all SPO triples where the entity appears as object.

---

[5]In DBpedia, the triple (e', `<dbo:wikiPageRedirects>`, e) records the fact that entity e' redirects to entity e. The name of e' is contained in a triple (e', `<foaf:name>`, e'_{name}) (or using the `<rdfs:label>` predicate). Putting these two together, e'_{name} will be a name variant of e. Similarly, name variants can also be inferred from disambiguations.

Table 3.5 Entity fields used in prior work

Field	[55]	[54]	[80]	[59]	[81]	[37]
Name	✓	✓	✓	✓	✓	✓
Name variants					✓	
Attributes		✓	✓		✓	
Types				✓	✓	✓
Outgoing relations		✓	✓	✓	✓	
Incoming relations		✓		✓		
Top predicates						✓
Catch-all	✓			✓		✓

- *Top predicates* may be considered as individual fields, e.g., Hasibi et al. [37] include the top-1000 most frequent DBpedia predicates as fields.
- *Catch-all* (or *content*) is a field that amasses all textual content related to the entity.

The example in Table 3.4 includes all these fields.

In Table 3.5, we summarize various field configurations that have been used in a selection of prior work. Notably, using as few as two fields, *name* and *content*, it is possible to achieve solid performance [55]. This setting, often referred to as "title+content," is also regarded as a common baseline [37, 81]. The importance of other fields can vary greatly depending on the type of queries [81].

3.2.3.2 From Triples to Text

Predicate folding governs how SPO triples in a knowledge base, $(s, p, o) \in \mathcal{K}$, should be mapped to entity description fields. It remains to be specified how to turn these triples into a text-based field representation. There are two triple patterns that contain information directly related to an entity e: (1) *outgoing relations*, where the entity stands as the subject of the triple, i.e., (e, p, o), and (2) *incoming relations*, where the entity stands as the object of the triple, i.e., (s, p, e). Let $\mathcal{M}_{out}(p)$ and $\mathcal{M}_{in}(p)$ be the mapping functions for outgoing and incoming relations labeled with predicate p; these return a set of target fields to which p is mapped (and they return $\emptyset$ if the predicate is not mapped to any field).

Using a bag-of-words representation, the frequency of a term in a given entity field is calculated by considering both the object values of outgoing relations and the subject values of incoming relations:

$$c(t; f_e) = \sum_{\substack{(e, p, o) \in \mathcal{K} \\ f \in \mathcal{M}_{out}(p)}} c(t; o) + \sum_{\substack{(s, p, e) \in \mathcal{K} \\ f \in \mathcal{M}_{in}(p)}} c(t; s) \,,$$

where $c(t; o)$ and $c(t; s)$ are the raw frequencies of term t in the object and subject, respectively.

To preserve term position information, essentially the subject/object values need to be concatenated across the set of triples that contain the entity. The sequence of terms within a given entity field may be written as:

$$f_{e_t} = \bigcup_{\substack{(e,p,o)\in\mathcal{K} \\ f\in\mathcal{M}_{out}(p)}} o_t \bigcup_{\substack{(s,p,e)\in\mathcal{K} \\ f\in\mathcal{M}_{in}(p)}} s_t \,,$$

where $\bigcup$ is the string concatenation operator.[6] The sequence of terms in the object and subject elements of SPO triples are denoted as o_t and s_t, respectively. Note that triples are a set, thus the order in which they are concatenated is of no importance. (That means that the order that is preserved here refers to the sequence of terms within each SPO element.)

Resolving URIs Recall that object values are either URIs or literals. Literals can be treated as regular text, so they do not need any further processing. URIs, however, are not suitable for text-based search. Consider for instance the full URI of our AUDI A4 example, which is http://dbpedia.org/resource/Audi_A4. Depending on the pre-processing and analysis steps applied during indexing, it may be tokenized as a single term, which will not match any query terms (other than that exact string). Or, it might be tokenized, e.g., as `http dbpedia org resource Audi A4`, which does contain the terms "Audi" and "A4," but also a lot of noise in addition. Mind that not all entity URIs are as friendly as the ones in DBpedia. The same entity in Freebase has the URI http://rdf.freebase.com/ns/m.030qmx, which is unfit for text-based search, no matter what pre-processing is applied. Since the names of entities (or of resources, in general) are contained in the knowledge base, we can replace URIs with the name of the entity (or resource) that they point to. This is what we refer to as *URI resolution*.

The specific predicate that holds the name of a resource depends on the RDF vocabulary used. Commonly,[7] `<foaf:name>` or `<rdfs:label>` are used. Let us take the following SPO triple as an example, which specifies the type of entity:

```
<dbr:Audi_A4> <rdf:type> <dbo:MeanOfTransportation>
```

The corresponding resource's name is contained in the object element of this triple:

```
<dbo:MeanOfTransportation> <rdfs:label> "mean of transportation"
```

Basically, we replace the boldfaced part of the first triple with the boldfaced part of the second triple, and we do this each time the object of a triple is a URI. (Additionally, we might also want to resolve certain subject URIs.) We note that there may be situations where the URI cannot be resolved, e.g., if it points to a

---

[6]This operator is assumed to insert a number of NIL-terms between subsequent strings, in order to delineate values originating from different triples.

[7]That is, using the FOAF or RDFS vocabularies.

resource outside the knowledge base. In that case, we need to fall back to some default URI tokenization scheme (e.g., remove `protocol://` prefixes and split on URI delimiters).

Let us take note of some of the prime limitations of term-based representations that are manifested here. By replacing unique entity identifiers with regular terms, much of the power of this semantically rich knowledge gets lost. For example, we can use only a single name for an entity, even though it might have multiple name variants or aliases. And that is just the name. There is a whole lot more information associated with an entity identifier. We will discuss models in Chap. 4 that preserve the semantics by going beyond a term-based representation.

3.2.3.3 Multiple Knowledge Bases

So far, we have considered a single knowledge base. The Web of Data, as we have explained in Sect. 2.3.6, enables different knowledge bases to be connected via the `<owl:sameAs>` predicate, which defines an equivalence relation between a pair of entities (resources). For a given entity e, we let E be the set of entity URIs such that each URI in E is connected to at least one other URI in E via a same-as link ($\forall e_i \in E : \exists e_j (e_i, \text{sameAs}, e_j) \lor (e_j, \text{sameAs}, e_i)$). During the construction of the fielded entity description of e, instead of only considering triples where e appears as subject or object, we now need to consider all triples where any of the URIs in E appears as subject or object.

3.3 Ranking Term-Based Entity Representations

Now that we have constructed term-based representations of entities, we turn to the task of ranking them with respect to their relevance to a search query q. This can be viewed as the problem of assigning a score to each entity in the entity catalog: $score(e; q)$. Entities are then sorted in descending order of their scores. We note that in practice we are typically interested only in returning the top-k results; for that, the calculation of scores (a procedure also known as *query processing*) can be optimized to significantly decrease response time. For example, entities that do not match any of the query terms can safely be ignored. Further performance optimization can be achieved via specialized indexing data structures, see, e.g., [18, 19, 22]. The description of these techniques is beyond the scope of this book.

Our focus in this section is on different ranking models and algorithms for scoring entities. Retrieval models are formal representations of the process of matching queries and entities. Since we are ranking term-based entity representations—or, as we call them, entity descriptions—the models we employ are well-known from document retrieval. While we will be referring to entities instead of documents, the only change in technical terms is replacing d with e in the equations. Readers already familiar with standard retrieval models may wish to skip this section. The reason we briefly discuss these models, apart from our general aim of being self-contained, is that we will present various extensions to them in Chap. 4. For a more in-depth treatment, we refer to an IR textbook, such as Manning et al. [49], Croft et al. [26], or Zhai and Massung [78].

3.3.1 Unstructured Retrieval Models

We present three effective and frequently used retrieval models: language models, BM25, and sequential dependence models.

3.3.1.1 Language Models

A probabilistic formulation of the ad hoc entity retrieval task is to estimate the probability of a particular entity e being relevant to the input query q, $P(e|q)$. Instead of estimating this probability directly, we apply Bayes's rule to rewrite it as follows:

$$P(e|q) = \frac{P(q|e)P(e)}{P(q)} \stackrel{\text{rank}}{=} P(q|e)P(e) . \tag{3.7}$$

For simplicity, we take the *entity prior*, $P(e)$, to be uniform,[8] which means that the ranking of entities boils down to the estimation of the *query likelihood*, $P(q|e)$. To avoid numerical underflows, the computation of this probability is performed in the log domain:[9]

$$score_{LM}(e;q) = \log P(q|e) = \sum_{t \in q} c(t;q) \log P(t|\theta_e) . \tag{3.8}$$

where $c(t;q)$ denotes the frequency of term t in query q and $P(t|\theta_e)$ is an entity language model. The model θ_e captures the language usage associated with the entity and represents it as a multinomial probability distribution over the vocabulary of terms. To avoid assigning zero probabilities to terms, the empirical entity language model, $P(t|e)$, is combined with a background (or collection) language model, $P(t|\mathcal{E})$. Using the *Jelinek-Mercer smoothing* method, this becomes the following linear interpolation:

$$P(t|\theta_e) = (1 - \lambda)P(t|e) + \lambda P(t|\mathcal{E}) , \tag{3.9}$$

---

[8]We shall present various entity priors in the next chapter, in Sect. 4.6.

[9]This means that retrieval scores will be negative numbers, where a higher (i.e., closer to zero) number means more relevant.

where the smoothing parameter λ controls the influence of the background model. Both components are taken to be *maximum likelihood* estimates:

$$P(t|e) = \frac{c(t;e)}{l_e} , \quad P(t|\mathcal{E}) = \frac{\sum_{e \in \mathcal{E}} c(t;e)}{\sum_{e \in \mathcal{E}} l_e} ,$$

where $c(t;e)$ is the frequency of term t in the entity's description and l_e is the length of the entity's description (measured in the number of terms).

Notice that according to Eq. (3.9), all entities receive the same amount of smoothing. Intuitively, entities with richer representation (i.e., longer descriptions, which means larger l_e) would require less smoothing. *Dirichlet prior smoothing* (or *Bayesian smoothing*) implements this idea by setting:

$$P(t|\theta_e) = \frac{c(t;e) + \mu P(t|\mathcal{E})}{l_e + \mu} . \tag{3.10}$$

Here, the amount of smoothing applied is controlled by the μ parameter. The choice of the smoothing method and smoothing parameter can have a considerable impact on retrieval performance [77]. Generally, a μ value between 1500 and 2500 is a good default setting.

3.3.1.2 BM25

Another popular and effective model is *BM25*. It is also a probabilistic model but differs considerably from language models. BM25 can be derived from the classical probability ranking principle [65], but the actual retrieval function may be seen as a variant of the vector space model. For its historical development and theoretical underpinnings, the interested reader is referred to Robertson and Zaragoza [63]. We use the following variant:[10]

$$score_{BM25}(e;q) = \sum_{t \in q} c(t;q) \frac{(k_1 + 1)\, c(t;e)}{k_1(1 - b + b\frac{l_e}{\bar{l}_e}) + c(t;e)} IEF(t) . \tag{3.11}$$

where $\bar{l}_e$ is the mean entity representation length across the entity catalog ($\bar{l}_e = \sum_{e \in \mathcal{E}} l_e / |\mathcal{E}|$). Further, k_1 and b are free parameters, where k_1 calibrates the term frequency saturation (typically chosen from the range $[1.2, 2]$) and $b \in [0, 1]$ controls the representation length normalization (commonly set to 0.75). It should be noted that using the common default parameter values can yield suboptimal

---

[10]Several dialects of BM25 have been used in the literature. We present what we believe is the most common variant, which includes an extra $(k_1 + 1)$ component to the numerator of the saturation function. As explained in [63], "this is the same for all terms, and therefore does not affect the ranking produced."

retrieval performance for BM25, especially when entity descriptions are created from a KB [38].

3.3.1.3 Sequential Dependence Models

The previous two models employ a bag-of-words representation of both entities and queries, which means that the order of terms is ignored. The *Markov random field* (MRF) model [52] provides a sound theoretical framework for modeling term dependence, thereby going beyond the bag-of-words assumption. MRF approaches are a family of undirected graphical models, which belong to the more general class of linear feature-based models [53]. A Markov random field is constructed from a graph G, which consists of an entity node e and query nodes q_i. The edges of the graph define the dependence semantics between nodes. The MRF ranking function is defined as the posterior probability of e given the query, parameterized by Λ. It is computed as a linear combination of feature functions over the set of cliques in G:

$$P_\Lambda(e|q) \overset{rank}{=} \sum_{c \in \mathcal{C}_G} \lambda_c f(c) .$$

The *sequential dependence model* (SDM) is one particular instantiation of the MRF model, which strikes a good balance between effectiveness and efficiency. SDM assumes dependence between neighboring query terms. The graphical representation is depicted in Fig. 3.6. Potential functions are defined for two types of cliques: (1) 2-cliques involving a query term and the entity, and (2) cliques containing two contiguous query terms and the entity. For the latter type of cliques, there are two further possibilities: either the query terms occur contiguously in the document (ordered match) or they do not (unordered match). The SDM ranking function is then given by a weighted combination of three feature functions, based on query terms (f_T), exact match of query bigrams (f_O), and unordered match of query bigrams (f_U):

$$score_{SDM}(e;q) = \lambda_T \sum_{i=1}^{n} f_T(q_i;e)$$

$$+ \lambda_O \sum_{i=1}^{n-1} f_O(q_i, q_{i+1};e)$$

$$+ \lambda_U \sum_{i=1}^{n-1} f_U(q_i, q_{i+1};e) .$$

Notice that the query is no longer represented as a bag of terms but as a sequence of terms $q = \langle q_1, \ldots, q_n \rangle$. The feature weights are subject to the constraint $\lambda_T + \lambda_O + \lambda_U = 1$. The recommended default setting is $\lambda_T = 0.85$, $\lambda_O = 0.1$,

Fig. 3.6 Sequential
dependence model (SDM) for
three query terms

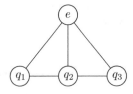

and $\lambda_U = 0.05$. In the presence of training data, the λ weights can be learned using grid search (exhaustive search over the parameter space) or the Coordinate Ascent algorithm (coordinate-level hill climbing); we refer to Metzler and Croft [53] for further details.

Below, we define the three feature functions based on language modeling estimates using Dirichlet prior smoothing.[11]

- *Unigram matches* are estimated using (smoothed) entity language models:

$$f_T(q_i; e) = \log P(q_i | \theta_e),$$

where $P(q_i | \theta_e)$ is given in Eq. (3.10).
- For *ordered bigram matches*, the feature function is defined as:

$$f_O(q_i, q_{i+1}; e) = \log \left(\frac{c_o(q_i, q_{i+1}; e) + \mu P_o(q_i, q_{i+1} | \mathcal{E})}{l_e + \mu} \right),$$

where $c_o(q_i, q_{i+1}; e)$ denotes the number of times the terms q_i, q_{i+1} occur in this exact order in the description of e. The background language model is a maximum likelihood estimate:

$$P_o(q_i, q_{i+1} | \mathcal{E}) = \frac{\sum_{e \in \mathcal{E}} c_o(q_i, q_{i+1}; e)}{\sum_{e \in \mathcal{E}} l_e}.$$

- For *unordered bigram matches* we set:

$$f_U(q_i, q_{i+1}; e) = \log \left(\frac{c_w(q_i, q_{i+1}; e) + \mu P_w(q_i, q_{i+1} | \mathcal{E})}{l_e + \mu} \right),$$

where $c_w(q_i, q_{i+1}; e)$ counts the co-occurrence of terms q_i and q_{i+1} in e, within an unordered window of w term positions. Typically, a window size of 8 is used, which corresponds roughly to sentence-level proximity [52]. The background language model is estimated as:

$$P_w(q_i, q_{i+1} | \mathcal{E}) = \frac{\sum_{e \in \mathcal{E}} c_w(q_i, q_{i+1}; e)}{\sum_{e \in \mathcal{E}} l_e}.$$

---

[11]Other weighting models may also be used with cliques; see Metzler [51] for BM25 weighting functions.

3.3.2 Fielded Retrieval Models

Next, we present extensions of the previously introduced models—language models, BM25, and sequential dependence models—that consider multiple fields.

3.3.2.1 Mixture of Language Models

According to the *mixture of language models* (MLM) [57] approach, a separate language model is estimated for each entity field f:

$$P(t|\theta_{f_e}) = (1 - \lambda_f)P(t|f_e) + \lambda_f P(t|f_{\mathcal{E}}),$$

where λ_f is a field-specific smoothing parameter. Both components are maximum likelihood estimates and are defined similarly as before, but with term counts limited to a certain field:

$$P(t|f_e) = \frac{c(t; f_e)}{l_{f_e}}, \quad P(t|f_{\mathcal{E}}) = \frac{\sum_{e \in \mathcal{E}} c(t; f_e)}{\sum_{e \in \mathcal{E}} l_{f_e}}.$$

For Dirichlet prior smoothing the computations follow similarly, by replacing e and $\mathcal{E}$ with f_e and $f_{\mathcal{E}}$, respectively, and making the smoothing parameter field-specific μ_f in Eq. (3.10).

Once field language models are created, they are combined together into a single entity-level model using a linear mixture:

$$P(t|\theta_e) = \sum_{f \in \mathcal{F}} \alpha_f P(t|\theta_{f_e}), \tag{3.12}$$

where α_f is the weight (or importance) of field f, such that $\sum_{f \in \mathcal{F}} \alpha_f = 1$. In the absence of training data, field weights can be set uniformly or following some heuristic; e.g., proportional to the field length (measured as the sum of field lengths of the given field type, i.e., $\alpha_f \propto \sum_e l_{f_e}$). When training data is available and the number of fields is small, grid search is possible (i.e., sweeping over the parameter space $[0, 1]$ for each α_f in, say, 0.1 steps). Alternatively, parameters may be trained using the Coordinate Ascent algorithm [53]. When the number of fields is large, one might perform retrieval using a single field at a time, then set α_f proportional to the retrieval performance of the corresponding field.

3.3.2.2 Probabilistic Retrieval Model for Semi-Structured Data

A different alternative for setting field weights is proposed by Kim et al. [40], called *probabilistic retrieval model for semi-structured data* (PRMS), which hinges on the

Table 3.6 Example mapping probabilities computed on the IMDb collection, taken from [40]

t = "Meg"		t = "Ryan"		t = "war"		t = "redemption"	
f	$P(f\|t)$	f	$P(f\|t)$	f	$P(f\|t)$	f	$P(f\|t)$
Cast	0.407	Cast	0.601	Genre	0.927	Title	0.983
Team	0.381	Team	0.381	Title	0.070	Location	0.017
Title	0.187	Title	0.017	Location	0.002	Year	0.000

following key points: (1) instead of using a fixed (static) field weight for all terms, field weights are determined dynamically on a term-by-term basis; and (2) field weights can be established based on the term distributions of the respective fields.

Formally, the static field weight α_f in Eq. (3.12) is replaced with the probability of mapping term t to the given entity field f, simply referred to as the *mapping probability*, $P(f|t)$:

$$P(t|\theta_e) = \sum_{f \in \mathcal{F}} P(f|t) P(t|\theta_{f_e}) ,$$

By applying Bayes' theorem and using the law of total probability we get:

$$P(f|t) = \frac{P(t|f)P(f)}{P(t)} = \frac{P(t|f)P(f)}{\sum_{f' \in \mathcal{F}} P(t|f')P(f')} . \tag{3.13}$$

The prior $P(f)$ can be used to incorporate, e.g., domain-specific background knowledge, or left to be uniform. The probability of a term given a field, $P(t|f)$, is conveniently estimated using the background language model of that field, i.e., $P(t|f) \cong P(t|f_\mathcal{E})$.

Table 3.6 presents an example of mapping probabilities computed using PRMS. It is important to note that PRMS assumes a homogeneous collection where fields can be characterized by distinctive term distributions, and works well only when these conditions are met [27].

3.3.2.3 BM25F

The fielded variant of BM25, commonly referred to as *BM25F*, uses a weighted variant of term frequencies:[12]

$$score_{BM25F}(e; q) = \sum_{t \in q} c(t; q) \frac{(k_1 + 1) \tilde{c}(t; e)}{k_1 + \tilde{c}(t; e)} IEF(t) . \tag{3.14}$$

---

[12]This version is according to Robertson and Zaragoza [63] and allows for field-specific b parameters, as opposed to the simpler version in [64]. We kept the $(k_1 + 1)$ term in the numerator to be consistent with Eq. (3.11).

The weighted term frequencies are calculated as a linear combination of term frequencies across the different fields, with field-specific normalization applied:

$$\tilde{c}(t;e) = \sum_{f \in \mathcal{F}} \alpha_f \frac{c(t;f_e)}{1 - b_f + b_f \frac{l_{f_e}}{\bar{l}_f}},$$

where $c(t;f_e)$ is the frequency of t in field f of entity e, α_f are the field weights, b_f are the field length normalization parameters, and $\bar{l}_f$ is the mean length of field f across the entity catalog. Note that the IEF component in Eq. (3.14) does not use field information.

3.3.2.4 Fielded Sequential Dependence Models

The main idea behind the fielded extension of SDM is to base the feature function estimates on term/bigram frequencies combined across multiple fields, much in the spirit of MLM and BM25F. Here, we present the *fielded sequential dependence model* (FSDM) by Zhiltsov et al. [81], which combines SDM and MLM. Similarly, SDM can also be combined with BM25F; for the BM25F-SD scoring function we refer to Broder et al. [14].

As before, we need to define three feature functions: for single terms, ordered bigrams, and unordered bigrams. For single term matches, we use the (log) MLM-estimated probability:

$$f_T(q_i;e) = \log \sum_{f \in \mathcal{F}} w_f^T P(t|\theta_{f_e}) .$$

The feature functions for unordered and ordered bigram matches are as follows:

$$f_O(q_i, q_{i+1};e) = \log \sum_{f \in \mathcal{F}} w_f^O \frac{c_o(q_i, q_{i+1}; f_e) + \mu_f P_o(q_i, q_{i+1}|f_\mathcal{E})}{l_{f_e} + \mu_f},$$

$$f_U(q_i, q_{i+1};e) = \log \sum_{f \in \mathcal{F}} w_f^U \frac{c_u^w(q_i, q_{i+1}; f_e) + \mu_f P_u^w(q_i, q_{i+1}|f_\mathcal{E})}{l_{f_e} + \mu_f} .$$

Notice that the background models and smoothing parameters have all been made field-specific. We can reasonably use the same smoothing parameter value for a given field for all types of matches (i.e., μ_f does not have a superscript). With that, FSDM will have $3 + 3 \times |\mathcal{F}|$ free parameters: $\lambda_T, \lambda_O, \lambda_U$, plus the field mapping weights (w_f^T, w_f^O, w_f^U) for each field. Even with using only a handful of fields, this results in a large number of free parameters.

Hasibi et al. [37] propose a parameter-free estimation of field mapping weights by using field mapping probability estimates from PRMS, cf. Eq. (3.13). The same estimation method can also be applied to ordered and unordered bigrams. In addition to reducing the number of free parameters to 3 (as only the λ weights will need to be set now), employing the mapping probability $P(f|.)$ comes with another important benefit. Instead of using the same (fixed) weight $w_f^{\{T,O,U\}}$ for all terms/bigrams within a field, the field mapping probability specifies field importance for each term/bigram individually.

3.3.3 Learning-to-Rank

Learning-to-rank (LTR) approaches represent the current state of the art in document retrieval (as well as for many other retrieval tasks) as they can effectively combine a large number of potentially useful signals from multiple sources [42]. Their downside is that performance is heavily dependent on the amount of training material available. The application of LTR to entity ranking goes as follows.

Each query-entity pair is represented as a feature vector, where features are designed to capture different relevance signals. The optimal way of combining these signals is then learned through discriminative training. The training data is composed of a set of (entity, query, relevance) triples. From this data, a set of training instances $\{(\mathbf{x}_{q,e}, r_{q,e})\}$ is created, where $\mathbf{x}_{q,e} \in \mathbb{R}^m$ is an m-dimensional feature vector representing the query-entity pair (q,e) and $r_{q,e} \in \mathbb{N}$ is the corresponding label indicating how relevant the entity is to the query. Relevance can be binary, i.e., $\{0,1\}$, or graded, e.g., $\{0,\ldots,4\}$, representing the scale from non-relevant to highly relevant. The objective is to learn a ranking model $h(q,e) = h(\mathbf{x})$ that gives a real-valued score to a given query-entity pair, or equivalently, to the corresponding feature vector $\mathbf{x}$. The scores assigned to each entity by the learned function then determine the ranking of entities for the query.

There is a separate feature function $\phi_i(q,e)$ corresponding to each element of the feature vector, i.e., $\mathbf{x}_{q,e} = \langle \phi_1(q,e), \ldots, \phi_m(q,e) \rangle$. The performance of learning highly depends on the choice of features (just like for any other machine learning task). In fact, much of the focus in IR revolves around designing and selecting effective features (a process commonly referred to as *feature engineering*), whereas machine learning algorithms are applied out-of-the-box.

3.3.3.1 Features

Generally, three groups of features are distinguished:

- *Query features* depend only on the query and have the same value across all entities for a given query. These are meant to help characterize the type of the query.
- *Entity features* depend only on the entity and capture some aspect of its general importance. Typically, these features are based on some indicator of "popularity," e.g., the total number of views, "likes," incoming/outgoing links (possibly

Table 3.7 Term-based features for entity ranking

Group	Feature	Description
Q	$|q|$	Length of the query
Q	$\sum_{q_i} IEF(q_i, f)$	Sum IEF of query terms, computed w.r.t. field f
E	l_{f_e}	Length of field f in the description of entity e
Q-E	$\sum_{q_i} TF(q_i, f_e)$	Sum TF of query terms for a given field
Q-E	$\sum_{q_i} TF(q_i, f_e)IEF(q_i, f)$	Sum TF-IEF of query terms for a given field
Q-E	$score_{LM}(q; f_e)$	LM score of query computed on field f
Q-E	$score_{MLM}(q; e)$	MLM score of query (considering all fields)
Q-E	$score_{BM25}(q; f_e)$	BM25 score of query computed on field f
Q-E	$score_{BM25F}(q; e)$	BM25F score of query (considering all fields)
Q-E	$score_{SDM}(q; f_e)$	SDM score of query computed on field f
Q-E	$score_{FSDM}(q; e)$	FSDM score of query (considering all fields)
Q-E	$QCE(q, e)$	Whether the query contains the name of the entity
Q-E	$ECQ(q, e)$	Whether the name of entity contains the query
Q-E	$EEQ(q, e)$	Whether the name of entity is equal to the query

Group is either of query features (Q), entity features (E), or query-entity features (Q-E)

involving some propagation, like PageRank), etc. This group also includes simple statistics about the entity's representation, such as the number of terms in various entity description fields. Notice that the values of these features would be the same for all queries, thus we can think of this group of features as an estimate of "how relevant the entity would be to any query."

- *Query-entity features* are the largest and most important type of features, as they capture the degree of matching between the query and the entity. Basic features include (1) simple term statistics of query terms with some aggregator function (min, max, or average) and (2) scores of unsupervised ranking algorithms applied to different entity representations (i.e., fields). More complex features would encompass the overall retrieval score computed by some entity retrieval method (e.g., MLM, BM25F, or FSDM). Retrieval models can be instantiated with different parameter settings to generate multiple features (e.g., different smoothing methods and parameters for language modeling approaches). Additionally, if user feedback data is available, various click-based statistics for the query-entity pair would also be incorporated under this feature group.

In this chapter we only consider term-based features, which are summarized in Table 3.7. This selection (which is by no means complete) includes features that have been proposed in the literature for document retrieval [20, 61] or entity ranking [21, 50]. Note that the choice of fields $\mathcal{F}$ depends on the actual entity representation (cf. Sect. 3.2); whenever we write f or f_e in Table 3.7 it actually implies $|\mathcal{F}|$ features, one for each field $f \in \mathcal{F}$. In Chaps. 4 and 7, we will introduce additional features that go beyond the term level.

Feature values are often normalized to be in the $[0, 1]$ range for a given query. One simple technique to achieve that, without making any assumptions about how the values are distributed, is called *min-max normalization*:

$$\tilde{x}_i = \frac{x_i - \min(x)}{\max(x) - \min(x)} \, ,$$

where $x_1, \ldots, x_n$ are the original values for a given feature and for a given query, and $\tilde{x}_i$ is the transformed feature value for the ith instance.

3.3.3.2 Learning Algorithms

When learning a ranking model from training data, the objective is to minimize the expected value of some loss function. Ranking algorithms can be divided into three categories based on the choice of loss function:

- *Pointwise* approaches consider a single entity at a time for a given query and attempt to approximate the relevance value or label, independent of the other candidate entities for that query. Many standard regression and classification algorithms can be directly used for pointwise learning-to-rank. Random Forests [13] and Gradient Boosted Regression Trees (GBRT, a.k.a. MART) [33] are among the most widely used approaches.
- *Pairwise* approaches look at a pair of entities and try to learn a binary classifier that can tell which of the two entities should be ranked first. The objective for the ranker is to minimize the number of inversions in the ranking compared to the ground truth. Commonly used pairwise algorithms include RankSVM [39], RankBoost [32], RankNet [15], and GBRank [79].
- *Listwise* approaches are similar to the pairwise approach, but they aim to optimize the ordering of the entire result list. Loss functions for listwise approaches tend to get more complex compared to pointwise or pairwise approaches. Representative listwise algorithms include Coordinate Ascent [53], ListNet [17], AdaRank [76], LambdaRank [16], and LambdaMART [75].

3.3.3.3 Practical Considerations

Note that computing features for all query-entity pairs is not only computationally prohibitive, but also unnecessary. For each query there is a small number of relevant entities (compared to the total number of entities in the catalog). Therefore, it is sufficient to consider only a *sample* of entities for each query. In practice, sampling is typically performed by using an unsupervised model (e.g., LM or BM25) to obtain an initial ranking and taking the top-k results. Features are computed only for the set of "possibly relevant" entities that are identified in this first-pass (or initial) retrieval round (both during training and when applying the model). The final ranking for an unseen query is determined in a second-pass retrieval round, by computing the features for the top-k initially retrieved entities and applying the model that was learned on the training data. This essentially means re-ranking the initial results. "Minimizing the number of documents [entities] for which features are calculated provides efficiency advantages, particularly if some features are expensive to compute" [48]. The optimal value of k can vary greatly depending on the task and the document collection; we refer to Macdonald et al. [48] for a study on document retrieval. For ad hoc entity retrieval, k is typically set between 100 and 1000 [21, 37].

3.4 Ranking Entities Without Direct Representations

Up until this point, we have ranked entities by creating term-based representations (entity descriptions) for them. We have further made the point, illustrated with examples, that ready-made descriptions are available for many entities and creating entity representations from these is straightforward. There are cases, however, when creating explicit entity representations is not desired or simply not possible. An example of such a scenario is an enterprise setting, where users may be permitted only to access a certain subset of documents, based on their access levels. Creating all entity descriptions for each individual user would be highly ineffective.

In this section, we will consider the retrieval process in a slightly different way, in which entities are not modeled directly. Instead, documents are modeled and queried, then entities associated with the top-ranked documents are considered (hence this strategy has been termed *document model* in [1]). As before, we shall assume that documents have been annotated with entities. Formally, the scoring of entities is performed according to the following equation:

$$score(e; q) = \sum_{d \in \mathcal{D}_q} score(d; q) \, w(e, d) \, ,$$

where $score(d; q)$ expresses the document's relevance to the query and can be computed using any existing document retrieval method (LM, BM25, SDM, etc.). As before, $w(e, d)$ is the association weight between entity e and document d (cf. Eq. (3.4)). The summation is done over the set $\mathcal{D}_q$, which contains all documents that bear any relevance to the query q (i.e., $score(d; q) > 0$). The efficiency of this approach can be further improved by restricting $\mathcal{D}_q$ to the top-k relevant documents. Finally, user permissions can be easily handled here by limiting $\mathcal{D}_q$ to documents that the user is allowed to access. This general formalism encompasses all methods that rank entities using documents returned by a document search engine, including, e.g., Voting models [47].

An appealing feature of the document-based entity ranking method is that it can be implemented with limited effort on top of an existing document search engine; see Algorithm 3.1. Provided that associated entities can effectively be looked up for each document (i.e., where $w(e, d) > 0$), the added overhead over standard document retrieval is minimal. That lookup can be conveniently performed

---

Algorithm 3.1: Document-based entity ranking

Input: query q, document-entity association weights w
Output: scoring of entities

1   $score(d;q) \leftarrow$ perform standard document retrieval
2   $\mathcal{D}_q \leftarrow \{d : score(d;q) > 0\}$ `/* may be further restricted */`
3   $score(e;q) \leftarrow 0$ for all entities   `/* initialize entity scores */`
4   **foreach** $d \in \mathcal{D}_q$ **do**
5   **foreach** $e : w(e,d) > 0$ **do**
6          $score(e;q) \leftarrow score(e;q) + score(d;q)\, w(e,d)$
7   **end**
8   **end**

---

using an inverted index structure (storing the associated entities, along with the corresponding weights, for each document).

With mention-level entity annotations, a proximity-based variant is also possible (even though an efficient implementation of that idea can be challenging). The interested reader is referred to Balog et al. [2] for details.

3.5 Evaluation

The evaluation of ad hoc entity retrieval is analogous to that of ad hoc document retrieval. Given a ranked list of items, the relevance of each item is judged with respect to the input query (and independently of all other items). Various rank-based measures can be used to measure how effective the system was at ranking results for a given query; we present a number of such measures in Sect. 3.5.1. We note that the main focus here is on *effectiveness* (i.e., the "goodness" of the ranking); *efficiency* (i.e., how "quickly" it is done) will not be discussed. Over the past decade, a number of evaluation campaigns have addressed the problem of entity ranking in various flavors. A major focus of these efforts is to build reusable test collections to facilitate further research and development. We review these benchmarking efforts in Sect. 3.5.2.

3.5.1 Evaluation Measures

Evaluation is performed by using rank-based measures: average precision and reciprocal rank in case of binary relevance assessments, and normalized discounted cumulative gain in case of graded relevance judgments. Set-based measures may also be used, specifically, precision at rank cutoff k, where k is typically small, i.e., P@10 or P@20. Below, we present the definitions of these measures. For an in-depth treatment of evaluation measures and for guidance on how to conduct statistical significance testing, the reader is encouraged to consult any standard IR textbook, e.g., Zhai and Massung [78, Chapter 9] or Croft et al. [26, Chapter 8].

Let us first consider the case of binary relevance, where Rel denotes the set of relevant entities for a given query (with all entities outside Rel being non-relevant). Let $L = \langle e_1, \ldots, e_n \rangle$ be the ranked list of entities returned for a given query. *Precision at rank k* (P@k) is the fraction of the top-k ranked results that are relevant:

$$P(k) = \frac{|\{e_1, \ldots, e_k\} \cap Rel|}{k}.$$

Average precision (AP) is computed by averaging the precision values from each rank position where a relevant result was retrieved:

$$AP(L) = \frac{1}{|Rel|} \sum_{\substack{i=1 \\ e_i \in Rel}}^{n} P(i).$$

There are situations when there is only a single relevant result (e.g., the user searches for a particular entity by its name). The *reciprocal rank* (RR) measure is defined as the reciprocal of the rank position r at which the first relevant result was found: $1/r$ (and 0 if no relevant result was retrieved).

In case of multi-level (or graded) relevance, we let r_i be the relevance level ("gain") corresponding to the ith ranked entity, e_i. The *discounted cumulative gain* (DCG) of a ranking is defined as:

$$DCG(L) = r_1 + \sum_{i=2}^{n} \frac{r_i}{\log_2 i}.$$

Each result's gain is "discounted" by dividing it with the log of the corresponding rank position. This way highly ranked results (which users are more likely to inspect) contribute more to the overall score than lowly ranked results. Note that DCG values vary depending on the number of relevant results. *Normalized discounted cumulative gain* (NDCG) scales the score to the range $[0, 1]$ by computing:

$$NDCG(L) = \frac{DCG(L)}{IDCG},$$

where IDCG is the *ideal* DCG for a particular query: sorting entities in decreasing order of relevance, and computing the DCG of that ranking. NDCG is often reported at a given rank cutoff, e.g., NDCG@10 or NDCG@20.

All the above measures are computed for a single query. To summarize overall system performance across a set of test queries, the average (arithmetic mean) over the individual query scores is taken. For average precision and reciprocal rank, these averaged measures are given a distinctive name, *mean average precision* (MAP) and *mean reciprocal rank* (MRR), respectively.

3.5.2 Test Collections

In IR, the use of test collections is the de facto standard of evaluation. A test collection consists of (1) a dataset, (2) a set of information needs (i.e., queries or topics), and (3) the corresponding relevance assessments (i.e., ground truth). Large-scale test collections are typically developed within the context of community benchmarking campaigns. The first of those benchmarks was the Text Retrieval Conference (TREC), organized by the US National Institute of Standards and Technology (NIST) in 1992 [72]. TREC operates on an annual cycle, completing the following sequence each year:

1. TREC provides a test dataset and information needs. Traditionally, each information need is provided as a *topic* definition, which consists of a *query*, which is the input to the search engine, and a more elaborate *description* and/or *narrative*, to guide human judges during the relevance assessment process.
2. Participants develop and run their retrieval systems on the test data and submit their rankings. Typically, participants can enter multiple submissions ("runs"), allowing for a comparison across different methods or parameter settings.
3. TREC pools the individual results, obtains relevance assessments, and evaluates the submitted systems.
4. TREC organizes a conference for participants to share their experiences.

TREC has had, and continues to have, a profound impact on the IR community [66]. Also, the TREC campaign style has been followed by several other benchmarking initiatives.

In this section, we present a number of benchmarking evaluation campaigns that have addressed the task of entity ranking in some form. Each of these has its peculiarities, and there are some fine details that we will omit here. As we go through these efforts in chronological order, we invite the reader to observe how the focus is shifting from unstructured to semi-structured and then to structured data sources. Table 3.8 provides an overview. At the end of this section, in Sect. 3.5.2.7, we introduce a derived test suite that combines queries and relevance assessments from multiple benchmarking campaigns.

3.5.2.1 TREC Enterprise

The TREC (Text Retrieval Conference) 2005–2008 Enterprise track [8, 25] featured an *expert finding* task, where a single type of entity was sought: people, who are experts on a given topic (specified by the query). The task is situated in a large knowledge-intensive organization, such as the World Wide Web Consortium (W3C) or the Commonwealth Scientific and Industrial Research Organisation (CSIRO). The document collection is the enterprise's intranet and people are uniquely identified by their email addresses. While a small number of people have personal homepages, the vast majority of persons are not represented as retrievable units.

Table 3.8 Entity retrieval test collections

Campaign, track/task	Entity ID	Data collection	Query type	#Queries
INEX XER 2007 [73]	Wikipedia page	Wikipedia	Keyword++	28+46
INEX XER 2008 [30]	Wikipedia page	Wikipedia	Keyword++	35
INEX XER 2009 [29]	Wikipedia page	Wikipedia	Keyword++	55
TREC Entity 2009 REF [9]	Homepage(s)	Web (ClueWeb09B)	Keyword++	20
TREC Entity 2010 REF [6]	Homepage(s)	Web (ClueWeb09)	Keyword++	47
TREC Entity 2010 ELC [6]	URI	Linked Data (BTC-2009)	Keyword++	14
TREC Entity 2011 REF [7]	Homepage(s)	Web (ClueWeb09)	Keyword++	50
TREC Entity 2011 ELC [7]	URI	Linked Data (Sindice-2011)	Keyword++	50
SemSearch 2009 ES [36]	URI	Linked Data (BTC-2009)	Keyword	92
SemSearch 2010 ES [10]	URI	Linked Data (BTC-2009)	Keyword	50
SemSearch 2010 LS [10]	URI	Linked Data (BTC-2009)	Keyword	50
INEX LD 2012 Ad Hoc [74]	Wikipedia page	Wikipedia-LOD (v1.1)	Keyword	100
INEX LD 2012 Ad Hoc [35]	Wikipedia page	Wikipedia-LOD (v2.0)	Keyword	144
QALD-1 [43]	URI	DBpedia (v3.6)	Natural lang.	50+50
QALD-2 [43]	URI	DBpedia (v3.7)	Natural lang.	100+100
QALD-3 [23]	URI	DBpedia (v3.8)	Natural lang.	100+99
QALD-4 [69]	URI	DBpedia (v3.9)	Natural lang.	200+50
QALD-5 [70]	URI	DBpedia (2014)	Natural lang.	340+59
QALD-6 [71]	URI	DBpedia (2015)	Natural lang.	400+150
DBpedia-Entity [4]	URI	DBpedia (v3.7)	Mixed	485
DBpedia-Entity v2 [38]	URI	DBpedia (2015-10)	Mixed	467

Keyword++ refers to queries that are enriched beyond the query string. The last column shows the number of test queries; when training queries are also provided, we use the syntax $x+y$, where x and y are the number of training and test queries, respectively

Two principal approaches to expert finding were proposed early on [1], which laid the foundations for much of the research we presented in this chapter: (1) *profile-based models* (building term-based entity representations, cf. Sect. 3.2) and (2) *document-based models* (ranking entities without building explicit representations, cf. Sect. 3.4). Given the specialized nature of the expert finding task (i.e., entities being restricted to a single type), we do not include the TREC Enterprise test collections in Table 3.8. For an overview of work on the broader subject of expertise retrieval, we refer to Balog et al. [3].

3.5.2.2 INEX Entity Ranking

The INEX (Initiative for the Evaluation of XML Retrieval) 2007–2009 Entity Ranking (XER) track [29, 30, 73] used Wikipedia as the data collection, where entities are represented and identified uniquely by their corresponding Wikipedia article. Two tasks are distinguished: *entity ranking* and *list completion*. Both tasks seek a ranked list of entities (e.g., "*olympic classes dinghy sailing*" or "*US presidents since 1960*"), but the input formulations differ. In addition to the free text query, the topic definition includes target entity types (Wikipedia categories) for the entity ranking task and a small set of example entities for the list completion task. As such, we have an enriched keyword++ query as input. Mind that in this chapter, we have only considered the keyword part of queries. Techniques for exploiting those additional query components (target types or example entities) will be discussed in Chap. 4.

The 2007 test collection [73] consists of 46 queries; additionally, a set of 28 training queries was also made available to participants. Twenty-five of the test queries are created specifically for XER. The rest (including all training queries) are derived from the INEX Ad Hoc track and have a different interpretation of relevance (e.g., for the query "*Bob Dylan songs,*" articles related to Bob Dylan, The Band, albums, cities where Bob Dylan lived, etc., are also considered relevant). Because of that, most (but not all) of these ad hoc topics were re-assessed as XER topics. As pointed out by de Vries et al. [73], "often surprisingly many articles that are *on topic* for the ad hoc track are not relevant entities." One particular example is list pages, which are not entities, and therefore are not considered as relevant entity results.

For the 2008 edition [30], 35 genuine XER topics were created. The 2009 track [29] switched to a newer Wikipedia dump and considered a set of 60 queries from the previous 2 years. All these queries were re-assessed, and the ones with too few (less than 7) or too many (more than 74) results were removed, leaving 55 queries in total.

3.5.2.3 TREC Entity

The TREC Entity track was launched in 2009 [9] with the aim to perform entity-oriented search tasks on the Web. The track introduced the *related entity finding* (REF) task, which asks for entities of a specified type that engage in a given relationship with a given source entity. This particular problem definition was motivated by the fact that many entity queries could have very large answer sets on the Web (e.g., "*actors playing in hollywood movies*"), which would render the assessment procedure problematic. In the first edition of the track, possible target entity types were limited to three: people, organizations, and products. An example query is "*airlines that currently use Boeing 747 planes,*" where Boeing 747 is the input entity and the target type is organization. Here, we consider only the free text part of the query; the input entity and target type may be viewed as semantic annotations (keyword++), which we will exploit in Chap. 4. The data collection is a web crawl (ClueWeb09B) and entities are identified by their homepages (URLs). An entity might have multiple homepages, including its Wikipedia page. Entity resolution, i.e., grouping homepages together that represent the same entity, was addressed at evaluation time.

In 2010 [6], a number of changes were implemented to the REF task. First, the document collection was extended to the English portion of ClueWeb09 (approx. 500 million pages). Second, Wikipedia pages were no longer accepted as entity homepages (to make the task more challenging), and the notion of entity homepage was made more precise by making a distinction between primary and relevant entity homepages. Third, a new task, *entity list completion* (ELC) was introduced. It addresses essentially the same problem as REF, but there are some important differences: (1) entities are represented by their URIs in a Linked Data crawl (BTC-2009, cf. Sect. 2.3.6.1), (2) a small number of example entities are made available, and (3) target types are mapped to the most specific class within the DBpedia Ontology. A subset of the 2009 REF topics were recycled, with previously identified relevant entities manually mapped to URIs in BTC-2009 and given out as examples.

The 2011, final edition of the track [7] made some further changes to the REF task, including that (1) only primary entity homepages are accepted (i.e., relevance is binary) and (2) target type is not limited anymore. For the ELC task a new larger Semantic Web crawl, Sindice-2011, was introduced (cf. Sect. 2.3.6.1), which replaced the BTC-2009 collection. The task definition remained the same as in 2010, but only ClueWeb homepages were provided for example entities and these were not mapped manually to Linked Data URIs. The REF 2010 topics were reused and known relevant answers were offered as example entities. Additionally, a Linked Data variant of the REF task (REF-LOD) was also proposed, using URIs instead of homepages for entity identification; this variant of the task, however, attracted little interest among participants.

3.5.2.4 Semantic Search Challenge

The Semantic Search Workshop series organized a challenge (SemSearch) in 2010 and 2011 [10, 36], sponsored by Yahoo!, where participants were required to answer ad hoc entity search queries over structured data collected from the Web (i.e., Linked Data). Entities are identified by their URIs. The dataset is BTC-2009, which combines the crawls of multiple semantic search engines (cf. Sect. 2.3.6.1). Two separate search tracks are distinguished. *Entity search* queries (2010 and 2011) seek information on one particular entity, e.g., "*YMCA Tampa*" or "*Hugh Downs.*" The target entity is often an ambiguous one, e.g., "*Ben Franklin,*" which may refer to the person or to the ship named after him. These queries were sampled from web search engine logs. *List search* queries (2011 only) describe sets of entities, where target entities are not named explicitly in the query. Examples include "*Arab states of the Persian Gulf*" and "*Matt Berry tv series.*" These queries were created by the challenge organizers.

Relevance assessments for both tracks were gathered via crowdsourcing. We note that unlike TREC and INEX practice, assessors for the entity search track were presented only with the keyword query (without a more detailed description of the underlying information need). For list search queries, crowd workers were additionally provided with the reference list of correct answers (obtained through manual searching by the organizers). In follow-up work, Blanco et al. [11] showed that crowdsourced judgments are repeatable (i.e., using a different pool of judges 6 months later led to the same evaluation results) and reliable (even though crowd workers mark somewhat relevant many of the items that expert judges would consider irrelevant, this does not change the relative ordering of systems).

3.5.2.5 INEX Linked Data

In 2012, INEX introduced the Linked Data track [74] with the aim to investigate retrieval techniques over a combination of textual and highly structured data. The data collection consists of Wikipedia articles, enriched with RDF properties from knowledge bases (DBpedia and YAGO2). Relevant to this chapter is the classic *ad hoc* task, where information needs are formulated using keyword queries. Results are Wikipedia articles, each uniquely identified by its page ID. Queries come from three main sources: (1) a selection of the worst performing topics from past editions of the INEX Ad Hoc track, (2) query suggestions related to some general concepts, obtained from Google (e.g., queries generated for "Vietnam" include *"vietnam war movie," "vietnam food recipes,"* and *"vietnam travel airports"*), and (3) natural language Jeopardy-style questions (e.g., *"Niagara Falls source lake"*). Assessments for (1) were taken from previous INEX editions; for (2) and (3) they were collected using crowdsourcing.

The 2013 edition of the track [35] changed the format of the reference collection and employed newer data dumps. The 2013 query set is a combination of (1) INEX 2009 and 2010 Ad Hoc topics and (2) natural language Jeopardy topics. As before, assessments were taken over from the previous years, for (1), and were collected using crowdsourcing, for (2).

3.5.2.6 Question Answering over Linked Data

Question Answering over Linked Data (QALD) is a series of evaluation campaigns, co-located with the European Semantic Web Conference (ESWC), on answering natural language questions over Linked Data [43]. Questions are of varying complexity and ask either for literal answers (Boolean, date, number, or string), e.g., *"How many students does the Free University in Amsterdam have?"* or for a list of resources, i.e., entities (identified by URIs), e.g., *"Which presidents of the United States had more than three children?"* Note that for us, questions of the latter type (seeking entities) are of particular interest. A gold standard SPARQL query was constructed manually for each question; the results of that query constitute the ground truth against which systems are evaluated.

The first instantiations of the challenge focused on answering English language questions from a single knowledge base, in particular DBpedia and MusicBrainz (a collaborative open-content music database). Later installments of the challenge focused on multilingual question answering (from QALD-3), extending the task to multiple, interlinked datasets (from QALD-4), hybrid question answering, which require information from both structured data and unstructured data (from QALD-4), and statistical question answering over RDF data cubes (QALD-6). Table 3.8

Table 3.9 Breakdown of the DBpedia-Entity v2 test collection [38] by query categories

Category	Type	#queries	R_1	R_2
SemSearch ES	Named entities	113	12.5	3.0
INEX-LD	Keyword queries	99	23.5	9.2
ListSearch	List of entities	115	18.1	12.7
QALD-2	NL questions	140	28.4	29.8
Total		467	21.0	14.7

R_1 and R_2 refer to the average number of relevant and highly relevant entities per query, respectively

shows, for each QALD edition, the total number of questions for all tasks (including hybrid QA), where DBpedia is used as the underlying knowledge base.

3.5.2.7 The DBpedia-Entity Test Collection

Balog and Neumayer [4] compiled the *DBpedia-Entity* test collection by synthesizing queries from a number of the above-presented benchmarking campaigns into a single query set and mapping known relevant answers to the DBpedia knowledge base. This amounts to a diverse query set ranging from short keyword queries to natural language questions. As part of the normalization, only the free text parts of queries are considered. That is, any additional markup, type information, example entities, etc., that may be available in the original task setup are ignored here. Further, relevance is taken to be binary. DBpedia-Entity has been used in recent work as a standard test collection for entity retrieval over knowledge bases [21, 44, 56, 81].

Recently, Hasibi et al. [38] introduced an updated and extended version, referred to as the *DBpedia-Entity v2* test collection. It uses a more recent version of DBpedia (2015-10) and comes with graded relevance judgments. Relevance assessments were collected via crowdsourcing and were then further curated manually by expert annotators, resulting in a high-quality dataset. Queries are further subdivided into four categories, with statistics presented in Table 3.9:

- *SemSearch ES* queries are from the Semantic Search Challenge, searching for a particular entity, like "*brooklyn bridge*" or "*08 toyota tundra.*"
- *INEX-LD* consists of IR-style keyword queries from the INEX 2012 Linked Data track, e.g., "*electronic music genres.*"
- *List Search* comprises queries from the Semantic Search Challenge (list search task), the INEX 2009 Entity Ranking track, and the TREC 2009 Entity track, seeking a list of entities, e.g., "*Professional sports teams in Philadelphia.*"
- *QALD-2* contains natural language queries that are from the Question Answering over Linked Data challenge, e.g., "*Who is the mayor of Berlin?*"

3.6 Summary

This chapter has introduced techniques for ranking entities in various datasets, ranging from unstructured documents to structured knowledge bases. Most of our effort has concentrated on constructing term-based representations of entities, which can then be ranked using traditional document retrieval techniques. Despite their simplicity, unstructured entity representations with bag-of-words retrieval models usually provide solid performance and a good starting point. The current state of the art is to employ fielded entity representations and supervised learning. According to a recent benchmark, this can yield a moderate, around 10%, relative improvement over unstructured and unsupervised models when entity retrieval is performed over a knowledge base [38]. Much larger differences can be observed across different query types. Queries seeking a particular entity are today tackled fairly successfully, while the same models perform 25–40% worse on more complex queries, such as relationship queries or natural language questions. Clearly, there is room for improvement, especially on the latter types of queries. Generally, careful data cleansing and pre-processing accounts for more than more sophisticated retrieval methods [55]. This is not surprising, as this likely applies to most information processing applications. There is no universal recipe—every collection has to be dealt with on a case-by-case basis.

Crucially, we have not really tapped into any specific characteristic of entities yet, such as types or relationships. That will follow in the next chapter.

3.7 Further Reading

In this chapter, we have limited ourselves to using "flat structures," i.e., entity fields are treated as a set, without taking into account the hierarchical relationships that may exist between them. *Hierarchical structures* have been studied in the context of element-level XML retrieval, see, e.g., [45, 58]. Neumayer et al. [54] consider hierarchical field structures for entity retrieval, but according to their experiments, these do not yield any performance improvements over flat structures. The different types of entity representation we have looked at in this chapter were built from a single type of data: unstructured, semi-structured, or structured. It is also possible to construct *hybrid representations*. For example, Graus et al. [34] combine various entity description sources, including a knowledge base, web anchors, social media, and search queries. Ranking without direct entity representations is also feasible, as we have discussed in Sect. 3.4. Schuhmacher et al. [68] implement this document-based strategy in a learning-to-rank framework and employ four types of features: mention, query-mention, query-entity, and entity-entity.

References

1. Balog, K., Azzopardi, L., de Rijke, M.: Formal models for expert finding in enterprise corpora. In: Proceedings of the 29th annual international ACM SIGIR conference on Research and development in information retrieval, SIGIR '06, pp. 43–50. ACM (2006). doi: 10.1145/1148170.1148181
2. Balog, K., Azzopardi, L., de Rijke, M.: A language modeling framework for expert finding. Inf. Process. Manage. **45**(1), 1–19 (2009a). doi: 10.1016/j.ipm.2008.06.003
3. Balog, K., Fang, Y., de Rijke, M., Serdyukov, P., Si, L.: Expertise retrieval. Found. Trends Inf. Retr. **6**(2-3), 127–256 (2012a). doi: 10.1561/1500000024
4. Balog, K., Neumayer, R.: A test collection for entity search in DBpedia. In: Proceedings of the 36th international ACM SIGIR conference on Research and development in Information Retrieval, SIGIR '13, pp. 737–740. ACM (2013). doi: 10.1145/2484028.2484165
5. Balog, K., de Rijke, M.: Associating people and documents. In: Proceedings of the IR Research, 30th European Conference on Advances in Information Retrieval, ECIR'08, pp. 296–308. Springer (2008). doi: 10.1007/978-3-540-78646-7_28
6. Balog, K., Serdyukov, P., de Vries, A.P.: Overview of the TREC 2010 Entity track. In: Proceedings of the Nineteenth Text REtrieval Conference, TREC '10. NIST (2011)
7. Balog, K., Serdyukov, P., de Vries, A.P.: Overview of the TREC 2011 Entity track. In: The Twentieth Text REtrieval Conference Proceedings, TREC '11. NIST (2012b)
8. Balog, K., Soboroff, I., Thomas, P., Craswell, N., de Vries, A.P., Bailey, P.: Overview of the TREC 2008 Enterprise track. In: Proceedings of the 17th Text REtrieval Conference, TREC '08. NIST (2009b)
9. Balog, K., de Vries, A.P., Serdyukov, P., Thomas, P., Westerveld, T.: Overview of the TREC 2009 Entity track. In: Proceedings of the Eighteenth Text REtrieval Conference, TREC '09. NIST (2010)
10. Blanco, R., Halpin, H., Herzig, D.M., Mika, P., Pound, J., Thompson, H.S., Duc, T.T.: Entity search evaluation over structured web data. In: Proceedings of the 1st International Workshop on Entity-Oriented Search, EOS '11, pp. 65–71 (2011a)
11. Blanco, R., Halpin, H., Herzig, D.M., Mika, P., Pound, J., Thompson, H.S., Tran, T.: Repeatable and reliable semantic search evaluation. Web Semant. **21**, 14–29 (2013)
12. Blanco, R., Mika, P., Vigna, S.: Effective and efficient entity search in RDF data. In: Proceedings of the 10th International Conference on The Semantic Web, ISWC '11, pp. 83–97. Springer (2011b). doi: 10.1007/978-3-642-25073-6_6
13. Breiman, L.: Random forests. Mach. Learn. **45**(1), 5–32 (2001). doi: 10.1023/A:1010933404324
14. Broder, A., Gabrilovich, E., Josifovski, V., Mavromatis, G., Metzler, D., Wang, J.: Exploiting site-level information to improve web search. In: Proceedings of the 19th ACM International Conference on Information and Knowledge Management, CIKM '10, pp. 1393–1396. ACM (2010). doi: 10.1145/1871437.1871630
15. Burges, C., Shaked, T., Renshaw, E., Lazier, A., Deeds, M., Hamilton, N., Hullender, G.: Learning to rank using gradient descent. In: Proceedings of the 22nd International Conference on Machine Learning, ICML '05, pp. 89–96. ACM (2005). doi: 10.1145/1102351.1102363
16. Burges, C.J.C., Ragno, R., Le, Q.V.: Learning to rank with nonsmooth cost functions. In: Proceedings of the 19th International Conference on Neural Information Processing Systems, NIPS '06, pp. 193–200. MIT Press (2006)
17. Cao, Z., Qin, T., Liu, T.Y., Tsai, M.F., Li, H.: Learning to rank: From pairwise approach to listwise approach. In: Proceedings of the 24th International Conference on Machine Learning, ICML '07, pp. 129–136. ACM (2007). doi: 10.1145/1273496.1273513

18. Chakrabarti, S., Kasturi, S., Balakrishnan, B., Ramakrishnan, G., Saraf, R.: Compressed data structures for annotated web search. In: Proceedings of the 21st International Conference on World Wide Web, WWW '12, pp. 121–130. ACM (2012). doi: 10.1145/2187836.2187854

19. Chakrabarti, S., Puniyani, K., Das, S.: Optimizing scoring functions and indexes for proximity search in type-annotated corpora. In: Proceedings of the 15th International Conference on World Wide Web, WWW '06, pp. 717–726. ACM (2006). doi: 10.1145/1135777.1135882

20. Chapelle, O., Chang, Y.: Yahoo! Learning to Rank Challenge overview. In: Proceedings of the Yahoo! Learning to Rank Challenge, pp. 1–24 (2011)

21. Chen, J., Xiong, C., Callan, J.: An empirical study of learning to rank for entity search. In: Proceedings of the 39th International ACM SIGIR Conference on Research and Development in Information Retrieval, SIGIR '16, pp. 737–740. ACM (2016). doi: 10.1145/2911451.2914725

22. Cheng, T., Chang, K.C.C.: Beyond pages: Supporting efficient, scalable entity search with dual-inversion index. In: Proceedings of the 13th International Conference on Extending Database Technology, EDBT '10, pp. 15–26. ACM (2010). doi: 10.1145/1739041.1739047

23. Cimiano, P., Lopez, V., Unger, C., Cabrio, E., Ngonga Ngomo, A.C., Walter, S.: Multilingual question answering over Linked Data (QALD-3): Lab overview. In: Information Access Evaluation. Multilinguality, Multimodality, and Visualization: 4th International Conference of the CLEF Initiative, CLEF 2013, Valencia, Spain, September 23–26, 2013. Proceedings, pp. 321–332. Springer (2013). doi: 10.1007/978-3-642-40802-1_30

24. Conrad, J.G., Utt, M.H.: A system for discovering relationships by feature extraction from text databases. In: Proceedings of the 17th Annual International ACM SIGIR Conference on Research and Development in Information Retrieval, SIGIR '94, pp. 260–270. Springer (1994)

25. Craswell, N., de Vries, A.P., Soboroff, I.: Overview of the TREC-2005 Enterprise track. In: Proceedings of the 14th Text REtrieval Conference, TREC '05. NIST (2006)

26. Croft, B., Metzler, D., Strohman, T.: Search Engines: Information Retrieval in Practice. 1st edn. Addison-Wesley Publishing Co. (2009)

27. Dalton, J., Huston, S.: Semantic entity retrieval using web queries over structured RDF data. In: Proceedings of the 3rd International Semantic Search Workshop, SEMSEARCH '10 (2010)

28. Dalvi, N., Kumar, R., Soliman, M.: Automatic wrappers for large scale web extraction. Proc. VLDB Endow. 4(4), 219–230 (2011). doi: 10.14778/1938545.1938547

29. Demartini, G., Iofciu, T., de Vries, A.: Overview of the INEX 2009 Entity Ranking track. In: Geva, S., Kamps, J., Trotman, A. (eds.) Focused Retrieval and Evaluation, *Lecture Notes in Computer Science*, vol. 6203, pp. 254–264. Springer (2010). doi: 10.1007/978-3-642-14556-8_26

30. Demartini, G., de Vries, A.P., Iofciu, T., Zhu, J.: Overview of the INEX 2008 Entity Ranking track. In: Advances in Focused Retrieval: 7th International Workshop of the Initiative for the Evaluation of XML Retrieval (INEX 2008), pp. 243–252 (2009). doi: 10.1007/978-3-642-03761-0_25

31. Firth, J.R.: A synopsis of linguistic theory 1930-55. Studies in Linguistic Analysis (special volume of the Philological Society) **1952-59**, 1–32 (1957)

32. Freund, Y., Iyer, R., Schapire, R.E., Singer, Y.: An efficient boosting algorithm for combining preferences. J. Mach. Learn. Res. **4**, 933–969 (2003)

33. Friedman, J.H.: Greedy function approximation: A gradient boosting machine. Annals of Statistics **29**, 1189–1232 (2000)

34. Graus, D., Tsagkias, M., Weerkamp, W., Meij, E., de Rijke, M.: Dynamic collective entity representations for entity ranking. In: Proceedings of the Ninth ACM International Conference on Web Search and Data Mining, WSDM '16, pp. 595–604. ACM (2016). doi: 10.1145/2835776.2835819

35. Gurajada, S., Kamps, J., Mishra, A., Schenkel, R., Theobald, M., Wang, Q.: Overview of the INEX 2013 Linked Data track. In: CLEF 2013 Evaluation Labs and Workshop, Online Working Notes (2013)

36. Halpin, H., Herzig, D.M., Mika, P., Blanco, R., Pound, J., Thompson, H.S., Tran, D.T.: Evaluating ad-hoc object retrieval. In: Proceedings of the International Workshop on Evaluation of Semantic Technologies, IWEST '10 (2010)

37. Hasibi, F., Balog, K., Bratsberg, S.E.: Exploiting entity linking in queries for entity retrieval. In: Proceedings of the 2016 ACM on International Conference on the Theory of Information Retrieval, ICTIR '16, pp. 209–218. ACM (2016). doi: 10.1145/2970398.2970406

38. Hasibi, F., Nikolaev, F., Xiong, C., Balog, K., Bratsberg, S.E., Kotov, A., Callan, J.: DBpedia-Entity v2: A test collection for entity search. In: Proceedings of the 40th International ACM SIGIR Conference on Research and Development in Information Retrieval, SIGIR '17, pp. 1265–1268. ACM (2017). doi: 10.1145/3077136.3080751

39. Joachims, T.: Optimizing search engines using clickthrough data. In: Proceedings of the Eighth ACM SIGKDD International Conference on Knowledge Discovery and Data Mining, KDD '02, pp. 133–142. ACM (2002). doi: 10.1145/775047.775067

40. Kim, J., Xue, X., Croft, W.B.: A probabilistic retrieval model for semistructured data. In: Proceedings of the 31th European Conference on IR Research on Advances in Information Retrieval, pp. 228–239. Springer (2009). doi: 10.1007/978-3-642-00958-7_22

41. Laender, A.H.F., Ribeiro-Neto, B.A., da Silva, A.S., Teixeira, J.S.: A brief survey of web data extraction tools. SIGMOD Rec. 31(2), 84–93 (2002). doi: 10.1145/565117.565137

42. Liu, T.Y.: Learning to Rank for Information Retrieval. Springer (2011)

43. Lopez, V., Unger, C., Cimiano, P., Motta, E.: Evaluating question answering over Linked Data. Web Semantics: Science, Services and Agents on the World Wide Web 21, 3–13 (2013). doi: 10.1016/j.websem.2013.05.006

44. Lu, C., Lam, W., Liao, Y.: Entity retrieval via entity factoid hierarchy. In: Proceedings of the 53rd Annual Meeting of the Association for Computational Linguistics and the 7th International Joint Conference on Natural Language Processing (Volume 1: Long Papers), ACL '15, pp. 514–523. Association for Computational Linguistics (2015). doi: 10.3115/v1/P15-1050

45. Lu, W., Robertson, S., MacFarlane, A.: Field-weighted XML retrieval based on BM25. In: Proceedings of the 4th International Conference on Initiative for the Evaluation of XML Retrieval, INEX '05, pp. 161–171 (2006). doi: 10.1007/11766278_12

46. Lv, Y., Zhai, C.: Positional language models for information retrieval. In: Proceedings of the 32nd International ACM SIGIR Conference on Research and Development in Information Retrieval, SIGIR '09, pp. 299–306. ACM (2009). doi: 10.1145/1571941.1571994

47. Macdonald, C., Ounis, I.: Voting for candidates: Adapting data fusion techniques for an expert search task. In: Proceedings of the 15th ACM international conference on Information and knowledge management, CIKM '06, pp. 387–396. ACM (2006). doi: 10.1145/1183614.1183671

48. Macdonald, C., Santos, R.L., Ounis, I.: The whens and hows of learning to rank for web search. Inf. Retr. 16(5), 584–628 (2013). doi: 10.1007/s10791-012-9209-9

49. Manning, C.D., Raghavan, P., Schütze, H.: Introduction to Information Retrieval. Cambridge University Press (2008)

50. Meij, E., Weerkamp, W., de Rijke, M.: Adding semantics to microblog posts. In: Proceedings of the Fifth ACM International Conference on Web Search and Data Mining, WSDM '12, pp. 563–572. ACM (2012). doi: 10.1145/2124295.2124364

51. Metzler, D.: A Feature-Centric View of Information Retrieval. Springer (2011)

52. Metzler, D., Croft, W.B.: A Markov random field model for term dependencies. In: Proceedings of the 28th Annual International ACM SIGIR Conference on Research and Development in Information Retrieval, SIGIR '05, pp. 472–479. ACM (2005). doi: 10.1145/1076034.1076115

53. Metzler, D., Croft, W.B.: Linear feature-based models for information retrieval. Inf. Retr. 10(3), 257–274 (2007). doi: 10.1007/s10791-006-9019-z

54. Neumayer, R., Balog, K., Nørvåg, K.: On the modeling of entities for ad-hoc entity search in the Web of Data. In: Proceedings of the 34th European conference on Advances in Information Retrieval, ECIR '12, pp. 133–145. Springer (2012a). doi: 10.1007/978-3-642-28997-2_12

55. Neumayer, R., Balog, K., Nørvåg, K.: When simple is (more than) good enough: Effective semantic search with (almost) no semantics. In: Proceedings of the 34th European conference on Advances in Information Retrieval, ECIR '12, pp. 540–543. Springer (2012b). doi: 10.1007/978-3-642-28997-2_59

56. Nikolaev, F., Kotov, A., Zhiltsov, N.: Parameterized fielded term dependence models for ad-hoc entity retrieval from knowledge graph. In: Proceedings of the 39th International ACM SIGIR Conference on Research and Development in Information Retrieval, SIGIR '16, pp. 435–444. ACM (2016). doi: 10.1145/2911451.2911545

57. Ogilvie, P., Callan, J.: Combining document representations for known-item search. In: Proceedings of the 26th Annual International ACM SIGIR Conference on Research and Development in Information Retrieval, SIGIR '03, pp. 143–150. ACM (2003). doi: 10.1145/860435.860463

58. Ogilvie, P., Callan, J.: Hierarchical language models for XML component retrieval. In: Fuhr, N., Lalmas, M., Malik, S., Szlávik, Z. (eds.) Advances in XML Information Retrieval, Third International Workshop of the Initiative for the Evaluation of XML Retrieval, INEX 2004, *Lecture Notes in Computer Science*, vol. 3493, pp. 224–237. Springer (2005). doi: 10.1007/11424550_18

59. Pérez-Agüera, J.R., Arroyo, J., Greenberg, J., Iglesias, J.P., Fresno, V.: Using BM25F for semantic search. In: Proceedings of the 3rd International Semantic Search Workshop, SEMSEARCH '10. ACM (2010). doi: 10.1145/1863879.1863881

60. Petkova, D., Croft, W.B.: Proximity-based document representation for named entity retrieval. In: Proceedings of the sixteenth ACM conference on Conference on information and knowledge management, CIKM '07, pp. 731–740. ACM (2007). doi: 10.1145/1321440.1321542

61. Qin, T., Liu, T.Y., Xu, J., Li, H.: LETOR: A benchmark collection for research on learning to rank for information retrieval. Inf. Retr. **13**(4), 346–374 (2010). doi: 10.1007/s10791-009-9123-y

62. Raghavan, H., Allan, J., Mccallum, A.: An exploration of entity models, collective classification and relation description. In: KDD Workshop on Link Analysis and Group Detection, LinkKDD '04, pp. 1–10 (2004)

63. Robertson, S., Zaragoza, H.: The probabilistic relevance framework: BM25 and beyond. Found. Trends Inf. Retr. **3**(4), 333–389 (2009). doi: 10.1561/1500000019

64. Robertson, S., Zaragoza, H., Taylor, M.: Simple BM25 extension to multiple weighted fields. In: Proceedings of the 13th ACM conference on Information and knowledge management, CIKM '04, pp. 42–49 (2004). doi: 10.1145/1031171.1031181

65. Robertson, S.E.: The probability ranking principle in information retrieval. Journal of Documentation **33**, 294–304 (1977)

66. Sanderson, M.: Test collection based evaluation of information retrieval systems. Found. Trends Inf. Retr. **4**(4), 247–375 (2010). doi: 10.1561/1500000009

67. Sandhaus, E.: The New York Times Annotated Corpus. Tech. rep. (2008)

68. Schuhmacher, M., Dietz, L., Paolo Ponzetto, S.: Ranking entities for web queries through text and knowledge. In: Proceedings of the 24th ACM International on Conference on Information and Knowledge Management, CIKM '15, pp. 1461–1470. ACM (2015). doi: 10.1145/2806416.2806480

69. Unger, C., Forascu, C., Lopez, V., Ngonga Ngomo, A.C., Cabrio, E., Cimiano, P., Walter, S.: Question answering over Linked Data (QALD-4). In: Cappellato, L., Ferro, N., Halvey, M., Kraaij, W. (eds.) Working Notes for CLEF 2014 Conference (2014)

70. Unger, C., Forascu, C., Lopez, V., Ngonga Ngomo, A.C., Cabrio, E., Cimiano, P., Walter, S.: Question answering over Linked Data (QALD-5). In: Cappellato, L., Ferro, N., Jones, G., San Juan, E. (eds.) Working Notes of CLEF 2015 - Conference and Labs of the Evaluation forum (2015)

71. Unger, C., Ngomo, A.C.N., Cabrio, E.: 6th Open Challenge on Question Answering over Linked Data (QALD-6). In: Sack, H., Dietze, S., Tordai, A., Lange, C. (eds.) Semantic Web Challenges: Third SemWebEval Challenge at ESWC 2016, Heraklion, Crete, Greece, May 29 - June 2, 2016, Revised Selected Papers, pp. 171–177. Springer (2016). doi: 10.1007/978-3-319-46565-4_13

72. Voorhees, E.M., Harman, D.K.: TREC: Experiment and Evaluation in Information Retrieval. The MIT Press (2005)

73. de Vries, A.P., Vercoustre, A.M., Thom, J.A., Craswell, N., Lalmas, M.: Overview of the INEX 2007 Entity Ranking track. In: Proceedings of the 6th Initiative on the Evaluation of XML Retrieval, INEX '07, pp. 245–251. Springer (2008). doi: 10.1007/978-3-540-85902-4_22

74. Wang, Q., Kamps, J., Camps, G.R., Marx, M., Schuth, A., Theobald, M., Gurajada, S., Mishra, A.: Overview of the INEX 2012 Linked Data track. In: CLEF 2012 Evaluation Labs and Workshop, Online Working Notes (2012)

75. Wu, Q., Burges, C.J., Svore, K.M., Gao, J.: Adapting boosting for information retrieval measures. Inf. Retr. **13**(3), 254–270 (2010). doi: 10.1007/s10791-009-9112-1

76. Xu, J., Li, H.: AdaRank: A boosting algorithm for information retrieval. In: Proceedings of the 30th Annual International ACM SIGIR Conference on Research and Development in Information Retrieval, SIGIR '07, pp. 391–398. ACM (2007). doi: 10.1145/1277741.1277809

77. Zhai, C., Lafferty, J.: A study of smoothing methods for language models applied to information retrieval. ACM Trans. Inf. Syst. **22**(2), 179–214 (2004). doi: 10.1145/984321.984322

78. Zhai, C., Massung, S.: Text Data Management and Analysis: A Practical Introduction to Information Retrieval and Text Mining. ACM and Morgan & Claypool (2016)

79. Zheng, Z., Zha, H., Zhang, T., Chapelle, O., Chen, K., Sun, G.: A general boosting method and its application to learning ranking functions for web search. In: Proceedings of the 20th International Conference on Neural Information Processing Systems, NIPS '07, pp. 1697–1704. Curran Associates Inc. (2007)

80. Zhiltsov, N., Agichtein, E.: Improving entity search over Linked Data by modeling latent semantics. In: Proceedings of the 22nd ACM International Conference on Information and Knowledge Management, CIKM '13, pp. 1253–1256. ACM (2013). doi: 10.1145/2505515.2507868

81. Zhiltsov, N., Kotov, A., Nikolaev, F.: Fielded sequential dependence model for ad-hoc entity retrieval in the Web of Data. In: Proceedings of the 38th International ACM SIGIR Conference on Research and Development in Information Retrieval, SIGIR '15, pp. 253–262. ACM (2015). doi: 10.1145/2766462.2767756

Chapter 4
Semantically Enriched Models for Entity Ranking

Most of our efforts in the previous chapter have revolved around constructing term-based representations of entities. These representations can then be ranked using direct adaptations of existing document retrieval models. On the one hand, the resulting approaches are robust and effective across a broad range of application scenarios. On the other hand, these term-based models have little awareness of what it takes to be an entity. Perhaps the most exciting challenge and opportunity in entity retrieval is how to leverage entity-specific properties—attributes, types, and relationships—to improve retrieval performance. This requires a departure from purely term-based approaches toward more semantically informed representations. This change of direction is supported by the emergence of knowledge bases over the past decade (cf. Sect. 2.3). Knowledge bases organize information about entities in a structured and semantically meaningful way. For us, semantics is taken to be synonymous with structure (more precisely, with references to meaningful structure). Our efforts in this chapter are driven by the following question: How can one leverage structured knowledge repositories in entity retrieval?

At its core, the entity ranking task (and most IR tasks for that matter) boils down to the problem of matching representations. That is, computing similarities between representations of queries (information needs) and those of entities (information objects). The question then becomes: How to preserve and represent structure associated with entities? Importantly, to be able to make use of richer (i.e., semantic) entity representations during matching, queries also need to have correspondingly enriched representations. For example, if types of entities are represented as semantic units, as opposed to sequences of words, then we also need to know the target types of the query. For now, we shall assume that we are provided with such enriched queries, which will be referred to as *keyword++ queries*.

In this chapter, we will look at semantically enriched entity retrieval in several flavors. Table 4.1 presents an overview of the different tasks, and serves as a roadmap to this chapter. Many of these tasks have been proposed and studied in the context of some benchmarking campaign. As we shall see, some of the most

© The Author(s) 2018
K. Balog, *Entity-Oriented Search*, The Information Retrieval Series 39,
https://doi.org/10.1007/978-3-319-93935-3_4

Table 4.1 Overview of various entity ranking tasks addressed in this chapter

Task	Query formulation	Section	Benchmark
Ad hoc entity retrieval	Keyword	4.2.1	
	Keyword	4.4.1	
	Keyword++ (query entities)	4.2.2	
	Keyword++ (target types)	4.3	INEX Entity Ranking
List search	Keyword	4.4.2	SemSearch Challenge
Related entity finding	Keyword++ (input entity, target type)	4.4.3	TREC Entity
Similar entity search	Keyword++ (example entities)[a]	4.5	INEX Entity Ranking

[a]The query may or may not have a keyword component (referred to as example-augmented vs. example-based search, respectively)

Table 4.2 Notation used in this chapter

Symbol	Meaning
e	Entity ($e \in \mathcal{E}$)
$\mathcal{E}$	Entity catalog (set of all entities)
$\mathcal{E}_q$	Set of query entities
$\mathcal{E}_q(k)$	Top-k ranked entities for query q
f	Field ($f \in \mathcal{F}$)
$\mathcal{F}$	Set of fields
f_e	Field f of entity e
$\mathcal{K}$	Knowledge base
$\mathcal{L}_e$	Links of entity e (i.e., set of nodes connected to e in the knowledge graph)
q	Keyword query
$\tilde{q}$	Keyword++ query ($\tilde{q} = (q, X_q, Y_q, \dots)$)
(s, p, o)	Subject-predicate-object (SPO) triple ($(s, p, o) \in \mathcal{K}$)
$\mathcal{T}$	Type taxonomy
$\mathcal{T}_e$	Set of types assigned to entity e
$\mathcal{T}_q$	Set of target types (a.k.a. query types)
y	Entity type ($y \in \mathcal{T}$)

effective approaches are tailor-made and highly specialized for the particular task. This chapter is mainly organized around the various aspects of entities that are utilized: properties generally (Sect. 4.2), then more specifically types (Sect. 4.3) and relationships (Sect. 4.4). In Sect. 4.5, we consider the task of similar entity search, which revolves around comparing representations of entities. Finally, in Sect. 4.6, we show that structure can also be exploited in a static (query-independent) fashion. Table 4.2 summarizes the notation used throughout this chapter.

4.1 Semantics Means Structure

Our objective in this chapter is to build *semantically enriched* entity retrieval models. We introduce the following working definition of semantics: "references to meaningful structures." What we mean by that is that specific entities, types, or relationships are recognized and identified uniquely, with references to an underlying knowledge repository, as opposed to being treated as mere strings. This makes it possible to search by *meaning* rather than just literal matches.

> Semantically enriched entity retrieval models extend the representation of entities from mere sequences of terms to include information about specific attributes, types, and relationships, and leverage this structured information when matching entities against information needs (queries).

This semantic enrichment needs to be woven through all elements of the retrieval process. In order to make use of rich entity representations, the retrieval model has to utilize these additional structures during matching. To be able to do that, queries also need to be enriched. See Fig. 4.1a vs. b for the illustration of the difference between term-based and semantic entity retrieval models.

The enrichment of queries may happen on the users' side or may be performed automatically. The former is typically facilitated through various query assistance services, such as facets or query auto-completion. For expressing complex information needs, custom user interfaces may also be built (see, e.g., [9, 36]). While such interactive query-builders offer the same expressivity as structured query languages (like SQL and SPARQL), they share the same disadvantages as well: users need to receive some training on how to use the tool. The other alternative is to rely on machine understanding of queries, i.e., to obtain semantic enrichments by automatic means. We will look at this direction in detail in Chap. 7. Finally, hybrid approaches that combine human and machine annotations are also possible.

Our focus in this chapter is not on the mechanics of query enrichment. We assume an enriched query as our input; how exactly it was obtained is presently immaterial for us. We make the following distinction for notational convenience.

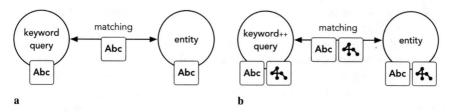

a   **b**

Fig. 4.1 (**a**) Ranking entities using term-based representations. (**b**) Semantically enriched entity ranking by incorporating structure

We use $\tilde{q}$ for a semantically enriched, i.e., keyword++, query. When referring to the keyword component of the query, we shall write q. In many cases, our input query will be a tuple $\tilde{q} = (q, X_q, Y_q, \dots)$, where X and Y are the additional query components or "enrichments" (e.g., target types or example entities). A typical approach, which we shall encounter several times throughout this chapter, is to build multiple representations of both entities and queries, in addition to the term-based one. Each of these additional "parallel representations" is designated to preserve the semantics associated with a specific entity property (e.g., types or relationships). Then, a given candidate entity is scored against the query based on each of these representations. Finally, these scores are combined, e.g., linearly:

$$score(e; \tilde{q}) = \lambda_t \, score_t(e; q) + \lambda_X \, score_X(e; X_q) + \lambda_Y \, score_Y(e; Y_q) + \dots ,$$

where $score_t(e; q)$ is the term-based retrieval score (using methods from the previous chapter), and the other score components correspond to the various other representations $(X, Y, \dots)$.

One important detail that needs attention when using the above formulation is that the different similarity scores need to be "compatible," i.e., must have the same magnitude. A simple technique to ensure compatibility is to normalize the top-k results (using the same k value across the different score components) by the sum of the scores for that query, and assign a zero score to results below rank k:

$$score'_x(e; q) = \begin{cases} \frac{1}{Z} \, score_x(e; q), & e \in \mathcal{E}_q(k) \\ 0, & \text{otherwise} , \end{cases}$$

where $\mathcal{E}_q(k)$ denotes the set of top-k results (entities) for query q and the normalization coefficient is set to $Z = \sum_{e \in \mathcal{E}_q(k)} score_x(e; q)$.

4.2 Preserving Structure

In this section, we look at how to preserve (and exploit) the rich structure associated with entities in a knowledge base. We will assume that each entity is stored as a set of SPO triples (cf. Sect. 2.3.1.2). Recall that the subject and predicate are always URIs, while the object can be a URI or a literal value. Previously, in Chap. 3, we have built fielded entity descriptions by grouping multiple predicates together into each field. The corresponding object values have been concatenated and set as that field's value. Further, we have replaced object URIs with the primary name (or label) of the given object (cf. Table 3.4). The resulting term-based entity representations are well suited for use with existing retrieval methods. On the other hand, most of the rich structure that is associated with an entity has been lost in the process.

We will now modify our approach in order to preserve the semantic information encapsulated in SPO triples. Two specific issues will be covered. The first is the case of multi-valued predicates (i.e., triples with the same subject and predicate, but with multiple object values). While these are implicitly handled to some extent when using proximity-aware retrieval models (cf. Sect. 3.3.2.4), we will address "multi-valuedness" more explicitly in Sect. 4.2.1. The second is the case of URI-valued objects, which are references to other entities. Instead of simply replacing these with the corresponding entity names and treating them as terms, we will consider them as first-class citizens and distinguish them from regular terms when constructing entity representations (Sect. 4.2.2).

4.2.1 Multi-Valued Predicates

For the issue of multi-valued predicates, we will continue to use a term-based entity representation. Accordingly, all entity properties may be seen as attributes (since types and related entities are also "just strings" in this case). The one adjustment we make is that we keep each predicate as a separate field, instead of folding predicates together. Thus, each field corresponds to a single predicate. Some of the predicates are multi-valued, i.e., have more than a single value. For example, a person may have multiple email addresses and most movies have multiple actors. Campinas et al. [14] present an extension to BM25F, called BM25MF, for dealing with multi-valued fields. Even though we focus exclusively on BM25F here, we note that similar extensions may be developed for other fielded retrieval models as well.

For convenience, we repeat how term frequencies are aggregated across different fields according to the original BM25F formula (cf. Sect. 3.3.2.3):

$$\tilde{c}(t; e) = \sum_{f \in \mathcal{F}} \alpha_f \frac{c(t; f_e)}{1 - b_f + b_f \frac{l_{f_e}}{\bar{l}_f}} .$$

Recall that for a given field f, α_f is the field's weight, b_f is a field length normalization parameter, and $\bar{l}_f$ is the average length of the field; these values are the same for all entities in the catalog. For a specific entity, e, $c(t; f_e)$ is the frequency of term t in field f that entity and l_{f_e} is the length of the field (number of terms).

According to the multi-valued extension BM25MF, the entity term frequency becomes:

$$\tilde{c}(t; e) = \sum_{f \in \mathcal{F}} \alpha_f \frac{\tilde{c}(t; f_e)}{1 - b_f + b_f \frac{|f_e|}{\overline{|f|}}} .$$

Here, $|f_e|$ is the cardinality of field f of e, while $\overline{|f|}$ denotes the average cardinality of field f across all entities in the catalog. Field cardinality refers to the number of distinct values in a field. Term frequencies are computed with respect to a given entity field according to:

$$\tilde{c}(t; f_e) = \sum_{v \in f_e} \alpha_v \frac{c(t; f_{e,v})}{1 - b_v + b_v \frac{l_v}{\bar{l}_f}} ,$$

where $c(t; f_{e,v})$ is the term frequency within the specific field value v. The length of the value, l_v, is measured in the number of terms. Finally, α_v and b_v are value-specific weights and normalization parameters, respectively.

4.2.1.1 Parameter Settings

For setting the field and value weight parameters (α_f and α_v), Campinas et al. [14] introduce a number of options, both query-independent and query-dependent. It is also possible to combine multiple weighting methods (by multiplying the parameters produced by each of the methods).

Recall that each field in the entity description corresponds to a unique predicate (URI). We write p_f to denote the predicate value that is assigned to field f. Field weights can be defined heuristically based on p_f, using a small set of regular expressions:[1]

$$\alpha_f = \begin{cases} 2.0, & p_f \text{ matches } .\star[\texttt{label|name|title|sameas}]\$ \\ 0.5, & p_f \text{ matches } .\star[\texttt{seealso|wikilinks}]\$ \\ 0.1, & p_f \text{ matches } \texttt{<rdf:\_[0-9]+>} \\ 1.0, & \text{otherwise} \, . \end{cases}$$

Alternatively, the field and value weight parameters may be estimated based on the portion of query terms covered. Note that this way α_f and α_v are set in a query-dependent manner. *Query coverage* measures the portion of query terms that are contained in the field or value. Additionally, it also considers the importance of terms, based on their IEF value. Formally:

$$\alpha_x(q) = \frac{\sum_{t \in x \cap q} IEF(t)^2}{\sum_{t \in q} IEF(t)^2} \, ,$$

where x stands for either f or v.

Another way of setting the value weight parameter is based on the notion of *value coverage*, which reflects the portion of terms for a given field value that match the query. To compensate for the differences in value lengths, the following formula is used:

$$\alpha_v(q) = \frac{\alpha}{1 + (\alpha - 1)\left[\frac{\sum_{t \in v \cap q} c(t; f_{e,v})}{l_v}\right]^B} \, ,$$

where $\alpha \in (0, 1)$ is a parameter that imposes a fixed lower bound, to prevent short values gaining a benefit over long values, and B is a parameter that controls the

---

[1] We use the prefixed version for the rdf namespace; see Table 2.4 for the full URL.

effect of coverage. "The higher B is, the higher the coverage needs to be for the value node [field value] to receive a weight higher than α" [14].

On top of field length normalization (b_f), the BM25MF ranking function offers an additional normalization on the field's cardinality (b_v). Based on the experiments in [14], $b_f \in [0.4, 0.7]$ and $b_v \in [0.5, 0.8]$ generally provide good overall performance (although there can be considerable differences across datasets).

4.2.2 References to Entities

Until this point, we have replaced the references to related entities (i.e., URI object values of SPO triples) with terms; specifically, with the primary names of the corresponding entities. This enhances the "findability" of entities by means of keyword queries. At the same time, a large part of the underlying structure (and hence semantics) gets lost in this translation. Consider in particular the issue of entity ambiguity. This has already been resolved in the underlying knowledge base, thanks to the presence of unique identifiers. Replacing those identifiers with the associated names re-introduces ambiguity.

Hasibi et al. [25] propose to preserve these entity references by employing a *dual entity representation*: On top of the traditional term-based representation, each entity is additionally represented by means of its associated entities. We will refer to the latter as *entity-based representation*. The idea is illustrated in Fig. 4.2. We will come back to the issue of field selection for building these fielded representations.

Crucially, to be able to make use of the proposed entity-based representation, we need to likewise enrich the term-based query with entity annotations. We shall assume that there is some query annotation mechanism in place that recognizes entities in the query and assigns unique identifiers to them (the dashed arrow in

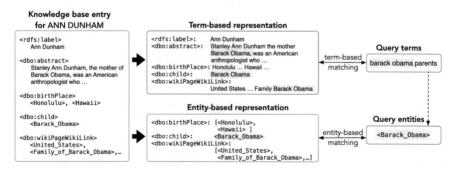

Fig. 4.2 Illustration of dual term-based and entity-based entity representations. On the left, predicate-object pairs are shown for a given subject entity (ANN DUNHAM). In the middle are the corresponding term-based and entity-based representations. On the right is the query, which also has a dual representation. Multi-valued fields are indicated using square brackets ([. . .]). URIs are typeset in monospace font. Matching query terms/entities are highlighted. Figure is based on [25]

Fig. 4.2). These query entities may be provided by the user (e.g., by using query auto-completion and clicking on a specific entity in the list of suggestions) or by some automatic entity linking method (see Sect. 7.3). The details of the query annotation process are not our focus here.

Formally, our input keyword++ query is a tuple $\tilde{q} = (q, \mathcal{E}_q)$, where $q = \langle q_1, \ldots, q_n \rangle$ is the keyword query (sequence of terms) and $\mathcal{E}_q = \{e_1, \ldots, e_m\}$ is a set of entities recognized in the query, referred to as *query entities* (possibly an empty set). Further, we assume that each of the query entities $e \in \mathcal{E}_q$ has an associated weight $w(e, q)$, reflecting the confidence in that annotation. These weights play a role when annotations are obtained automatically; if query entities are provided by the user then they would all be assigned $w(e, q) = 1$.

There are several possibilities to combine term-based and entity-based representations during matching. Hasibi et al. [25] propose a theoretically sound solution, referred to as ELR (which stands for *entity linking incorporated retrieval*). It is based on the Markov random field (MRF) model [40], and thus may be applied on top of any term-based retrieval model that can be instantiated in the MRF framework. In our discussion, we will focus on the sequential dependence model (SDM) variant (we refer back to Sect. 3.3.1.3 for details). ELR extends the underlying graph representation of SDM with query entity nodes; see the shaded circles in Fig. 4.3. Notice that query entities are independent of each other. This introduces a new type of clique: 2-cliques between the given entity (that is being scored) and the query entities. Denoting the corresponding feature function as $f_{\mathcal{E}}(e_i; e)$ and the associated weight as $\lambda_{\mathcal{E}}$, the MRF ranking function is defined as:

$$
P_\Lambda(e|\hat{q}) \stackrel{\text{rank}}{=} \lambda_T \sum_{i=1}^{n} f_T(q_i; e)
$$

$$
+ \lambda_O \sum_{i=1}^{n-1} f_O(q_i, q_{i+1}; e)
$$

$$
+ \lambda_U \sum_{i=1}^{n-1} f_U(q_i, q_{i+1}; e)
$$

$$
+ \lambda_{\mathcal{E}} \sum_{i=1}^{m} f_{\mathcal{E}}(e_i; e) .
$$

Fig. 4.3 Graphical representation of the ELR model [25] for a query with three terms and two query entities

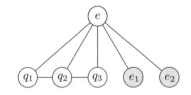

There is, however, a crucial difference between term-based and entity-based matches, which we need to take account of. As explained in Hasibi et al. [25], "the number of cliques for term-based matches is proportional to the length of the query (n for unigrams and $n - 1$ for ordered and unordered bigrams), which makes them compatible (directly comparable) with each other, irrespective of the length of the query." This is why the SDM parameters are outside the summations and can be trained without having to deal with query length normalization. The same cannot be done with $\lambda_{\mathcal{E}}$, for two reasons. First, the number of query entities varies and is independent of the length of the query (e.g., a short but ambiguous keyword query may have several query entities, while a long natural language query might have only a single one). Second, query entities have different weights (confidence scores) associated with them, and these should be taken into consideration (this is of particular importance when automatic query annotation is used). To overcome the above issues, the λ parameters are rewritten as parameterized functions over each clique:

$$\lambda_T(q_i) = \lambda_T \frac{1}{n},$$

$$\lambda_O(q_i, q_{i+1}) = \lambda_O \frac{1}{n - 1},$$

$$\lambda_U(q_i, q_{i+1}) = \lambda_U \frac{1}{n - 1},$$

$$\lambda_{\mathcal{E}}(e_i) = \lambda_{\mathcal{E}} \frac{w(e_i, q)}{\sum_{j=1}^{m} w(e_j, q)}.$$

Using these parametrized λ functions, and factoring constants out of the summations, the final ranking function becomes:

$$
\begin{aligned}
P_\Lambda(e|\hat{q}) \overset{\text{rank}}{=} & \frac{\lambda_T}{n} \sum_{i=1}^{n} f_T(q_i; e) \\
& + \frac{\lambda_O}{n - 1} \sum_{i=1}^{n-1} f_O(q_i, q_{i+1}; e) \\
& + \frac{\lambda_U}{n - 1} \sum_{i=1}^{n-1} f_U(q_i, q_{i+1}; e) \\
& + \frac{\lambda_{\mathcal{E}}}{\sum_{j=1}^{m} w(e_j, q)} \sum_{i=1}^{m} w(e_i, q) \, f_{\mathcal{E}}(e_i; e).
\end{aligned}
$$

We subject the free parameters to the constraint $\lambda_T + \lambda_O + \lambda_U + \lambda_{\mathcal{E}} = 1$. Notice that by setting λ_O and λ_U to zero, the above model is an extension of unigram language models, LM, MLM, and PRMS. Otherwise, it extends SDM and FSDM. The default parameter values in [25] are (1) $\lambda_T = 0.9$ and $\lambda_{\mathcal{E}} = 0.1$ for unigram language models and (2) $\lambda_T = 0.8$, $\lambda_O = 0.05$, $\lambda_U = 0.05$, and $\lambda_{\mathcal{E}} = 0.1$ for sequential dependence models (SDM and FSDM).

The last component of the model that remains to be defined is the feature function $f_{\mathcal{E}}(e_i; e)$. Before that, let us briefly discuss the construction of entity-based representations, as the choices we make there will have an influence on the feature function. Let $\tilde{e}$ denote the entity-based representation of entity e. Each unique predicate from the set of SPO triples belonging to e corresponds to a separate field in $\tilde{e}$, and $\tilde{\mathcal{F}}$ denotes the set of fields across the entity catalog. Unlike for terms-based representations, we do not fold predicates together here. When computing the degree of match between a candidate entity e that is being scored and a query entity e_i, we take a weighted combination of field-level scores (for each field in $\tilde{e}$). There are two important differences compared to a traditional term-based representation. Firstly, each entity appears at most once in each field (because of how we mapped SPO triples to fields). Secondly, if a query entity appears in a field, then it shall be regarded as a "perfect match," independent of what other entities are present in the same field. Using our example from Fig. 4.2, if the query entity <dbp:Barack_Obama> is contained in the <dbo:child> field of the entity ANN DUNHAM, then it is treated as a perfect match (since it should not matter how many other children she has). Driven by the above considerations the feature function is defined as:

$$f_{\mathcal{E}}(e_i; e) = \log \sum_{f \in \tilde{\mathcal{F}}} w_f^{\mathcal{E}} \left((1 - \lambda)\, \mathbb{1}(e_i, f_{\tilde{e}}) + \lambda \frac{\sum_{e' \in \mathcal{E}} \mathbb{1}(e_i, f_{\tilde{e}'})}{|\{e' \in \mathcal{E} : f_{\tilde{e}'} \neq \emptyset\}|} \right),$$

where the linear interpolation implements the Jelinek-Mercer smoothing method (using $\lambda = 0.1$ in [25]), and $\mathbb{1}(e, f_{\tilde{e}})$ is a binary indicator function, which is 1 if e_i is present in the entity field $f_{\tilde{e}}$ and otherwise 0. The background model part of the interpolation employs a notion of (fielded) entity frequency; the number of entities in the catalog that contain e_i in field f is divided by the total number of entities for which that field is non-empty.

Finally, for setting the field weights $w_f^{\mathcal{E}}$, we employ dynamic mapping using PRMS, exactly the same way as we did for terms (cf. Eq. (3.13)), but using entity identifier (URI) tokens instead of terms. Setting the field weights in this manner has a number of advantages: (1) there are no additional free parameters to be trained and (2) the importance of fields is chosen dynamically for each query entity (depending on which fields it typically occurs in).

We note that the ELR model we discussed here may be applied to entity types as well (as those are also represented with a URI value as object in SPO triples). However, since types are conceptually different from related entities, they shall receive special treatment in the next section.

4.3 Entity Types

One distinctive characteristic of entities is that they are typed. Each entity in a knowledge base in principle has a type, or multiple types, assigned to it. These types are often organized hierarchically in a *type taxonomy*. Types may also be referred to as *categories*, e.g., in an e-commerce context.

In this section, we will assume a keyword++ input query that takes the form $\tilde{q} = (q, \mathcal{T}_q)$, where q is the keyword query and $\mathcal{T}_q$ is a set of *target types* (also referred to as *query types*). Like with query entities in the previous section, these types may be provided by the user (e.g., through the use of faceted search interfaces, see Fig. 4.4) or identified automatically (cf. Sect. 7.2). An abstraction of this very scenario has been studied at the INEX Entity Ranking track [19, 20, 54], where a keyword query is complemented with a small number of target types. There, Wikipedia categories were used as the type taxonomy. An example INEX topic is shown in Listing 4.1.

Having knowledge of the target types of the query, retrieval results may be filtered or re-ranked based on how well they match these target types. There are, however, a number of complicating factors, which we shall elaborate upon below.

4.3.1 Type Taxonomies and Challenges

When the type taxonomy is flat and comprises of only a handful of types (such as in the Airbnb example in Fig. 4.4 (left)), the usage of type information is rather straightforward: A strict type filter may be employed to return only entities of the desired type(s). Many type taxonomies, however, are not like that. They consist of a large number of types that are organized in a multi-layered hierarchy. The type hierarchy is transitive, i.e., if entity e is of type y and y is a subtype of z, then e is

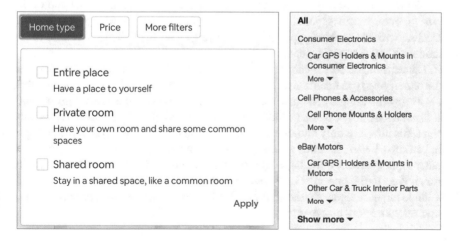

Fig. 4.4 Entity type filters on the SERP in AirBnb (Left) and eBay (Right)

Table 4.3 Overview of type taxonomies corresponding to different knowledge repositories

	DBpedia[a]	Freebase[b]	Wikipedia[c]	YAGO
#types	713	1719	1,333,352	568,672
Depth	7	2	N/A	19
#top-level types	22	92	N/A	61
#leaf-level types	561	1626	662,609	549,775
Avg. #types/entity	2.8	4.4	4.5	12.2

[a]Version 2015-10
[b]Latest available dump (2015-03-31)
[c]Corresponding to DBpedia version 2015-10

```
<inex_topic topic_id="132">
  <title>living nordic classical composers</title>
  <description>I want a list of living classical composers, who are born in
nordic countries.</description>
  <narrative>Iceland, Denmark, Sweden, Norway and Finland are the Nordic
countries. They share quite a similar musical heritage. Therefore, a set of
contemporary living Nordic composers are sought.</narrative>
  <categories>
    <category id="47342">21st century classical composers</category>
    <category id="39380">finnish composers</category>
    <category id="37202">living classical composers</category>
  </categories>
</inex_topic>
```

Listing 4.1 Example topic definition from the INEX 2008 Entity Ranking track. Systems only use the contents of the `<title>` and `<categories>` tags, `<description>` and `<narrative>` are meant for human relevance assessors to clarify the query intent

also of type z. The root node of the hierarchy is often a generic concept, such as "object" or "thing." There might be less "well-behaved" type systems (Wikipedia categories being a prime example), where types do not form a well-defined "is-a" hierarchy (i.e., the type taxonomy is a graph, not a tree). Table 4.3 shows the type taxonomies corresponding to four popular large-scale knowledge repositories. Dealing with hierarchical type taxonomies brings a set of challenges, related to the modeling and usage of type information.

Concerning the user's side, in many cases the user has no extensive knowledge of the underlying type taxonomy. One real-life example is provided in Fig. 4.4 (right), displaying the category filters shown on eBay in response to the query *"gps mount."* It is not necessarily obvious which of these categories should be picked (especially since this interface only allows for selecting a single one). The INEX topic example in Listing 4.1 also identifies a very specific (albeit incomplete) set of categories; for other topics, however, the target type(s) may be more wide-ranging. For example, for another INEX query, *"Nordic authors who are known for children's literature,"* a single target category "writers" is provided. Thus, the target types provided by the user might be very broad or very narrow.

There are also issues on the data side. Type assignments of entities, particularly in large entity repositories, are imperfect. Types associated with an entity may be incomplete or missing altogether, wrong types may be attributed, and type assignments may be done inconsistently across entities. The type taxonomy may suffer from various coverage and quality problems too. It is often the case that certain branches of the taxonomy are very detailed, while for specific entities there is no matching category other than the root node of the hierarchy, which is overly generic to be useful. The upshot is as follows:

> In many application scenarios, which involve a type system (taxonomy) with more than a handful of possible categories, target type information should be treated as "hints" rather than as strict filters.

4.3.2 Type-Aware Entity Ranking

We model type-based similarity as a separate retrieval component. The type-aware scoring formula can be written in the form of a linear mixture [3, 32, 45, 47]:

$$score(e; \tilde{q}) = \lambda \, score_t(e; q) + (1 - \lambda) \, score_T(e; \mathcal{T}_q) \, ,$$

where the first component, $score_t(e; q)$, is the term-based similarity between entity e and the keyword part of the query. This score may be computed using any of the methods from the previous chapter. The second component, $score_T(e; \mathcal{T}_q)$, expresses the type-based similarity between the entity and the set of target types $\mathcal{T}_q$. The interpolation parameter λ is chosen empirically.

Alternatively, the type-aware score may also be written as a multiplication of the components:

$$score(e; \tilde{q}) = score_t(e; q) \times score_T(e; \mathcal{T}_q) \, .$$

Since the relative influence of the individual components cannot be adjusted, this formulation is primarily used for type-based filtering of search results. We shall see an example for this sort of usage in Sect. 4.4.3.

4.3.3 Estimating Type-Based Similarity

Let us next consider a number of different ways of establishing the similarity between an entity and a set of target types, that is, estimating $score_T(e; \mathcal{T}_q)$. Note that relevant entities may not be associated with the provided target types. This is alleviated by leveraging hierarchical relationships of types in the taxonomy (either explicitly in the scoring formula or by expanding the types of the entity and/or the query). The choice of method to use depends on the particular application scenario and type taxonomy.

Term-Based Similarity One simple solution is to join the labels of the corresponding entity types in a separate field and measure the similarity between the labels of target types and this field using any term-based retrieval model; see, e.g., [18]. Considering the example in Listing 4.1, the bag-of-words type query becomes "21st[1] century[1] classical[2] composers[3] finnish[1] living[1]," where the numbers in the superscript denote the query term frequency. The advantage of this method is that a separate type field is often distinguished anyway in the fielded entity description (cf. Table 3.5), thus no extra implementation effort is involved on the entity side. Further, the hierarchical nature of the type taxonomy can be easily exploited by expanding the type query with labels of subcategories, siblings, etc. The disadvantage is that we are limiting ourselves to surface-level matches.

Another variant of this idea is to represent types in terms of their "contents" (not just by their labels). This content-based representation can be obtained by concatenating the descriptions of entities that belong to that type; see, e.g., [32]. A term-based type query is formulated the same way as before, and is scored against the content-based representation.

Set-Based Similarity Since both the target types and the types of an entity are sets (of type identifiers), it is natural to consider the similarity of those sets. Pehcevski et al. [45] measure the *ratio of common types* between the set of types associated with an entity ($\mathcal{T}_e$) and the set of target types ($\mathcal{T}_q$):

$$score_T(e; \mathcal{T}_q) = \frac{|\mathcal{T}_e \cap \mathcal{T}_q|}{|\mathcal{T}_q|} .$$

Both the entity and target types may be expanded using ancestor and descendant categories in the type taxonomy.

Graph-Based Distance Raviv et al. [47] propose to base the type score on the distance of the entity and query types in the taxonomy:

$$score_T(e; \mathcal{T}_q) = e^{-\alpha d(\mathcal{T}_q, \mathcal{T}_e)} ,$$

where e stands for the mathematical constant that is the base of the natural logarithm (not to be confused with entity e), α is a decay coefficient (set to 3 in [47]), and $d(\mathcal{T}_q, \mathcal{T}_e)$ is the distance between the types of the query and the types of entity e, computed as follows:

$$d(\mathcal{T}_q, \mathcal{T}_e) = \begin{cases} 0, & \mathcal{T}_q \cap \mathcal{T}_e \neq \emptyset \\ \min\left(d_{max}, \min_{y \in \mathcal{T}_q, z \in \mathcal{T}_e} d(y, z)\right), & \text{otherwise} . \end{cases}$$

In words, if the query and entity share any types then their distance is taken to be zero; otherwise, their distance is defined to be the minimal path length between all pairs of query and entity types (denoted as $d(y, z)$ for types y and z). Additionally, there is a threshold on the maximum distance allowed, in case the query and entity types are too far apart in the taxonomy (d_{max}, set to 5 in [47]).

Probability Distributions Balog et al. [3] model type information as probability distributions over types, analogously to the idea of language modeling (which is about representing documents/entities as probability distributions over terms). Let θ_{T_q} and θ_{T_e} be the type models (i.e., probability distributions) corresponding to the query and entity, respectively. The type-based similarity then is measured in terms of the distance between the two distributions:

$$score_T(e; T_q) = \max_{e' \in \mathcal{E}} KL(\theta_{T_q} || \theta_{T_{e'}}) - KL(\theta_{T_q} || \theta_{T_e}) \,.$$

The distance function employed is the Kullback–Leibler (KL) divergence; the maximum distance is used for turning this into a similarity function.

Type models are represented as multinomial distributions.[2] The probability of a type y given an entity is estimated analogously to unigram language models employing Dirichlet prior smoothing. Mind that we denote individual types as y (so as not to be confused with terms):

$$P(y|\theta_{T_e}) = \frac{\mathbb{1}(y \in T_e) + \mu_T P(y|\mathcal{E})}{|T_e| + \mu_T} \,,$$

where $\mathbb{1}(y \in T_e)$ takes the value 1 if y is one of the types assigned to entity e, and otherwise equals to 0. The total number of types assigned to e is denoted as $|T_e|$. The smoothing parameter μ_T is set to the average number of types assigned to an entity across the catalog. Finally, the background (catalog-level) type model is the relative frequency of the type across all entities:

$$P(y|\mathcal{E}) = \frac{\sum_{e \in \mathcal{E}} \mathbb{1}(y \in T_e)}{\sum_{e \in \mathcal{E}} |T_e|} \,.$$

The types of the query are also modeled as a probability distribution. In the simplest case, we can set it according to the relative type frequency in the query:

$$P(y|\theta_{T_q}) = \frac{\mathbb{1}(y \in T_q)}{|T_q|} \,.$$

---

[2]Since types are binary assignments, the attentive reader might wonder why not use a multivariate Bernoulli distribution instead. That is certainly a possibility, which the reader is invited to explore. General reasons for using a multinomial distribution include that it is simpler to understand, easier to implement, and appears to be more efficient [38]. Our specific reason is that the same statistical distribution is used for modeling the query; when employing query expansion techniques, we need to be able to capture the importance of a given type (as opposed to its mere presence/absence).

Since each type appears at most once in the query, this basically means distributing the probability mass uniformly across the query types. Considering that input type information may be very sparse, it makes sense to enrich the type query model ($\theta_{\mathcal{T}_q}$) by (1) considering other types that are relevant to the keyword query, or (2) applying (pseudo) relevance feedback techniques (analogously to the term-based case) [3].

4.4   Entity Relationships

Relationships, or "typed links," are another unique characteristic of entities. Many information needs involve searching for entities based on the relationships between them. Consider, e.g., the queries *"teammates of Michael Schumacher," "wives of Tom Cruise,"* or *"astronauts who landed on the Moon."* In this section, we discuss approaches for utilizing relationship information for entity retrieval. We look at three particular variants of the entity ranking task. We start with classical *ad hoc search* using keyword queries (Sect. 4.4.1). Next, we consider *list search*, where we still use a keyword query, but we have an additional piece of information, namely, that the query seeks a list of entities (Sect. 4.4.2). Finally, we discuss *related entity finding*, where the input is a keyword++ query, which includes an input entity and a target type (Sect. 4.4.3).

Depending on the task, we may view the knowledge repository as a *knowledge graph*. There, each entity is a node that is connected to other resources[3] via labeled directed edges (i.e., predicates).

4.4.1   Ad Hoc Entity Retrieval

To begin with, we consider the standard ad hoc entity retrieval task (using conventional keyword queries). However, instead of relying only on term-based ranking, we will additionally exploit the structure of the knowledge graph. Specifically, we present the approach proposed by Tonon et al. [52], where (1) a set of seed entities are identified using term-based entity retrieval, and then (2) edges of these seed entities are traversed in the graph in order to identify potential additional results.

Let $\mathcal{E}_q(k)$ denote the set of top-k entities identified using the term-based retrieval method; these are used as *seed entities*. Let $\hat{\mathcal{E}}_q$ denote the set of *candidate entities* that may be reached from the seed entities. Tonon et al. [52] define a number of graph patterns, which are expressed as SPARQL queries. *Scope one* structured queries look for candidate entities that have direct links to the seed entities, i.e., follow the pattern $e' \overset{p}{\rightleftharpoons} e$, where e' is a seed entity, e is a candidate entity, and the

---

[3]We say resources because in addition to entities and types, the graph may contain other types of URI-nodes as well (e.g., general concepts or disambiguations).

two are connected by predicate p. Based on the predicate, four different variations are considered, such that each extends the set of predicates from the one above:

- *Same-as links.* The `<owl:sameAs>` predicate connects identifiers that refer to the same real-world entity.
- *Disambiguations and redirects.* Disambiguations (`<dbo:wikiPageDisambi-guates>`) and redirects (`<dbo:wikiPageRedirects>`) to other entities are also incorporated.
- *Properties specific to user queries.* An additional set of predicates that connect seed and candidate entities is selected empirically using a training dataset. These include generic properties, such as Wikipedia links (`<dbp:wikilink>`) and categories (`<dct:subject>`), as well as some more specific predicates, like `<dbo:artist>` or `<dbp:region>`.
- *More general concepts.* On top of the predicates considered by the previous methods, links to more general concepts (predicate `<skos:broader>`) are also included.

An extension of the above approach is to look for candidate entities multiple hops away in the graph. Mind that the number of candidate entities reached potentially grows exponentially with distance from the seed entity. Therefore, only *scope two* queries are used in [52]. These queries search for patterns in the form of $e' \overset{p_1}{\rightleftharpoons} x \overset{p_2}{\rightleftharpoons} e$, where x can be an entity or a type standing in between the seed entity e' and the candidate entity e. The connecting graph edges (predicates p_1 and p_2) are selected from the most frequent predicate pairs.

According to the results in [52], using the second type of scope one queries (i.e., same-as links plus Wikipedia redirects and disambiguations) and retrieving $k = 3$ seed entities performs best. The resulting set of candidate entities may be further filtered based on a pre-defined set of predicates [52]. For each of the candidate entities $e \in \hat{\mathcal{E}}_q$, it is kept track of which seed entity it was reached from. Noting that there may be multiple such seed entities, we let $\mathcal{E}_{e',e}$ denote the set of seed entities that led to e.

Finally, retrieval scores are computed according to the following formula:

$$score(e; q) = \lambda\, score_t(e; q) + (1 - \lambda) \sum_{e' \in \mathcal{E}_{e',e}} score_t(e; e', q) .$$

The first component is the term-based retrieval score of the entity. The second component is the sum of the retrieval scores of the seed entities e' that led to entity e. The interpolation parameter λ is set to 0.5 in [52].

4.4.2 List Search

Next, we consider a specific flavor of ad hoc entity retrieval, where the user is seeking a list of entities. That is, we are supplied with an extra bit of information concerning the intent of the query. We are not concerned with how this knowledge is obtained; it could be an automatic query intent classifier or it could be the user indicating it somehow. This scenario was addressed by the *list search* task of the Semantic Search Challenge in 2011 [10] (cf. Sect. 3.5.2.4). Examples of such queries include "*Apollo astronauts who walked on the Moon*" and "*Arab states of the Persian Gulf.*" Notice that queries are still regular keyword queries, which means that existing term-based entity ranking approaches are applicable (which is indeed what most challenge participants did). Below, we discuss a tailored solution that performs substantially better than conventional entity ranking methods.

The *SemSets* model by Ciglan et al. [15] is a three-component retrieval model, where entities are ranked according to:

$$score(e;q) = score_C(e;q) \times score_S(e;q) \times score_P(e;q) . \qquad (4.1)$$

We detail each of the three score components below.

Candidate Entity Score To identify candidate entities that are possible answers to the query, the process starts with a standard (term-based) entity ranking step (using any model of choice from Sect. 3.3). Let $\mathcal{E}_q(k)$ denote the set of top-k ranked entities and let $rank(e,q) \in [0..k-1]$ indicate the rank position of these entities (lower rank means higher relevance). The "base" entity score is set inversely proportional to the rank position:

$$score_B(e;q) = \begin{cases} 1 - \frac{rank(e,q)}{k} , & e \in \mathcal{E}_q(k) \\ 0, & \text{otherwise} . \end{cases}$$

The base scores are then propagated in the knowledge graph, following the principle of the activation spreading. Ciglan et al. [15] restrict the spreading to only one hop from the given vertices (and use $k = 12$ base entities). Accordingly, the candidate score becomes:

$$score_C(e;q) = score_B(e;q) + \sum_{\substack{(s,p,o)\in\mathcal{K} \\ s=e',o=e}} score_B(e';q) .$$

Thus, each entity receives, in addition to its base score, the sum of base scores of all entities that link to it. Optionally, spreading may be restricted to a selected subset of predicates. The candidate set comprises of entities with a non-zero candidate score: $\mathcal{E}_C = \{e : score_C(e;q) > 0\}$. Entities outside this set would receive a final score of zero because of the multiplication of score components in Eq. (4.1), and therefore are not considered further.

Semantic Set Score A key idea in this model is the concept of *semantic sets* (SemSets): "sets of semantically related entities from the underlying knowledge base" [15]. Intuitively, members of a music band or companies headquartered in the same city would constitute SemSets. We shall first explain how SemSets are used for scoring entities, and then detail how they can be obtained.

The SemSets score of an entity is calculated as a sum of the relevance scores of all SemSets it belongs to:

$$score_S(e; q) = 1 + b \left(\sum_{S \in S_q} \sum_{e \in S} score(S; q) \right),$$

where S_q are the candidate semantic sets for the query and b is a boost parameter (set to 100 in [15]). For a given SemSet S, the relevance score $score(S; q)$ is established based on its similarity to the query. A term-based representation of the set is built by concatenating the descriptions of all entities that belong to that set: $S_t = \bigcup_{e \in S} e_t$, where e_t is the term-based representation of e. This representation can then be scored against the query using any standard text retrieval model.

The selection of candidate semantic sets is based on their overlap with the set of candidate entities. That is, a certain fraction of the SemSet's member entities must also be in the candidate entity set identified for the query. Denoting the set of all possible SemSets as S, the set S_q of candidate SemSets for the query is given by:

$$S_q = \left\{ S \in S : \frac{|S \cap \mathcal{E}_C|}{|S|} \geq \gamma \right\},$$

where γ is a threshold parameter (set to 0.7 in [15]).

The construction of possible SemSets S is governed by two graph patterns, illustrated in Fig. 4.5:

(a) Vertices (entities) with outgoing edges, labeled with the same predicate, to the same object (or, in terms of SPO triples: $(\star, p, o)$). For instance, Wikipedia categories are examples of sets of this type, i.e., the predicate is <dct:subject> and the object is a given category (e.g., <dbc:People_who_have_walked-_on_the Moon>).

Fig. 4.5 Two graph patterns for forming semantic sets (SemSets) of entities. Figure is based on [15]

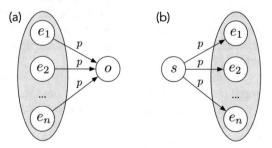

(b) Vertices (entities) with incoming edges, labeled with the same predicate, from the same subject (SPO pattern $(s, p, \star)$). For example, members of some music band (predicate `<dbo:bandMember>`) constitute a semantic set of this type.

A problem with the above constructions is that the number of possible SemSets is huge and becomes impractical to handle. Arguably, not all types of edges (predicates) are equally useful. Thus, set formation may be restricted to specific predicates. It is shown in [15] that using only two types of predicates, Wikipedia category memberships (`<dct:subject>`) and Wikipedia templates (`<dbp:wikiPageUses-Template>`), provides solid performance. This is not surprising, since Wikipedia categories and templates are complementary, manually curated lists of semantically related entities (cf. Sect. 2.2). Alternatively, sets may be filtered automatically, using graph structural measures [15].

Principal Entity Relatedness The third score component considers the distance of a given entity from the *principal entity* of the query. Entities in the query are recognized using a dictionary-based approach (cf. Sect. 7.3). Then, the entity with the highest confidence score is selected as the principal entity of the query. (It might also happen that no entities are recognized in the query, in which case this score component is ignored.) The principal entity relatedness score is defined as:

$$score_P(e; q) = 1 + c \times sim\left(e, e_q\right) ,$$

where e_q denotes the principal entity of the query, c is a boost parameter (set to 100 in [15]), and $sim(e, e')$ is a graph structural similarity measure. Specifically, entities are represented in terms of their neighbors in the knowledge graph and the cosine similarity of the corresponding vectors is computed. Other possible measures of pairwise entity similarity will be discussed in Sect. 4.5.1.

4.4.3 Related Entity Finding

Taking the previous task one step further, one may explicitly target a class of queries that mention a focal entity and seek for entities that are related to that entity. The TREC Entity track in 2009 introduced the *related entity finding* (REF) task as follows: "Given an *input entity*, by its name and homepage, the *type of the target entity*, as well as the *nature of their relation*, described in free text, find related entities that are of target type, standing in the required relation to the input entity" [7].

Formally, the input keyword++ query is $\tilde{q} = (q, e_q, y_q)$, where q is the keyword query (describing the relation), e_q is the input entity, and y_q is the target type. An example REF query is show in Listing 4.2. There are several possibilities for defining the input entity and target type. At the first edition of the TREC Entity track, homepages were used as entity identifiers and the target type could be either person, organization, or product. Later editions of the track also experimented,

```
<query>
   <num>7</num>
   <entity_name>Boeing 747</entity_name>
   <entity_URL>clueweb09-en0005-75-02292</entity_URL>
   <target_entity>organization</target_entity>
   <narrative>Airlines that currently use Boeing 747 planes.</narrative>
</query>
```

Listing 4.2 Example topic definition from the TREC 2009 Entity track. Entities are identified by their homepages in a web crawl. The `narrative` tag holds the keyword query q, the input entity e_q and the target entity type y_q are specified by the `entity_URL` and `target_entity` tags, respectively

Fig. 4.6 Related entity finding pipeline

among other things, with using a Linked Data crawl for entity identification and lifting the restrictions on target types [5, 6]. Here, we will consider a simplified version of the task assuming that (1) entities are equipped with unique identifiers and come from a given catalog (knowledge repository) and (2) the target type is from a handful of possible coarse-grained categories, such as person, organization, product, or location.

Even though we use a knowledge repository for identifying entities, that repository, as a single source of data, is often insufficient for answering entity relationship queries. For the ease of argument, let us assume that the said repository is a general-purpose knowledge base (like DBpedia or Freebase). First, the number of distinct predicates in a KB is very small compared to the wide range of possible relationships between entities. For instance, the "[airline] uses [aircraft]" relationship from our topic example is not recognized in DBpedia. Second, even if the given relationship is recognized in the KB, the KB may be incomplete with regards to a specific entity. Therefore, for the REF task, we will complement the knowledge base with a large unstructured data collection: a web corpus. We shall assume that this web corpus has been annotated with entity identifiers from the underlying entity catalog.

Commonly, the REF task is tackled using a pipeline of three steps: (1) identifying candidate entities, (2) filtering entities that are of incorrect type, and (3) computing the relevance of the (remaining) candidates with respect to the input entity and relation. This pipeline is shown in Fig. 4.6.

Bron et al. [12] address the REF task using a generative probabilistic model. Entities are ranked according to the probability $P(e|q,e_q,y_q)$ of entity e being relevant. Using probability algebra (Bayes' theorem) and making certain independence assumptions, the following ranking formula is derived:

$$P(e|q,e_q,y_q) \stackrel{\text{rank}}{=} P(e|e_q)P(y_q|e)P(q|e,e_q), \qquad (4.2)$$

Fig. 4.7 Generative model
for related entity finding
by Bron et al. [12]

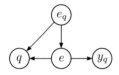

where $P(e|e_q)$, $P(y_q|e)$, and $P(q|e,e_q)$ correspond to the candidate selection, type filtering, and entity relevance steps of the pipeline, respectively. The graphical representation of the model is shown in Fig. 4.7. We note that this is only one of the many possible ways to go about modeling this task. Nevertheless, the components that make up the scoring formula in Eq. (4.2) are rather typical.

4.4.3.1 Candidate Selection

The first step of the pipeline is concerned with the identification of candidate entities. At this stage the focus is on achieving high recall, in order to capture all entities that are possible answers to the query. In Bron et al. [12], this is done through a so-called co-occurrence model, $P(e|e_q)$, which reflects the degree of association between a candidate entity e and the input entity e_q. Let $a(e,e_q)$ be a function (to be defined) that expresses the strength of association between a pair of entities. The co-occurrence probability is then estimated according to:

$$P(e|e_q) = \frac{a(e,e_q)}{\sum_{e' \in \mathcal{E}} a(e',e_q)} .$$

In order to arrive at a reliable estimate, the function $a(e,e_q)$ is based on co-occurrence statistics of the two entities in a large web corpus. Let $c(e)$ denote the number of documents in which e occurs and let $c(e,e_q)$ denote the number of documents in which e and e_q co-occur.

There are many possible ways to set the association function. One of the simplest options is to compute the *maximum likelihood estimate*:

$$a_{MLE}(e,e_q) = \frac{c(e,e_q)}{c(e_q)} .$$

Another alternative, which was reported to perform best empirically in [12], is using the χ^2 hypothesis test (to determine "if the co-occurrence of two entities is more likely than just by chance" [12]):

$$a_{\chi^2}(e,e_q) = \frac{|\mathcal{D}| \left(c(e,e_q)c(\bar{e},\bar{q}_e) - c(e,\bar{q}_e)c(\bar{e},e_q) \right)^2}{c(e)c(e_q) \left(|\mathcal{D}| - c(e) \right) \left(|\mathcal{D}| - c(e_q) \right)} ,$$

where $|\mathcal{D}|$ is the total number of documents in the collection, and $\bar{e}$ and $\bar{q}_e$ indicate that e and e_q do not occur.

Rather than relying on the mere co-occurrence of two entities in documents, one might want to consider "stronger evidence." Requiring that the two entities cross-link to each other constitutes one particular solution. Specifically, the *anchor-based co-occurrence* method in [12] takes account of how many times one entity appears in the anchor text of the other entity's description (e.g., the Wikipedia page of that entity). The co-occurrence probability in this case is estimated as:

$$P(e|e_q) = \frac{1}{2}\frac{c(e,e_q)}{\sum_{e'\in\mathcal{E}}c(e',e_q)} + \frac{1}{2}\frac{c(e_q,e)}{\sum_{e'\in\mathcal{E}}c(e',e)},$$

where $c(e,e_q)$ is the number of times entity e occurs in the anchor text in the description of e_q. For the sake of simplicity, both linking directions are taken into consideration with the same weight.

At the end of the candidate selection step, entities with a non-zero $P(e|e_q)$ value are considered for downstream processing. Commonly, this set is further restricted to the top-k entities with the highest probability.

4.4.3.2 Type Filtering

Earlier, in Sect. 4.3, we have discussed type-aware entity retrieval and have presented various ways of comparing target types of the query with types of entities. The very same methods can be used here as well, for estimating the type component, $P(y_q|e)$, which expresses the probability that entity e is of type y_q. In our earlier scenario, however, it was assumed that target types from the corresponding taxonomy are provided explicitly. Here, the target types are only given as coarse-grained categories (such as person, organization, product, or location). Selecting the appropriate types from the type taxonomy is part of the task.

One strategy for dealing with this is to establish a mapping from each possible coarse-grained input type to multiple categories in the type taxonomy. Such mapping may be constructed using a handful of simple rules. For example, using Wikipedia categories as the underlying type taxonomy, Kaptein et al. [33] map the *person* target type to categories that (1) start with "People," (2) end with "births" or "deaths," and (3) the category "Living People." This initial mapping may be further expanded by adding descendent types until a certain depth according to the type taxonomy [12].

Another strategy is to infer target types from the keyword query directly [34]. We will detail these methods in Sect. 7.2.

4.4.3.3 Entity Relevance

The last part of the pipeline is responsible for determining the relevance of entities. In Bron et al. [12], it is expressed as $P(q|e, e_q)$, the likelihood that the relation contained in the keyword query is "observable" in the context of an input and candidate entity pair. This context is represented as the entity co-occurrence language model, θ_{e, e_q}. The query is scored against this model by taking the product over individual query terms:

$$P(q|\theta_{e, e_q}) = \prod_{t \in q} P(t|\theta_{e, e_q})^{c(t; q)} .$$

The probability of a term given the entity co-occurrence language model is estimated using the following aggregation:

$$P(t|\theta_{e, e_q}) = \frac{1}{|\mathcal{D}_{e, e_q}|} \sum_{d \in \mathcal{D}_{e, e_q}} P(t|\theta_d) ,$$

where $\mathcal{D}_{e, e_q}$ is the set of documents, or document snippets, in which e and e_q co-occur, and θ_d is the (smoothed) language model corresponding to document d.

As a matter of choice, the conditional dependence of the query on the input entity may be dropped when computing entity relevance, like it is done in [22]. Thus, $P(q|e, e_q) \cong P(q|e)$. Then, the query likelihood $P(q|e)$ may be estimated using entity language models, which we have already discussed in Chap. 3 (cf. Sect. 3.3.1.1).

4.5 Similar Entity Search

Similar entity search is the task of ranking entities given a small set of example entities. One might imagine a specialized search interface that allows the user to explicitly provide a set of example entities. A more practical scenario is to construct the set of entities via implicit user interactions, by taking the entities that the user viewed or "dwelled on" during the course of her search session. These examples may complement the keyword query, which we shall call *example-augmented search*, or serve on their own as the expression of the user's information need, which will be referred to as *example-based search*. The keyword++ query may be written as $\tilde{q} = (q, \mathcal{E}_q)$, where $\mathcal{E}_q$ is a set of example entities. Note that q may be empty.

Example-augmented search has been studied at the INEX 2007–2009 Entity Ranking track [19, 20, 54] under the name *list completion*. Listing 4.3 shows an example INEX list completion topic. Example-based search (also known as *set expansion*) is considerably harder than example-augmented search, because of the inherent ambiguity. In the lack of a keyword query, there are typically multiple

```
<inex_topic topic_id="88">
  <title>Nordic authors who are known for children's literature</title>
  <description>I want a list of Nordic authors who are known for children's
literature.</description>
  <narrative>Each answer should be the article about a Danish, Finnish,
Icelandic, Norwegian or Swedish author that has distinguished himself or
herself among others by writing stories or fiction for children. (A possible
query in a library setting.)</narrative>
  <entities>
    <entity id="13550">Hans Christian Andersen</entity>
    <entity id="37413">Astrid Lindgren</entity>
    <entity id="49274">Tove Jansson</entity>
  </entities>
</inex_topic>
```

Listing 4.3 Example topic definition from the INEX 2009 Entity Ranking track. Systems only use the contents of the `<title>` and `<entities>` tags, `<description>` and `<narrative>` are meant for human relevance assessors to clarify the query intent

possible interpretations. Take, for instance, the following three example entities: CANON, SONY, and NIKON. The underlying information need may be "camera brands" or "multinational corporations headquartered in Japan" or something else. Domain-specific applications of example-based search include, for instance, finding people with expertise similar to that of others within an organization [4, 28]. A more generic application area is *concept expansion* for knowledge base population (cf. Chap. 6).

As we have already grown accustomed to it in this chapter, we will employ a two-component mixture model:

$$score(e; \tilde{q}) = \lambda\, score_t(e; q) + (1 - \lambda)\, score_\mathcal{E}(e; \mathcal{E}_q) \,,$$

where $score_t(e; q)$ is the standard text-based retrieval score and $score_\mathcal{E}(e; \mathcal{E}_q)$ is the example-based similarity. This general formula can encompass both flavors of similar entity search. In the case of example-based search, where the input comprises only of $\mathcal{E}_q$, the interpolation parameter λ is set to 0. For example-augmented search, $0 \leq \lambda \leq 1$, with the exact value typically set empirically. Bron et al. [13] adjust λ on a per-query basis, depending on which of the text-based and example-based score components is more effective in retrieving the example entities.

Our main focus in this section is on estimating the example-based similarity component, $score_\mathcal{E}(e; \mathcal{E}_q)$. Before we delve in, a quick note on terminology. To make the distinction clear, we will refer to the entity e that is being scored as the *candidate entity*, and the set $\mathcal{E}_q$ of examples complementing the keyword query as *example entities* (or *seed entities*).

Next, we discuss two families of approaches. *Pairwise methods* consider the similarity between the candidate entity and each of the example entities (Sect. 4.5.1).

Collective methods, on the other hand, treat the entire set of examples as a whole (Sect. 4.5.2).

4.5.1 Pairwise Entity Similarity

A simple and intuitive method is to take the average pairwise similarity between the candidate entity e and each of the example entities $e' \in \mathcal{E}_q$:

$$score_{\mathcal{E}}(e; \mathcal{E}_q) = \frac{1}{|\mathcal{E}_q|} \sum_{e' \in \mathcal{E}_q} sim(e, e') \ .$$

This approach is rather universal, as it boils down to the pairwise entity similarity function, *sim*(). This similarity is of fundamental importance, which extends well beyond this specific task. The choice of the entity similarity measure is closely tied to, and constrained by, the entity representation used. We discuss a range of options below, organized by the type of entity representation employed. Note that the combination of multiple similarity measures is also possible.

4.5.1.1 Term-Based Similarity

Perhaps the most conventional method is to compare term-based representations of entities (i.e., entity descriptions, cf. Chap. 3). Sometimes, this is referred to as *topical similarity*. Let $\mathbf{e}$ denote the term vector corresponding to entity e:

$$\mathbf{e} = \langle w(t_1, e), \dots, w(t_m, e) \rangle \ , \tag{4.3}$$

where m is the size of the vocabulary, $t_1 \dots t_m$ are the distinct terms in the vocabulary, and $w(t_j, e)$ is the weighted term frequency of t_j. Typically, a TF-IDF weighting scheme is used. A standard way of comparing two term vectors is using the *cosine similarity*:

$$sim_{cos}(e, e') = \frac{\mathbf{e} \cdot \mathbf{e}'}{\| \mathbf{e} \| \| \mathbf{e}' \|} = \frac{\sum_{i=1}^{m} w(t_i, e)\, w(t_i, e')}{\sqrt{\sum_{i=1}^{m} w(t_i, e)^2} \sqrt{\sum_{i=1}^{m} w(t_i, e')^2}} \ .$$

Instead of using individual terms (unigrams), the vector representation may also be made up of n-grams or keyphrases [27, 28]. Specifically, Hoffart et al. [27] introduce the *keyphrase overlap relatedness* (KORE) measure, and present approximation techniques, based on min-hash sketches and locality-sensitive hashing, for efficient computation.

4.5.1.2 Corpus-Based Similarity

Entity similarity may be established based on co-occurrence statistics in some corpus of data. This corpus may be a collection of documents, in which case entities are represented by the set of documents mentioning them. Letting $\mathcal{D}_e$ and $\mathcal{D}_{e'}$ denote the set of documents in which entities e and e' occur, respectively, the similarity of the two document sets can be measured, e.g, using the Jaccard coefficient:

$$sim_{Jac}(e,e') = \frac{|\mathcal{D}_e \cap \mathcal{D}_{e'}|}{|\mathcal{D}_e \cup \mathcal{D}_{e'}|} .$$

Other co-occurrence-based similarity functions include the maximum likelihood estimate and the χ^2 hypothesis test, which we have already discussed in Sect. 4.4.3.1.

It is also possible to consider co-occurrences on the sub-document level, e.g., in lists [26, 48] or tables [55]. As another option, a corpus of query log data may also be utilized for the same purpose [26].

4.5.1.3 Distributional Similarity

The conventional term-based representation (also called *one-hot representation*) allows only for exact word matches. Consider the following oversimplified example, for the sake of illustration. Let us assume that entity A has a single term in its representations, "apple," and entity B also has a single term in its representation, "orange." These two entities would have a similarity of 0 according to any standard term-based similarity measure (like the Jaccard coefficient or cosine similarity). Yet, arguably, the similarity of "apple" to "orange" should be higher than, say, that of "apple" to "chess." With the traditional term-based representation, it is not possible to make this distinction. This, in fact, is one of the fundamental challenges in IR, known as the *vocabulary mismatch* problem. The overall idea behind *distributed representations* (or *distributional semantics*) is to represent each word as a "pattern." Words are embedded into a continuous vector space such that semantically related words are close to each other.[4]

Word vector representations of terms, a.k.a. *word embeddings*, are obtained from large text corpora using neural networks. The embedding (or latent factor) vector space has low dimensionality, typically in the order of 200–500. At the time of writing, the two dominating methods for computing word vector representations are

---

[4]*Distributional representations* and *distributed representations* are both used, and in this case, both are correct. The former has to do with the linguistic aspect, "meaning is context," i.e., items with similar distributions (i.e., context words) have similar meanings. The latter refers to the idea of having a compact, dense, and low dimensional representation; a single component of a vector representation does not have a meaning on its own, i.e., it is distributed among multiple vector components. Word embeddings have both these properties.

Word2vec [41] and Glove [46]. Both these implementations are publicly available, and may be run on any text corpus. A variety of pre-trained word vectors are also available for download; these may be used off-the-shelf.[5, 6]

Let us now see how these word embeddings can be used for entities. One way to compute the distributed representation of an entity is to take the weighted sum of distributed word vectors for each term in the entity's term-based representation. Variants of this idea may be found in recent literature, see, e.g., [49, 53, 56]. Formally, let n be the dimensionality of the embedding vector space. We write $\bar{\mathbf{e}} = \langle \bar{e}_1, \ldots, \bar{e}_n \rangle$ to denote the distributed representation of entity e—this is what we wish to compute. Let $\mathbf{T}$ be an $m \times n$ dimensional matrix, where m is the size of the vocabulary; row j of the matrix corresponds to the distributed vector representation of term t_j (i.e., is an n-dimensional vector). The distributed representation of entity e can then be computed simply as:

$$\bar{\mathbf{e}} = \mathbf{eT} \,,$$

where $\mathbf{e}$ denotes the term-based entity representation, cf. Eq. (4.3). The above equation may be expressed on the element level as:

$$\bar{e}_i = \sum_{j=1}^{m} \mathbf{e}_j \mathbf{T}_{ji} \,.$$

One way to look at this transformation is that an m-dimensional representation is compressed into an n-dimensional one (where $m \gg n$), in such a way that entities that are more similar become closer in this lower dimensional space. The similarity of two entities in the embedding vector space is computed using $sim_{cos}(\bar{\mathbf{e}}, \bar{\mathbf{e}}')$.

4.5.1.4 Graph-Based Similarity

Viewing entities as nodes in a knowledge graph gives rise to another family of similarity functions. One way to establish similarity between two entities is based on the set of other nodes that they are connected to, the idea being that "two objects are similar if they are related to similar objects" [30]. Let $\mathcal{L}_e$ denote the set of nodes connected to e, where connectedness may be interpreted as (1) incoming links (nodes that link to e), (2) outgoing links (nodes linked by e), or (3) the union of incoming and outgoing links of e. For entity e', $\mathcal{L}_{e'}$ is defined analogously. Then, it is possible to measure the similarity of the two link sets, $\mathcal{L}_e$ and $\mathcal{L}_{e'}$, e.g., using the Jaccard coefficient [26]. The *Wikipedia link-based measure* (WLM) [42] gives

---

[5]https://code.google.com/archive/p/word2vec/.
[6]http://nlp.stanford.edu/projects/glove/.

another quantification of semantic relatedness of two entities, based on the same idea of the overlap between links; see Eq. (5.4) in Sect. 5.6.1.3 for details.

Instead of using only the direct neighbors of entities, like the above measures do, one might consider their distance in the knowledge graph. One way of establishing similarity is by setting it inversely proportional to the minimum (weighted) distance of the two entities [50]. Alternatively, the problem may be approached as that of propagating similarity from an entity node through graph edges, using some variant of graph walk [30, 43]. Graph edge weights can be set uniformly, manually, or automatically (using some weighting function [50] or a learning procedure [43]).

4.5.1.5   Property-Specific Similarity

In addition to the above methods, which are generally applicable, similarity measures may be tailored to individual properties. This may be imagined as having a distinct entity representation corresponding to each particular property, such as entity name or type. These property-level similarities are then combined, e.g., as:

$$sim(e, e') = \sum_i \lambda_i sim_i(e, e'),$$

where $sim_i(e, e')$ is a similarity function for property i and λ_i is the corresponding weight (importance of that property). Weights may be set manually (based on domain knowledge) or learned from training data. We note that any of the similarity functions from above may be used as $sim_i(e, e')$.

A distinguished property is entity type (or category), which naturally provides a grouping of similar entities; here, similarity is understood in an ontological sense. We refer back to Sect. 4.3.3 for various ways of establishing type similarity (where $\mathcal{T}_q$ is to be replaced with $\mathcal{T}_{e'}$). Note that type-based similarity is only effective in telling apart entities that are of a different kind, e.g., people vs. products. It needs to be combined with other similarity measures when the two entities belong to the same semantic category (e.g., racing drivers). Effectiveness further depends on the granularity of type information, i.e., how detailed the type taxonomy is.

A range of options exists for particular domains or applications. For example, Balog [2] introduces similarity measures for specific product attributes. Product names are compared using various string distance measures, both character-based (e.g., Levenshtein or Jaro-Winkler distance) and term-based (e.g., Jaccard or Dice coefficient). For product prices, the relative value difference is used. Another example, for geospatial entities, is given in [51], for matching the location coordinates of entities.

4.5.2   Collective Entity Similarity

We now switch to collective methods, which consider the set of example entities as a whole. One simple solution, which works for some but not all kinds of entity representations, is to take the centroid of example entities, and compare it against the candidate entity (using a similarity function corresponding to that kind of representation). Here, we introduce two more advanced methods, which make explicit use of the structure associated with entities.

4.5.2.1   Structure-Based Method

Bron et al. [13] employ a structured entity representation, which is comprised of the set of properties of the entity. Given an entity e, and the set of SPO triples that contain e, each triple yields a property r by removing the entity in question from that triple. (If the data is viewed as a knowledge graph, these properties correspond to the nodes adjacent to the entity, along with the connecting edges.) Formally, the structured representation of a given entity e is defined as:

$$\tilde{e} = \{r = (p,o) : (e,p,o) \in \mathcal{K}\} \cup \{r = (s,p) : (s,p,e) \in \mathcal{K}\} \ .$$

For example, given the entity MICHAEL SCHUMACHER, with a set of RDF triples from Sect. 2.3.1.2, the structured representation of this entity becomes:

$$\tilde{e} = \big\{(\texttt{<foaf:name>, "Michael Schumacher"}),$$
$$(\texttt{<dbo:birthPlace>, <dbr:West\_Germany>}),$$
$$(\texttt{<dbr:1996\_Spanish\_Grand\_Prix>, <dbp:firstDriver>}),$$
$$\dots \big\}.$$

Under this representation, the set of example entities $\mathcal{E}_q$ becomes a set of set of properties: $\tilde{\mathcal{E}}_q = \{\tilde{e}_1, \dots, \tilde{e}_{|\mathcal{E}_q|}\}$. The similarity between the entity e and the set of examples is estimated by marginalizing over all properties of the entity and considering whether they are observed with the examples:

$$score_{\mathcal{E}}(e; \mathcal{E}_q) = P(\tilde{e}|\tilde{\mathcal{E}}_q) = \sum_{r \in \tilde{e}} P(r|\theta_{\mathcal{E}_q}) = \sum_{r \in \tilde{e}} \frac{\sum_{i=1}^{|\mathcal{E}_q|} \mathbb{1}(r, \tilde{e}_i)}{\sum_{i=1}^{|\mathcal{E}_q|} \sum_{r \in \tilde{e}_i} \mathbb{1}(r, \tilde{e}_i)} \ ,$$

where $\mathbb{1}(r, \tilde{e}_i)$ is a binary indicator function which is 1 if r occurs in the representation of $\tilde{e}_i$ and is 0 otherwise. The denominator is the representation length of the seed entities, i.e., the total number of relations of all seed entities.

4.5.2.2 Aspect-Based Method

In their approach called QBEES, Metzger et al. [39] aim to explicitly capture the different potential user interests behind the provided examples. Each entity e is characterized based on three kinds of aspects:

- *Type aspects*, $\mathcal{A}_T(e)$, include the set of types that the entity is an instance of, e.g., {<dbo:Person>, <dbo:RacingDriver>, <dbo:FormulaOneRacer>,...}.
- *Relational aspects*, $\mathcal{A}_R(e)$, are the predicates (edge labels) connecting the entity to other non-type nodes (irrespective of the direction of the edge), e.g., {<dbo:birthPlace>, <dbp:firstDriver>, <dbo:championships>,...}.
- *Factual aspects*, $\mathcal{A}_F(e)$, are relationships to other entities, i.e., other entity nodes incident to the entity, along with the connecting edges,[7] e.g., {(<dbo:birthPlace>, <dbr:West_Germany>), (<dbr:1996_Spanish_Grand_Prix>, <dbp:firstDriver>),...}.

The *basic aspects* of an entity is the union of the above three kinds of aspects: $\mathcal{A}(e) = \mathcal{A}_T(e) \cup \mathcal{A}_R(e) \cup \mathcal{A}_F(e)$. A *compound aspect* of an entity is any subset of its basic aspects: $A_e \subseteq \mathcal{A}(e)$.

For each basic aspect a, the *entity set of an aspect*, $\mathcal{E}_a$, is a set of entities that have this aspect: $\mathcal{E}_a = \{e \in \mathcal{E} : a \in \mathcal{A}(e)\}$. This concept can easily be extended to compound aspects. Let $\mathcal{E}_A$ be the set of entities that share *all* basic aspects in A: $\mathcal{E}_A = \{e \in \mathcal{E} : a \in \mathcal{A}(e) \, \forall a \in A\}$.

A compound aspect A is *maximal aspect of an entity* e iff

1. $\mathcal{E}_A$ contains at least one entity other than e;
2. A is maximal w.r.t. inclusion, i.e., extending this set with any other basic aspect of e would violate the first condition.

We write $\mathcal{M}(e)$ to denote the family (i.e., collection) of *all maximal aspects of e*.

Next, we extend all the above concepts from a single entity to the set of example entities $\mathcal{E}_q$. Let $\mathcal{A}(\mathcal{E}_q)$ denote the set of basic aspects that are shared by all example entities: $\mathcal{A}(\mathcal{E}_q) = \cap_{e \in \mathcal{E}_q} \mathcal{A}(e)$. A compound aspect A is said to be a *maximal aspect of the set of example entities $\mathcal{E}_q$* iff

1. $\mathcal{E}_A$ contains at least one entity that is not in $\mathcal{E}_q$;
2. A is maximal w.r.t. inclusion.

The family of all maximal aspects of $\mathcal{E}_q$ is denoted as $\mathcal{M}(\mathcal{E}_q)$.

The ranking of entities is based on the fundamental observation that for a given set of example entities $\mathcal{E}_q$, the most similar entities will be found in entity sets of maximal aspects of $\mathcal{E}_q$. One of the main challenges in similar entity search, i.e., in computing $score_{\mathcal{E}}(e; \mathcal{E}_q)$, is the inherent ambiguity because of the lack of an explicit user query. Such situations are typically addressed using *diversity-aware*

---

[7]This is similar to the structured entity representation in the previous subsection, the difference being that only entity nodes are considered (type and literal nodes are not).

Algorithm 4.1: QBEES [39]

Input: set of example entities, $\mathcal{E}_q$
Output: top-k similar entities, $\mathcal{E}_R$

1 $\mathcal{E}_R \leftarrow \emptyset$
2 $\mathcal{M}(\mathcal{E}_q) \leftarrow$ maximal aspects of $\mathcal{E}_q$
3 Filter $\mathcal{M}(\mathcal{E}_q)$ by type constraints
4 Rank $\mathcal{M}(\mathcal{E}_q)$
5 **while** $|\mathcal{E}_R| < k$ *and* $|\mathcal{M}(\mathcal{E}_q)| > 0$ **do**
6 $\quad$ $A^* \leftarrow$ top ranked aspect $(A^* \in \mathcal{M}(\mathcal{E}_q))$
7 $\quad$ $\mathcal{E}^* \leftarrow \{e \in \mathcal{E} : e \in \mathcal{E}_{A^*}, e \notin \mathcal{E}_q \cup \mathcal{E}_R\}$
8 $\quad$ $e^* \leftarrow \arg\max_{e \in \mathcal{E}^*} score(e)$
9 $\quad$ $\mathcal{E}_R \leftarrow \mathcal{E}_R \cup \{e^*\}$
10 $\quad$ update ranking of $\mathcal{M}(\mathcal{E}_q)$
11 **end**

---

approaches. That is, constructing the result set in a way that it covers multiple possible interpretations of the user's underlying information need.

In QBEES, these possible interpretations are covered by the compound aspects. Further, the concept of maximal aspect is designed such that it naturally provides diversity-awareness. This is articulated in the following theorem.

Theorem 4.1 *Let $A_1(\mathcal{E}_q)$ and $A_2(\mathcal{E}_q)$ be two different maximal aspects of $\mathcal{E}_q$. In this case, $A_1(\mathcal{E}_q)$ and $A_2(\mathcal{E}_q)$ do not share any entities, except those in $\mathcal{E}_q$: $A_1(\mathcal{E}_q) \cap A_2(\mathcal{E}_q) \setminus \mathcal{E}_q = \emptyset$.*

We omit the proof here and refer to [39] for details. According to the above theorem, the partitioning provided by the maximal aspects of $\mathcal{E}_q$ can guide the process of selecting a diverse set of result entities $\mathcal{E}_R$ that have the highest similarity to $\mathcal{E}_q$.

We detail the key steps of the procedure used for selecting top-k similar entities, shown in Algorithm 4.1.

- **Finding the maximal aspects of the example entities.** From the aspects shared by all example entities, $\mathcal{A}(\mathcal{E}_q)$, the corresponding family of maximal aspects, $\mathcal{M}(\mathcal{E}_q)$, is computed. For that, all aspects that are subsets of $\mathcal{A}(\mathcal{E}_q)$ need to be considered and checked for the two criteria defining a maximal aspect. For further details on how to efficiently perform this, we refer to [39].
- **Constraining maximal aspects on typical types.** It can reasonably be assumed that the set of example entities is to be completed with entities of similar types. Therefore, the set of *typical types* of the example entities, $\mathcal{T}_{\mathcal{E}_q}$. To determine $\mathcal{T}_{\mathcal{E}_q}$, first, the types that are shared by all example entities are identified (according to a predefined type granularity). Then, only the most specific types in this set are kept (by filtering out all types that are super-types of another type in the set). Maximal aspects that do not contain at least one of the typical types (or their subtypes) as a basic aspect are removed.
- **Ranking maximal aspects.** "The resulting maximal aspects are of different specificity and thus quality" [39]. Therefore, a ranking needs to be established,

and maximal aspects will be considered in that order. Metzger et al. [39] discuss various ranking functions. We present one particular ranker (*cost*), which combines the (normalized) "worth" of aspects with their specificity:

$$cost(A) = \frac{1}{\mathcal{E}_A} \times \frac{val(A)}{\sum_{A' \in \mathcal{M}(\mathcal{E}_q)} val(A')} .$$

Recall that $\mathcal{E}_A$ is the set of entities that share all basic aspects in A. Thus, the first term expresses the specificity of an aspect (the larger the set of entities $\mathcal{E}_A$, the less specific aspect A is). The value of a compound aspect is estimated as a sum of the value of its basic aspects. The value of a basic aspect is its "inverse selectivity," i.e., aspects with many entities are preferred:

$$val(A) = \sum_{a \in A} (1 - \frac{1}{\mathcal{E}_a}) .$$

- **Selecting an entity.** An entity is picked from the top ranked aspect's entity set. That is, if A^* is the top ranked aspect, then an entity is selected from $\mathcal{E}_{A^*}$. Entities in $\mathcal{E}_{A^*}$ are ranked according to their (static) importance, $score(e)$, which can be based on, e.g., popularity. (We discuss various query-independent entity ranking measures in Sect. 4.6.) Aspects that cannot contribute more entities to the result list (i.e., $\mathcal{E}_A \setminus \mathcal{E}_q \cup \mathcal{E}_R = \emptyset$) are removed. If the objective is to provide diversification, the aspects are re-ranked after each entity selection. Metzger et al. [39] further suggest a relaxation heuristic; instead of removing a now empty maximal aspect, it may be relaxed to cover more entities from "the same branch of similar entities." A simple relaxation strategy is to consider each of the basic aspects a and either dropping it or replacing it with the parent type, if a is a type aspect.

4.6 Query-Independent Ranking

Up until this point, we have focused on ranking entities with respect to a query. In a sense, all entities in the catalog have had an equal chance when competing for the top positions in the ranked result list. There are, however, additional factors playing a role, which are independent of the actual query.

For the sake of illustration, let us consider the task of searching on an e-commerce site. Let us imagine that for some search query there are two or more products that have the same retrieval score, i.e., they are equally good matches. This could happen, for instance, because those products have very similar names and descriptions. It might also happen that the user is not issuing a keyword query, but is browsing the product catalog by product category or brand instead. In this case, the selected category or brand is the query, for which a complete and perfect set of results may be returned (i.e., all entities that belong to the selected category/brand are equally relevant). In both the above scenarios, the following question arises: How can we determine the relative ranking of entities that have the same relevance to the query?

Intuitively, the ranking of products in the above example could be based on which (1) has been clicked on more, (2) has been purchased more, (3) has a better consumer rating, (4) has a better price, (5) is a newer model, etc., or some combination of these. Notice that all these signals are independent of the actual query.

> Query-independent (or "static") entity scores capture the notion of *entity importance*. That is, they reflect the extent with which a given entity is relevant to *any* query.

Let us refer back to the language modeling approach from the previous chapter (Sect. 3.3.1.1). There, we rewrote the probability $P(e|q)$ using Bayes' theorem. We repeat that equation (Eq. (3.7)) here for convenience:

$$P(e|q) = \frac{P(q|e)P(e)}{P(q)} \stackrel{\text{rank}}{=} P(q|e)P(e) \,.$$

Back then, we ignored the second component, $P(e)$, which is the *prior* probability of the entity. As we can see, the generative formulation of the language modeling framework offers a theoretically principled way of incorporating the query-independent entity score.

Generally, the query-independent entity score may be incorporated into any retrieval method by way of multiplication of the (original) query relevance score:

$$score'(e; q) = score(e) \times score(e; q) \,.$$

Since $score(e)$ is independent of the actual query, it can be computed offline and stored in the entity index (hence then name *static* score).

In this section, we present two main groups of query-independent measures, which revolve around the notions of *popularity* and *centrality*. Multiple query-independent features may be combined using a learning-to-rank approach [16].

4.6.1   Popularity

Intuitively, the popularity of an entity is an indicator of its importance. It may be expressed as a probability:

$$P(e) = \frac{c(e)}{\sum_{e' \in \mathcal{E}} c(e')} \,,$$

where $c(e)$ is some measure of popularity. Options for estimating $c(e)$ include the following:

- Aggregated *click or view statistics* over a certain period of time [39]. One useful resource is the Wikipedia page view statistics, which is made publicly available.[8]
- The *frequency of the entity* in a large corpus, like the Web. This may be approximated by using the entity's name as a query and obtaining the number of hits from a web search engine API [16]. Alternatively, the name of the entity can be checked against a large and representative n-gram corpus to see how often it was used [16]. One such resource is the Google Books Ngram Corpus [37].[9] When the web corpus is annotated with specific entities, it is possible to search using entity IDs, instead of entity names; see Sect. 5.9.2 for web corpora annotated with entities from Freebase.

In addition to the above "universal" signals, there are many possibilities in specific domains, e.g., in an e-commerce setting, the number of purchases of a product; in academic search, the number of citations of an article; on a social media platform, the number of likes/shares, etc.

4.6.2   Centrality

Entity *centrality* can be derived from the underlying graph structure using (adaptations of) link analysis algorithms, like PageRank [11] and HITS [35]. In this section, we focus on the PageRank algorithm, due to its popularity, and discuss its application to entities. For similar work using the HITS algorithm, see, e.g., [8, 23].

4.6.2.1   PageRank

Web pages can be represented as a directed graph, where nodes are pages and edges are hyperlinks connecting these pages to each other. *PageRank* assigns a numerical score to each page that reflects its importance. The main idea is that it is not only the number of incoming links that matters but also the quality of those links. Links that originate from more important sources (i.e., pages with high PageRank score) weigh more than unimportant pages. Another way to understand PageRank is that it measures the likelihood of a random surfer landing on a given web page. The random surfer is assumed to navigate on the Web as follows. In each step, the user either (1) moves to one of the pages linked from the current page or (2) jumps to a random web page. The random jump also ensures that the user does not get stuck on a page that has no outgoing links.

---

[8]http://dumps.wikimedia.org/other/pagecounts-raw/.

[9]http://books.google.com/ngrams.

The PageRank score of an entity is computed as follows:

$$PR(e) = \frac{\alpha}{|\mathcal{E}|} + (1 - \alpha) \sum_{i=1}^{n} \frac{PR(e_i)}{|\mathcal{L}_{e_i}|} , \tag{4.4}$$

where α is the probability of a random jump (typically set to 0.15 [11]), $e_1 \ldots e_n$ are the entities linking to e, and $|\mathcal{L}_{e_i}|$ is the number of outgoing links of e_i. Notice that PageRank is defined recursively, thus needs to be computed iteratively. In each successive iteration, the PageRank score is determined using the PageRank values from the previous iteration. Traditionally, all nodes are initialized with an equal score (i.e., $1/|\mathcal{E}|$). The final values are approximated fairly accurately even after a few iterations. Notice that the PageRank scores form a probability distribution ($\sum_{e \in \mathcal{E}} PR(e) = 1$). Therefore, the iterative computation process may also be interpreted as propagating a probability mass across the graph. It is common to use quadratic extrapolation (e.g., between every fifth iteration) to speed up convergence [31].

4.6.2.2 PageRank for Entities

A number of variations and extensions of PageRank have been proposed for entities. We shall take look at a selection of them below. The first important question that needs to be addressed is: How to construct the entity graph?

- Using *unstructured* entity descriptions, references to other entities need to be recognized and disambiguated (see Chap. 5). Directed edges are added from each entity to all the other entities that are mentioned in its description.
- In a *semi-structured* setting, e.g., Wikipedia, links to other entities might be explicitly provided.
- When working with *structured* data, RDF triples define a graph (i.e., the knowledge graph). Specifically, subject and object resources (URIs) are nodes and predicates are edges.

In the remainder of this section, we shall concentrate on the structured data setting. Recall that nodes in an RDF graph include not only entities but other kinds of resources as well (in particular entity types, cf. Fig. 2.3). Since it is not pages but resources that are being scored, Hogan et al. [29] refer to the computed quantity as *ResourceRank*. Instead of computing static scores on the entire graph, Resource-Rank computes PageRank scores over topical subgraphs (resources matching the keyword query and their surrounding resources).

It can be argued that the traditional PageRank model is unsuitable for entity popularity calculation because of the heterogeneity of entity relationships [44]. *ObjectRank* [1] extends PageRank to weighted link analysis, applied to the problem of keyword search in databases. ObjectRank, however, relies on manually assigned link weights. This makes the approach applicable only in restricted domains (e.g., academic search). Nie et al. [44] introduce *PopRank*, to rank entities within a specific domain. A separate *popularity propagation factor* (i.e., weight) is assigned to each link depending on the type of entity relationship. For example, in their case study on academic search, the types of entities involved are authors, papers, and conferences/journals. The types of relationships between entities include cited-by, authored-by, and published-by. It is problematic to manually decide the propagation factors for each type of relationship. Instead, partial entity rankings are collected from domain experts for some subsets of entities. The setting of propagation factor then becomes a parameter estimation problem using the partial rankings as training data. This method is applicable in any vertical search domain, where the number of entity types and relationships is sufficiently small (product search, music search, people search, etc.).

The quality of ranking may be improved by taking the context of data into account. When working with RDF, this context is the provenance or source of data. Hogan et al. [29] extract a context graph from the RDF graph and compute PageRank scores on this graph, referred to as *ContextRank*. Further, by inferring links between contexts and resources (as well as between contexts), a combined graph can be created. This graph contains both resource and context nodes. *ReConRank* refers to the PageRank scores computed on this unified graph. ReConRank captures the "symbiotic relationship between context authority and resource authority" [29].

4.6.2.3 A Two-Layered Extension of PageRank for the Web of Data

The heterogeneity of data is further increased when moving from a single knowledge base to the Web of Data, which is comprised of multiple datasets. Instead of considering only the ranks of entities, the dataset where that entity originates from should be also taken into account. Additionally, computing popularity scores on a graph of that scale brings challenges.

Delbru et al. [17] propose a two-layered extension of PageRank. Their method, called *DING*, operates in a hierarchical fashion between the dataset and entity layers. The top layer is comprised of a collection of inter-connected datasets, whereas the bottom layer is composed of (independent) graphs of entities; see Fig. 4.8 for an illustration. The hierarchical random surfer model works as follows:

1. The user randomly selects a dataset.
2. Then, the user may choose and perform one of these actions:

    (a) Select randomly an entity in the current dataset.
    (b) Jump to another dataset by following a link from the current dataset.
    (c) End the browsing.

According to the above model, the computation is performed in two stages. First, the importance of the top level dataset nodes (*DatasetRank*) are calculated. The rank score of a dataset is composed of two parts: (1) the contribution from the other datasets via incoming links, and (2) the probability of selecting the dataset

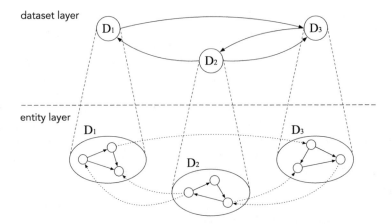

Fig. 4.8 A two-layer model for computing entity importance in the Web of Data. Dashed edges on the entity layer represent inter-dataset links; these are aggregated to form weighted links on the dataset layer. Figure is based on [17]

during a random jump (which is set proportional to its size). Then, in a second stage, the importance of entities within a given dataset (*local entity rank*) are computed. Local entity rank is PageRank applied on the intra-links of the dataset. Delbru et al. [17] further introduce an unsupervised link weighting scheme (LF-IDF, motivated by TF-IDF), which assigns high importance to links with high frequency within a given dataset and low dataset frequency across the collection of datasets. Mind that the computation of the local entity ranks can be performed independently for each dataset (and thus can easily be parallelized). The two scores are combined using the following equation:

$$score_{DING}(e) = r(D)\, r(e)\, \frac{|D|}{\sum_{D'} |D'|} \, ,$$

where $r(D)$ is DatasetRank and $r(e)$ is the local entity rank. The last term in the equation serves as a normalization factor. Since the magnitude of local entity ranks depends on the size of the dataset (recall that $\sum_{e \in \mathcal{E}} r(e) = 1$), entities in small datasets will receive higher scores than in large datasets. This is compensated by taking the dataset size into account, where $|D|$ denotes the size of dataset D (measured in the number of entities contained).

4.6.3 Other Methods

Without detailed elaboration, we mention a number of other query-independent methods that have been proposed in the literature. Instead of centrality, more simple frequency-based measures may be used, which can be extracted from the knowledge graph. These include the number of subjects, objects, distinct predicates, etc., for at most k steps away from the entity node [16]. Alternatively, one may not even need to consider the graph structure but can obtain statistics from the set of RDF triples that contain a given entity. For example, how many times does the entity occur as a subject in an RDF triple where the object is a literal [16]?

4.7 Summary

This chapter has introduced semantically enriched models for entity retrieval. In the various entity retrieval tasks we have looked at, the common theme has been to exploit specific properties of entities: attributes, types, and relationships. Typically, this semantic enrichment is operationalized by combining the term-based retrieval score with one or multiple property-specific score components. Utilizing entity attributes and relationships is shown to yield relative improvements in the 5–20% range [14, 25, 52]. Exploiting entity type information is even more rewarding, with relative improvements ranging from 25% to well over 100% [3, 15, 47].

While these results are certainly encouraging, a lot remains to be done. First, most of the approaches we have presented in this chapter were tailored to a rather specific entity retrieval task, making assumptions about the format/characteristics of the query and of the underlying user intent. It is an open question whether a "one size fits all" model can be developed, or if effort might be better spent by designing task-specific approaches and automatically deciding which one of these should be used for answering a given input query. Second, we have assumed a semantically enriched query as input, referred to as *keyword++ query*, which explicitly provides query entities, types, etc. Obtaining these "enrichments" automatically is an area of active research (which we will be discussing in Chap. 7). Third, the potential of combining structured and unstructured data has not been fully explored and realized. Instead of performing this combination inside the retrieval model, an alternative is to enrich the underlying collections. Annotating documents with entities can bring structure to unstructured documents, which in turn can help populate knowledge bases with new information about entities. In Part II, we will discuss exactly how this strategy of annotating documents, and consequently populating knowledge bases, can be accomplished.

4.8 Further Reading

It is worth pointing out that there are many additional factors beyond topical relevance to consider in real-life applications. One particular example is e-commerce search, where the quality of images has shown to have an impact on which results get clicked [21, 24]. Another example scenario is that of expert search, where it has been shown that contextual factors (e.g., availability, experience, physical proximity) also play a role [28].

References

1. Balmin, A., Hristidis, V., Papakonstantinou, Y.: Objectrank: Authority-based keyword search in databases. In: Proceedings of the Thirtieth International Conference on Very Large Data Bases - Volume 30, VLDB '04, pp. 564–575. VLDB Endowment (2004)
2. Balog, K.: On the investigation of similarity measures for product resolution. In: Proceedings of the Workshop on Discovering Meaning On the Go in Large Heterogeneous Data, LHD-11 (2011)
3. Balog, K., Bron, M., De Rijke, M.: Query modeling for entity search based on terms, categories, and examples. ACM Trans. Inf. Syst. **29**(4), 22:1–22:31 (2011a)
4. Balog, K., de Rijke, M.: Finding similar experts. In: Proceedings of the 30th annual international ACM SIGIR conference on Research and development in information retrieval, SIGIR '07, pp. 821–822. ACM (2007). doi: 10.1145/1277741.1277926
5. Balog, K., Serdyukov, P., de Vries, A.P.: Overview of the TREC 2010 Entity track. In: Proceedings of the Nineteenth Text REtrieval Conference, TREC '10. NIST (2011b)
6. Balog, K., Serdyukov, P., de Vries, A.P.: Overview of the TREC 2011 Entity track. In: The Twentieth Text REtrieval Conference Proceedings, TREC '11. NIST (2012)
7. Balog, K., de Vries, A.P., Serdyukov, P., Thomas, P., Westerveld, T.: Overview of the TREC 2009 Entity track. In: Proceedings of the Eighteenth Text REtrieval Conference, TREC '09. NIST (2010)
8. Bamba, B., Mukherjea, S.: Utilizing resource importance for ranking semantic web query results. In: Proceedings of the Second International Conference on Semantic Web and Databases, SWDB '04, pp. 185–198. Springer (2005). doi: 10.1007/978-3-540-31839-2_14
9. Bast, H., Bäurle, F., Buchhold, B., Haussmann, E.: Semantic full-text search with Broccoli. In: Proceedings of the 37th International ACM SIGIR Conference on Research and Development in Information Retrieval, SIGIR '14, pp. 1265–1266. ACM (2014). doi: 10.1145/2600428.2611186
10. Blanco, R., Halpin, H., Herzig, D.M., Mika, P., Pound, J., Thompson, H.S., Duc, T.T.: Entity search evaluation over structured web data. In: Proceedings of the 1st International Workshop on Entity-Oriented Search, EOS '11, pp. 65–71 (2011)
11. Brin, S., Page, L.: The anatomy of a large-scale hypertextual web search engine. In: Seventh International World-Wide Web Conference, WWW '98 (1998)
12. Bron, M., Balog, K., de Rijke, M.: Ranking related entities: Components and analyses. In: Proceedings of the 19th ACM International Conference on Information and Knowledge Management, CIKM '10, pp. 1079–1088 (2010). doi: 10.1145/1871437.1871574
13. Bron, M., Balog, K., de Rijke, M.: Example-based entity search in the web of data. In: Proceedings of the 35th European conference on Advances in Information Retrieval, ECIR '13, pp. 392–403. Springer (2013). doi: 10.1007/978-3-642-36973-5_33
14. Campinas, S., Delbru, R., Tummarello, G.: Effective retrieval model for entity with multi-valued attributes: BM25MF and beyond. In: Proceedings of the 18th International Conference on Knowledge Engineering and Knowledge Management, EKAW '12, pp. 200–215. Springer (2012). doi: 10.1007/978-3-642-33876-2_19
15. Ciglan, M., Nørvåg, K., Hluchý, L.: The SemSets model for ad-hoc semantic list search. In: Proceedings of the 21st International Conference on World Wide Web, WWW '12, pp. 131–140. ACM (2012). doi: 10.1145/2187836.2187855

16. Dali, L., Fortuna, B., Duc, T.T., Mladenić, D.: Query-independent learning to rank for RDF entity search. In: Proceedings of the 9th International Conference on The Semantic Web: Research and Applications, ESWC'12, pp. 484–498. Springer (2012). doi: 10.1007/978-3-642-30284-8_39

17. Delbru, R., Toupikov, N., Catasta, M., Tummarello, G., Decker, S.: Hierarchical link analysis for ranking web data. In: Proceedings of the 7th International Conference on The Semantic Web: Research and Applications - Volume Part II, ESWC'10, pp. 225–239. Springer (2010). doi: 10.1007/978-3-642-13489-0_16

18. Demartini, G., Firan, C.S., Iofciu, T., Krestel, R., Nejdl, W.: Why finding entities in Wikipedia is difficult, sometimes. Information Retrieval **13**(5), 534–567 (2010a). doi: 10.1007/s10791-010-9135-7

19. Demartini, G., Iofciu, T., de Vries, A.: Overview of the INEX 2009 Entity Ranking track. In: Geva, S., Kamps, J., Trotman, A. (eds.) Focused Retrieval and Evaluation, *Lecture Notes in Computer Science*, vol. 6203, pp. 254–264. Springer (2010b). doi: 10.1007/978-3-642-14556-8_26

20. Demartini, G., de Vries, A.P., Iofciu, T., Zhu, J.: Overview of the INEX 2008 Entity Ranking track. In: Advances in Focused Retrieval: 7th International Workshop of the Initiative for the Evaluation of XML Retrieval (INEX 2008), pp. 243–252 (2009). doi: 10.1007/978-3-642-03761-0_25

21. Di, W., Sundaresan, N., Piramuthu, R., Bhardwaj, A.: Is a picture really worth a thousand words? - on the role of images in e-commerce. In: Proceedings of the 7th ACM International Conference on Web Search and Data Mining, WSDM '14, pp. 633–642. ACM (2014). doi: 10.1145/2556195.2556226

22. Fang, Y., Si, L.: Related entity finding by unified probabilistic models. World Wide Web **18**(3), 521–543 (2015). doi: 10.1007/s11280-013-0267-8

23. Franz, T., Schultz, A., Sizov, S., Staab, S.: TripleRank: Ranking semantic web data by tensor decomposition. In: Proceedings of the 8th International Semantic Web Conference, ISWC '09, pp. 213–228. Springer (2009). doi: 10.1007/978-3-642-04930-9_14

24. Goswami, A., Chittar, N., Sung, C.H.: A study on the impact of product images on user clicks for online shopping. In: Proceedings of the 20th International Conference Companion on World Wide Web, WWW '11, pp. 45–46. ACM (2011). doi: 10.1145/1963192.1963216

25. Hasibi, F., Balog, K., Bratsberg, S.E.: Exploiting entity linking in queries for entity retrieval. In: Proceedings of the 2016 ACM on International Conference on the Theory of Information Retrieval, ICTIR '16, pp. 209–218. ACM (2016). doi: 10.1145/2970398.2970406

26. He, Y., Xin, D.: SEISA: Set expansion by iterative similarity aggregation. In: Proceedings of the 20th International Conference on World Wide Web, WWW '11. ACM (2011). doi: 10.1145/1963405.1963467

27. Hoffart, J., Seufert, S., Nguyen, D.B., Theobald, M., Weikum, G.: KORE: Keyphrase overlap relatedness for entity disambiguation. In: Proceedings of the 21st ACM International Conference on Information and Knowledge Management, CIKM '12, pp. 545–554. ACM (2012). doi: 10.1145/2396761.2396832

28. Hofmann, K., Balog, K., Bogers, T., de Rijke, M.: Contextual factors for finding similar experts. Journal of the American Society for Information Science and Technology **61**(5), 994–1014 (2010). doi: 10.1002/asi.v61:5

29. Hogan, A., Harth, A., Decker, S.: ReConRank: A scalable ranking method for semantic web data with context. In: 2nd Workshop on Scalable Semantic Web Knowledge Base Systems (2006)

30. Jeh, G., Widom, J.: SimRank: A measure of structural-context similarity. In: Proceedings of the Eighth ACM SIGKDD International Conference on Knowledge Discovery and Data Mining, KDD '02, pp. 538–543. ACM (2002). doi: 10.1145/775047.775126

31. Kamvar, S.D., Haveliwala, T.H., Manning, C.D., Golub, G.H.: Extrapolation methods for accelerating pagerank computations. In: Proceedings of the 12th International Conference on World Wide Web, WWW '03, pp. 261–270. ACM (2003). doi: 10.1145/775152.775190

32. Kaptein, R., Kamps, J.: Exploiting the category structure of Wikipedia for entity ranking. Artificial Intelligence **194**, 111–129 (2013). doi: 10.1016/j.artint.2012.06.003

33. Kaptein, R., Koolen, M., Kamps, J.: Result diversity and entity ranking experiments: anchors, links, text and Wikipedia. In: Proceedings of the Eighteenth Text REtrieval Conference, TREC '09. NIST (2010a)

34. Kaptein, R., Serdyukov, P., De Vries, A., Kamps, J.: Entity ranking using Wikipedia as a pivot. In: Proceedings of the 19th ACM international conference on Information and knowledge management, CIKM '10, pp. 69–78. ACM (2010b). doi: 10.1145/1871437.1871451

35. Kleinberg, J.M.: Authoritative sources in a hyperlinked environment. J. ACM **46**(5), 604–632 (1999). doi: 10.1145/324133.324140

36. Li, X., Li, C., Yu, C.: EntityEngine: Answering entity-relationship queries using shallow semantics. In: Proceedings of the 19th ACM International Conference on Information and Knowledge Management, CIKM '10, pp. 1925–1926. ACM (2010). doi: 10.1145/1871437.1871766

37. Lin, Y., Michel, J.B., Aiden, E.L., Orwant, J., Brockman, W., Petrov, S.: Syntactic annotations for the Google books Ngram corpus. In: Proceedings of the ACL 2012 System Demonstrations, ACL '12, pp. 169–174. Association for Computational Linguistics (2012)

38. Losada, D.E., Azzopardi, L.: Assessing multivariate Bernoulli models for information retrieval. ACM Trans. Inf. Syst. **26**(3), 17:1–17:46 (2008). doi: 10.1145/1361684.1361690

39. Metzger, S., Schenkel, R., Sydow, M.: QBEES: Query-by-example entity search in semantic knowledge graphs based on maximal aspects, diversity-awareness and relaxation. J. Intell. Inf. Syst. **49**(3), 333–366 (2017). doi: 10.1007/s10844-017-0443-x

40. Metzler, D., Croft, W.B.: A Markov Random Field model for term dependencies. In: Proceedings of the 28th Annual International ACM SIGIR Conference on Research and Development in Information Retrieval, SIGIR '05, pp. 472–479. ACM (2005). doi: 10.1145/1076034.1076115

41. Mikolov, T., Sutskever, I., Chen, K., Corrado, G., Dean, J.: Distributed representations of words and phrases and their compositionality. In: Proceedings of the 26th International Conference on Neural Information Processing Systems, NIPS'13, pp. 3111–3119. Curran Associates Inc. (2013)

42. Milne, D., Witten, I.H.: Learning to link with Wikipedia. In: Proceedings of the 17th ACM Conference on Information and Knowledge Management, CIKM '08, pp. 509–518 (2008). doi: 10.1145/1458082.1458150

43. Minkov, E., Cohen, W.W.: Improving graph-walk-based similarity with reranking: Case studies for personal information management. ACM Trans. Inf. Syst. **29**(1), 4:1–4:52 (2010). doi: 10.1145/1877766.1877770

44. Nie, Z., Zhang, Y., Wen, J.R., Ma, W.Y.: Object-level ranking: Bringing order to web objects. In: Proceedings of the 14th International Conference on World Wide Web, WWW '05, pp. 567–574. ACM (2005). doi: 10.1145/1060745.1060828

45. Pehcevski, J., Thom, J.A., Vercoustre, A.M., Naumovski, V.: Entity ranking in Wikipedia: utilising categories, links and topic difficulty prediction. Information Retrieval **13**(5), 568–600 (2010). doi: 10.1007/s10791-009-9125-9

46. Pennington, J., Socher, R., Manning, C.D.: Glove: Global vectors for word representation. In: Empirical Methods in Natural Language Processing, EMNLP '14, pp. 1532–1543 (2014)

47. Raviv, H., Carmel, D., Kurland, O.: A ranking framework for entity oriented search using Markov Random Fields. In: Proceedings of the 1st Joint International Workshop on Entity-Oriented and Semantic Search, JIWES '12, pp. 1:1–1:6. ACM (2012). doi: 10.1145/2379307.2379308

48. Sarmento, L., Jijkuon, V., de Rijke, M., Oliveira, E.: "More like these": Growing entity classes from seeds. In: Proceedings of the Sixteenth ACM Conference on Conference on Information and Knowledge Management, CIKM '07, pp. 959–962. ACM (2007). doi: 10.1145/1321440.1321585

49. Schuhmacher, M., Dietz, L., Paolo Ponzetto, S.: Ranking entities for web queries through text and knowledge. In: Proceedings of the 24th ACM International on Conference on

Information and Knowledge Management, CIKM '15, pp. 1461–1470. ACM (2015). doi: 10.1145/2806416.2806480

50. Schuhmacher, M., Ponzetto, S.P.: Knowledge-based graph document modeling. In: Proceedings of the 7th ACM International Conference on Web Search and Data Mining, WSDM '14, pp. 543–552 (2014). doi: 10.1145/2556195.2556250

51. Sehgal, V., Getoor, L., Viechnicki, P.D.: Entity resolution in geospatial data integration. In: Proceedings of the 14th Annual ACM International Symposium on Advances in Geographic Information Systems, GIS '06, pp. 83–90. ACM (2006). doi: 10.1145/1183471.1183486

52. Tonon, A., Demartini, G., Cudré-Mauroux, P.: Combining inverted indices and structured search for ad-hoc object retrieval. In: Proceedings of the 35th International ACM SIGIR Conference on Research and Development in Information Retrieval, SIGIR '12, pp. 125–134. ACM (2012). doi: 10.1145/2348283.2348304

53. Voskarides, N., Meij, E., Tsagkias, M., de Rijke, M., Weerkamp, W.: Learning to explain entity relationships in knowledge graphs. In: Proceedings of the 53rd Annual Meeting of the Association for Computational Linguistics and the 7th International Joint Conference on Natural Language Processing (Volume 1: Long Papers), pp. 564–574. Association for Computational Linguistics (2015)

54. de Vries, A.P., Vercoustre, A.M., Thom, J.A., Craswell, N., Lalmas, M.: Overview of the INEX 2007 Entity Ranking track. In: Proceedings of the 6th Initiative on the Evaluation of XML Retrieval, INEX '07, pp. 245–251. Springer (2008). doi: 10.1007/978-3-540-85902-4_22

55. Wang, C., Chakrabarti, K., He, Y., Ganjam, K., Chen, Z., Bernstein, P.A.: Concept expansion using web tables. In: Proceedings of the 24th International Conference on World Wide Web, WWW '15, pp. 1198–1208. International World Wide Web Conferences Steering Committee (2015). doi: 10.1145/2736277.2741644

56. Yang, L., Guo, Q., Song, Y., Meng, S., Shokouhi, M., McDonald, K., Croft, W.B.: Modeling user interests for zero-query ranking. In: Proceedings of the 38th European Conference on IR Research, ECIR '16, pp. 171–184. Springer (2016). doi: 10.1007/978-3-319-30671-1_13

Part II
Bridging Text and Structure

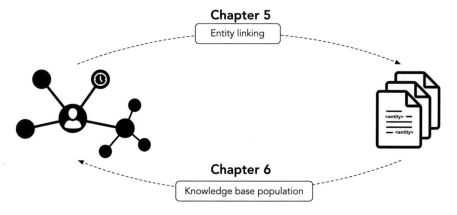

Combining structured and unstructured information is key to improving search breadth and quality. In Part I, we have seen how unstructured and structured data representations can be blended on the retrieval level. In this part, our focus is on the interconnectivity between unstructured and structured data on the dataset level. Entities play a central role in this process, as they help to bridge the gap between the world of unstructured data and the world of structured data. On the one hand, textual content can be enriched by annotating mentions of specific entities with unique identifiers from an underlying knowledge repository, a process commonly known as *entity linking*. On the other hand, knowledge bases can be extended by extracting structured information from unstructured and semi-structured sources, an activity often referred to as *knowledge base population*. Ultimately, the two complementary tasks of entity linking and knowledge base population can together be used to set up the necessary resources for the approaches discussed in the final part of the book.

Chapter 5
Entity Linking

Machine-understanding of text is an extremely challenging problem. The importance of named entities in this regard has been acknowledged early on in natural language processing research; being able to identify entities in a document is a key step towards understanding what the document is about. Like words, entity names can be ambiguous and the same entity may be referred to by many different names. Human readers can use their prior knowledge in combination with the context of a particular *entity mention* (i.e., a text span referring to an entity) to make a decision between the possible choices; for machines, the automatic disambiguation of entity mentions presents many difficulties and challenges. A key enabling component in this process is the availability of large-scale knowledge repositories (such as Wikipedia and various knowledge bases). Having a reference catalog of entities, which are equipped with unique identifiers, the ambiguity of the recognized entity mentions can be resolved by assigning ("linking") them to the corresponding entries in the entity catalog. For instance, there are at least three different Freebase IDs that may be assigned to the mention "Ferrari," depending on whether it refers to the Italian sports car manufacturer (`/m/02_kt`), their racing division that competes in Formula One (`/m/0179v6`), or the founding father Enzo Ferrari (`/m/0gc0s`). The topic of this chapter, *entity linking*, is the task of annotating an input text with entity identifiers from a reference knowledge repository (KR). The output of this annotation process is illustrated in Fig. 5.1.

Linking entities in unstructured text to a structured knowledge repository can greatly empower users in their information consumption activities. For instance, readers of a document can acquire contextual or background information with a single click or can gain easy access to related entities. Entity annotations can also be used in downstream processing to improve retrieval performance or to facilitate better user interaction with search results. We shall look at some of these usages in detail in Part III. Finally, semantic enrichment of documents with entities can prove useful in a number of other text processing tasks as well, including summarization,

© The Author(s) 2018
K. Balog, *Entity-Oriented Search*, The Information Retrieval Series 39,
https://doi.org/10.1007/978-3-319-93935-3_5

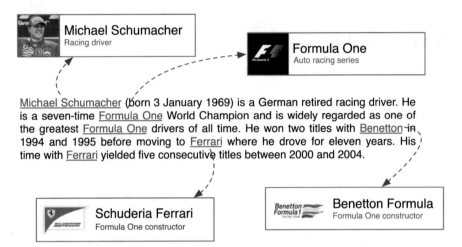

Fig. 5.1 Example of text annotated with entities from Wikipedia. Blue underlined text indicates the linked entity mentions. Text is taken from https://en.wikipedia.org/wiki/Michael_Schumacher

text categorization, topic detection and tracking, knowledge base population, and question answering.

In this chapter our focus is on long text, where it is implicitly assumed that there is in principle always enough context to resolve all entity mentions unambiguously. We discuss the case of short text, such as tweets and search queries, that can possibly have multiple interpretations, in Chap. 7.

The remainder of this chapter is organized as follows. We begin by situating entity linking in the broader context of entity annotation problems in Sect. 5.1. Next, Sect. 5.2 presents an overview of the entity linking task, which is commonly approached as a pipeline of three components. The following sections elaborate on these components: mention detection (Sect. 5.4), candidate selection (Sect. 5.5), and disambiguation (Sect. 5.6). Section 5.7 provides a selection of prominent, publicly available entity linking systems. Evaluation measures and test collections are introduced in Sect. 5.8. We list some useful large-scale resources in Sect. 5.9.

5.1 From Named Entity Recognition Toward Entity Linking

The importance of named entities has long been recognized in natural language processing [64]. Before discussing various approaches to solving the entity linking task, this introductory section gives a brief overview of a range of related entity annotation tasks that have been studied in the past. Table 5.1 provides an overview.

Table 5.1 Overview of named entity recognition and disambiguation tasks

Task	Recognition	Assignment
Named entity recognition	Entities	Entity type
Named entity disambiguation	Entities	Entity identifier/NIL
Wikification	Entities and concepts	Entity identifier/NIL
Entity linking	Entities	Entity identifier

```
<LOC>Silicon Valley</LOC> venture capitalist <PER>Michael Moritz</PER>
said that today's billion-dollar "unicorn" startups can learn from
<ORG>Apple</ORG> founder <PER>Steve Jobs</PER>.
```

Listing 5.1 Text annotated with ENAMEX entity types

5.1.1 Named Entity Recognition

The task of *named entity recognition* (NER) (also known as *entity identification, entity extraction*, and *entity chunking*) is concerned with detecting mentions of entities in text and labeling them each with one of the possible entity types. Listing 5.1 shows an example.

Traditionally, NER has focused on three specific types of proper names: person (PER), organization (ORG), and location (LOC). These are collectively known as ENAMEX types [78]. Proper names falling outside the standard ENAMEX types are sometimes considered under an additional fourth type, miscellaneous (MISC). From an information extraction point of view, temporal expressions (TIMEX) and certain types of numerical expressions (NUMEX) (such as currency and percentages) may also be considered as named entities [78] (primarily because the techniques used to recognize them can be similar). The ENAMEX types only allow for a coarse distinction, whereas for certain applications a more fine-grained classification of entities may be desired. Question answering, in particular, has been a driving problem for the development of type taxonomies. Sekine et al. [71] developed an extended named entity hierarchy, with 150 entity types organized in a tree structure. In follow-up work, they extended their hierarchy to 200 types [70] and defined popular attributes for each category to make their type taxonomy an ontology [69]. This approach, however, "relies heavily on an encyclopedia and manual labor" [69]. That is, why in recent years, (existing) type systems of large-scale knowledge bases have been leveraged for NER. For instance, Ling and Weld [53] introduced a (flat) set of 112 types manually curated from Freebase types, while Yosef et al. [83] derived a fine-grained taxonomy with 505 types, organized in a 9 levels deep hierarchy, from YAGO.

NER is approached as a sequence labeling problem, where a categorical label (entity type or not-an-entity) is to be assigned to each term. The dominant technique is to train a machine-learned model on a large collection of annotated documents. Widely used sequence labeling models are *hidden Markov models* [86] and *conditional random fields* [23]. Commonly used features include word-level features (word case, punctuation, special characters, word suffixes and prefixes, etc.), character-level n-grams, part-of-speech tags, dictionary lookup features (whether the term is present in a dictionary, often referred to as gazetteer list), and document and corpus features (other entities, position in document, meta information, corpus frequency, etc.). We refer to Nadeau and Sekine [64] for a detailed overview of traditional NER techniques. Recently, neural networks have been shown to achieve state-of-the-art performance without resorting to hand-engineered features [50, 54].

Named entity recognition is a basic functionality in most NLP toolkits (e.g., GATE,[1] Stanford CoreNLP,[2] NLTK,[3] or Apache OpenNLP[4]). NER techniques have been evaluated at the MUC, IREX, CoNLL, and ACE conferences.

5.1.2 Named Entity Disambiguation

Named entity disambiguation (NED), also called *named entity normalization* or *named entity resolution*, is the task of disambiguating entity mentions by assigning entity identifiers to them from some catalog. It is usually assumed that entity mentions have already been detected in the input text (i.e., it has been processed by a NER system). NER is closely related to *word sense disambiguation* (WSD), which is one of the earliest problems in natural language processing. WSD is the process of identifying in what sense (meaning) a word is being used in the given context, when the word has multiple meanings [65]. The possible senses are assigned from some dictionary or thesaurus (typically, WordNet [59]). This way, one can decide, e.g., "whether the word 'church' refers to a building or an institution in a given context" [12]. WSD evaluations exclude proper noun disambiguation (that is addressed separately in NED). It is easy to see that NED and WSD share similarities: they attempt to resolve language ambiguity by mapping words or phrases to unique identifiers. However, there are at least two key differences. First, the input in WSD is a single token (e.g., "church"), while in NED it may be a sequence of tokens (e.g., "Church of England") or an abbreviation (e.g., "CofE"). Second, WSD assumes that each possible word sense has an entry in the dictionary and candidate senses are provided directly; since "named entity mentions vary more than lexical mentions in WSD" [32], candidate entity generation (i.e., identifying the set of entities that the mention possibly refers to) is a critical step in NED. Furthermore, entity mentions without a corresponding catalog entry need are annotated with a special NIL identifier. Nevertheless, the two tasks may be seen as analogous, and early NED approaches were indeed inspired by WSD research [58].

---

[1] https://gate.ac.uk/.

[2] http://stanfordnlp.github.io/CoreNLP/.

[3] http://www.nltk.org/.

[4] https://opennlp.apache.org/.

The advent of Wikipedia has facilitated large-scale entity recognition and disambiguation by providing a comprehensive catalog of entities along with other invaluable resources (specifically, hyperlinks, categories, and redirection and disambiguation pages; cf. Sect. 2.2). The work by Bunescu and Paşca [2] was the first to perform named entity disambiguation using Wikipedia and was soon followed by others [12, 58].

Within the general problem area of named entity disambiguation, a number of more specific tasks can be distinguished, cf. Table 5.1. Mihalcea and Csomai [58] define *wikification* as "the task of automatically extracting the most important words and phrases in the document, and identifying for each such keyword the appropriate link to a Wikipedia article." The *entity linking* task is to assign mentions of entities in a document to entity identifiers in a reference knowledge repository. We make a conscious distinction between wikification and entity linking, emphasizing that the latter considers only proper names, while the former includes concepts too. Nevertheless, the techniques for the two are essentially the same. We also wish to point out that named entity disambiguation and entity linking are often considered to be synonymous in the NLP community; we make a distinction between the two because of the following important differences:

- Most NED datasets mark up entity mentions explicitly and supply these as part of the input; entity linking is also concerned with the detection of these mentions in the input text.
- Recognizing out-of-KR entities and marking them as NIL is an important subproblem within NED; in entity linking a "closed world" assumption is typically made, i.e., all "possible meanings of a name are known upfront" [37].

We present evaluation methodology and resources in Sect. 5.8.

5.1.3   Entity Coreference Resolution

Another task related to but different from named entity disambiguation is *entity coreference resolution*. Here, entity mentions are to be clustered "such that two mentions belong to the same cluster if and only if they refer to the same entity" [75]. In this task, "there is no explicit mapping onto entities in a knowledge base" [36]. The task is addressed in two flavors: within-document and cross-document coreference resolution. Coreference resolution has been evaluated at the MUC and ACE conferences. We refer to Ng [66] for a survey of approaches.

5.2   The Entity Linking Task

Definition 5.1 *Entity linking* is the task of recognizing entity mentions in text
and linking them to the corresponding entries in a knowledge repository.

For simplicity, we will refer to the input text as a *document*. Consider, e.g., the
mention "Ferrari" that can refer to any of the entities FERRARI (the Italian sports
car manufacturer), SCUDERIA FERRARI (the racing division), FERRARI F2007
(a particular model with which Ferrari competed during the 2007 Formula One
season), or ENZO FERRARI (the founder), among others. Based on the context
in which the mention occurs (i.e., the document's content), a single one of these
candidate entities is selected and linked to the corresponding entry in a knowledge
repository. Our current task, therefore, is limited to recognizing entities for which a
target entry exists in the reference knowledge repository. Further, it is assumed that
the document provides sufficient context for disambiguating entities.

Formally, given an input document d, the task is to generate entity annotations for
the document, denoted by $\mathcal{A}_d$, where each annotation $a \in \mathcal{A}_d$ is given as a triple $a =
(e, m_i, m_t)$: e is an entity (reference to an entry in the knowledge repository), and m_i
and m_t denote the initial and terminal character offsets of the entity's mention in d,
respectively. The linked entity mentions in $\mathcal{A}_d$ must not overlap.

Unless pointed out explicitly, the techniques presented below rely on a rather
broad definition of a knowledge repository: It provides a catalog of entities, each
with one or more names (surface forms), links to other entities, and, optionally,
a textual description. The attentive reader might have noticed that we are here
using the term knowledge repository as opposed to knowledge base. This is on
purpose. The reference knowledge repository that is most commonly used for entity
linking is Wikipedia, which is not a knowledge base (cf. Sect. 2.3). General-purpose
knowledge bases—DBpedia, Freebase, and YAGO—are also frequently used, since
these provide sufficient coverage for most tasks and applications. Also, mapping
between their entries and Wikipedia is straightforward. Alternatively, domain-
specific resources may also be used, such as the Medical Subject Headings (MeSH)
controlled vocabulary.[5]

We refer to Table 5.2 for the notation used throughout this chapter.

5.3   The Anatomy of an Entity Linking System

Over the years, a canonical approach to entity linking has emerged that consists of
a pipeline of three components [4, 32], as shown in Fig. 5.2.

---

[5]https://www.nlm.nih.gov/mesh/.

Table 5.2 Notation used in this chapter

Symbol	Meaning
a	Annotation ($a = (e, m_i, m_t) \in \mathcal{A}_d$)
$\mathcal{A}_d$	Entity annotations for document d
d	Document
d_e	Textual representation (entity description) of entity e
e	Entity ($e \in \mathcal{E}$)
$\mathcal{E}$	Entity catalog (set of all entities)
$\mathcal{E}_d$	Set of all candidate entities in the document d
$\mathcal{E}_m$	Set of candidate entities for mention m
$\mathcal{E}_s$	Set of entities denoted by the surface form s
$\mathcal{L}_e$	Set of links of an entity e
m	Mention (text span) ($m \in \mathcal{M}_d$)
$\mathcal{M}_d$	Set of mentions for document d
$n(m, e)$	Number of times e is a link target of m
s	Surface form ($s \in \mathcal{S}$)
$\mathcal{S}$	Surface form dictionary

Fig. 5.2 Entity linking pipeline

Mention detection   The first component, also known as *extractor* or *"spotter,"* is
responsible for the identification of text snippets that can potentially be linked
to entities. Commonly, mention detection is based on an extensive dictionary of
entity names and variations thereof, which we will refer to as (entity) *surface
forms*. Mention detection is closely related to the problem of named entity
recognition (cf. Sect. 5.1.1) and can indeed be performed with the help of NER
techniques. Since only mentions detected by the extractor are considered for
subsequent processing in the pipeline, the emphasis here is on achieving high
recall.

Candidate selection   Next, a set (or ranked list) of candidate entities is generated
for each mention. This component is sometimes referred to as the *searcher*.
Given that the next step (disambiguation) is typically the computationally most
expensive one of all, "an ideal searcher should balance precision and recall to
capture the correct entity [for each mention] while maintaining a small set of
candidates" [32].

Disambiguation   Finally, in the disambiguation step, a single best entity (or none)
is selected for each mention, based on the context. This task can be framed as a
ranking problem: Given a mention along with the set of candidate entities for that
mention, rank candidates based on their likelihood of being the correct referent
for the mention. The assigned score can be interpreted as the confidence in the
linking, and the annotation (mention-entity pair) may be rejected if its score falls

below a certain (user-defined or machine-learned) threshold. This threshold may also be used to balance the trade-off between precision and recall. Alternatively, disambiguation may be approached as an inference problem, with the objective of optimizing the coherence among all entity linking decisions in the document.

In the following three sections, we look at each of the processes corresponding to the components in Fig. 5.2 in detail.

Before we continue, we note that the organization of the entity linking task along these steps is the most commonly used, but certainly not the only possibility. One particular alternative is where only two stages are distinguished: entity detection and disambiguation [2]. With this approach, mention detection and candidate selection are essentially performed jointly in a single step—a reasonable choice when detection is performed using dictionary-based methods.

5.4   Mention Detection

The first component in the entity linking pipeline is responsible for the detection of entity mentions in the document.

Definition 5.2 A *mention* is a text span (contiguous sequence of terms) in the document that refers to a particular entity. The referred entity may or may not exist in the reference knowledge repository.

Formally, for an input document d, the set of mentions $\mathcal{M}_d$ is to be identified, where each mention $m \in \mathcal{M}_d$ is defined by its initial and terminal character offsets. Bear in mind that the scope of this task is restricted to entities that are contained in the knowledge repository. For that reason, virtually all modern entity linking systems rely on a dictionary of known surface forms to detect mentions; see, e.g., [2, 12, 22, 37, 49, 57, 68]. In a sense, we work under a controlled vocabulary setting; if the text span under consideration does not match any entry in the dictionary then it will not be recognized as a mention, and, consequently, will not be linked to any entity. Therefore, it is vitally important for the dictionary of surface forms to be extensive, including common variations, nicknames, abbreviations, etc. We detail the construction of the surface form dictionary in Sect. 5.4.1.

Assuming that this surface form dictionary $\mathcal{S}$ has been constructed, mention detection works as follows. The input document is parsed and all possible text spans are checked if they are present in $\mathcal{S}$. Text spans are typically token n-grams at length $\leq n$, with n set between 6 and 8. Figure 5.3 illustrates the process. This kind of lexicon-based string matching can be performed efficiently using, e.g., the Aho–Corasick algorithm [1]. To reduce the number of unnecessary dictionary lookups, and thereby increase the efficiency and throughput of mention detection, certain snippets may be disregarded. For example, a system might be instructed not to annotate common words or text spans that are only composed of verbs, adjectives,

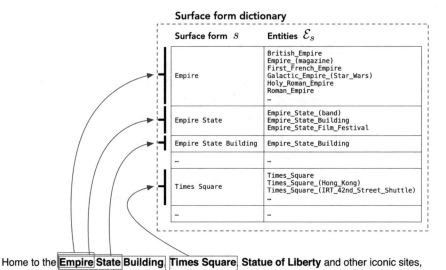

Fig. 5.3 Illustration of dictionary-based mention detection. Detected mentions are boldfaced. The boxes show some of the mentions being looked up (indicated by arrows) in the surface form dictionary. Note the overlaps between mentions

adverbs, and prepositions [57]. Moreover, one might employ simple heuristics, for instance, restrict detection to words that have at least one capitalized letter [16].

Another approach to mention detection is to use NER techniques from natural language processing (cf. Sect. 5.1.1) to identify text spans (typically noun phrases) that refer to named entities; see, e.g., [9, 37, 68]. In this case, an additional string comparison step is involved, where the detected mentions are to be matched against known entity surface forms using some string similarity measure, e.g., edit distance [85], character Dice score, skip bigram Dice score, or Hamming distance [14]. Mentions that do not match any of the dictionary entries, even under a relaxed matching criteria, are likely to denote new, out-of-KR entities.

In practice, it is often desirable that mention detection works directly on the raw text, before any of the standard pre-processing steps, such as tokenization, stopword removal, case-folding, etc., would take place. Sentence boundaries and capitalization can provide cues for recognizing named entities.

5.4.1 Surface Form Dictionary Construction

Dictionary-based mention detection relies on known *surface forms* of entities. These surface forms, also known as *name variants* or *aliases*, are organized in a dictionary structure (map), $\mathcal{S} : s \rightarrow \mathcal{E}_s$, where the surface form s is the key and it is mapped to the set $\mathcal{E}_s$ of entities.

The reference knowledge repository that entity linking is performed against might already contain a list of name variants for each entity. Below, we focus on the scenario where such lists of aliases are either unavailable or need to be expanded, and discuss how entity surface forms may be obtained from a variety of sources.

Collecting Surface Forms from Wikipedia   Wikipedia is a rich resource that has been heavily utilized for extracting name variants. For a given entity, represented by a Wikipedia article, the following sources may be used for collecting aliases:

- *Page title* is the canonical (most common) name for the entity (cf. Sect. 2.2.1.1).
- *Redirect pages* exist for alternative names (including spelling variations and abbreviations) that are frequently used to refer to an entity (cf. Sect. 2.2.3.1).
- *Disambiguation pages* contain a list of entities that share the same name (cf. Sect. 2.2.3.2).
- *Anchor texts* of links pointing to the article can be regarded as aliases of the linked entity (cf. Sect. 2.2.2).
- *Bold texts from first paragraph* generally denote other name variants of the entity.

Recall that not all Wikipedia pages represent entities. With the help of a small set of heuristic rules, it is possible to retain only those Wikipedia articles that refer to named entities (i.e., entities with a proper name title) [2].

Collecting Surface Forms from Other sources   The idea of using anchor texts may be generalized from inter-Wikipedia links to links from (external) web pages pointing to Wikipedia articles; one such dictionary resource is presented in Sect. 5.9.1.

The task of identifying name variants is also known as the problem of *entity synonym discovery*. Synonyms might be identified by expanding acronyms [84], or leveraging search results [7, 14] or query click logs [6, 8] from a web search engine.

5.4.2   Filtering Mentions

The surface form dictionary can easily grow (too) large, since, in principle, it contains all strings as keys that have ever been used as anchor text for a link pointing to an entity. While our main focus is on recall, it is still desirable to filter out mentions that are unlikely to be linked to any entity. In this subsection we present two Wikipedia-based measures that may be used for that. Notice that we intentionally call this procedure "filtering mentions," as opposed to "filtering surface forms:" it may be performed early on in the pipeline (i.e., even during the construction of the surface form dictionary) or later, as part of candidate selection or disambiguation.

In their seminal work, Mihalcea and Csomai [58] introduce the concept of *keyphraseness*, which is an estimate of how likely it is that a given text span will be linked to an entity:

$$P(\text{keyphrase}|m) = \frac{|\mathcal{D}_{link}(m)|}{|\mathcal{D}(m)|} , \qquad (5.1)$$

where $|\mathcal{D}_{link}(m)|$ is the number of Wikipedia articles where m appears as an anchor text of a link, and $|\mathcal{D}(m)|$ is the number of Wikipedia articles that contain m.

It is essentially the same idea that is captured under the notion of *link probability* in [22]:

$$P(\text{link}|m) = \frac{n_{link}(m)}{n(m)} , \qquad (5.2)$$

where $n_{link}(m)$ is the number of times mention m appears as an anchor text of a link, and $n(m)$ denotes the total number of times mention m occurs in Wikipedia (as a link or not).

The main difference between keyphraseness and link probability is that the former considers at most one occurrence (and linking) of a mention per document, while the latter counts all occurrences. (An analogy can be drawn to document frequency vs. term frequency in term importance weighting.) To get a more reliable estimate, it is common to discard mentions that are composed of a single character, made up of only numbers, appear too infrequently in Wikipedia (e.g., less than five times [58]), or have too low relative frequency (e.g., $P(\text{link}|m) < 0.001$ [22]).

5.4.3 Overlapping Mentions

It should be pointed out that the recognized mentions may be overlapping (cf. Fig. 5.3), while the final entity annotations must not overlap. To deal with this, either of two main strategies is employed: (1) containment mentions are dealt with in the mention detection phase, e.g., by dropping a mention if it is subsumed by another mention [29] or by selecting the mention with the highest link probability [22], or (2) overlapping mentions are kept and the decision is postponed to a later stage (candidate selection or disambiguation).

5.5 Candidate Selection

The detection of entity mentions is followed by the selection of candidates for each mention. Let $\mathcal{E}_m$ denote the set of candidate entities for mention m. Potentially, all entities with surface forms matching the mention are candidates: $\mathcal{E}_m = \{e : m \in \mathcal{S}_e\}$. However, as Mendes et al. [57] point out, "candidate selection offers a chance to narrow down the space of disambiguation possibilities." Selecting fewer

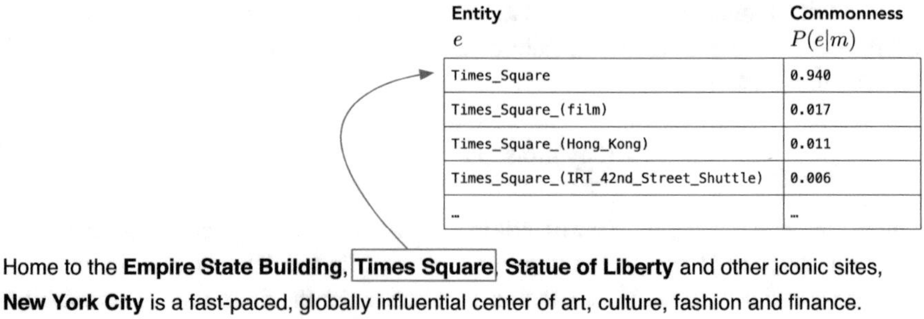

Entity e	Commonness $P(e\|m)$
Times_Square	0.940
Times_Square_(film)	0.017
Times_Square_(Hong_Kong)	0.011
Times_Square_(IRT_42nd_Street_Shuttle)	0.006
...	...

Home to the **Empire State Building,** Times Square **Statue of Liberty** and other iconic sites, **New York City** is a fast-paced, globally influential center of art, culture, fashion and finance.

Fig. 5.4 Ranking candidate entities based on commonness

candidates can greatly reduce computation time, but it may hurt recall if performed too aggressively. In the process of entity linking, candidate selection plays a crucial role in balancing the trade-off between effectiveness and efficiency. Therefore, candidate selection is often approached as a ranking problem: Given a mention m, determine the prior probability of an entity e being the link target for m: $P(e|m)$. The probabilistic interpretation comes naturally here as it emphasizes the fact that this estimate is based only on the mention, a priori to observing its context. We note that this estimate does not have to be an actual probability; any monotonic scoring function may be used. The top ranked candidate entities, based on a score or rank threshold, are then selected to form $\mathcal{E}_m$.

A highly influential idea by Medelyan et al. [56] is to take into account the overall popularity of entities as targets for a given mention m in Wikipedia. The *commonness* of an entity e is defined as the number of times it is used as a link destination for m divided by the total number of times m appears as a link. In other words, commonness is the maximum-likelihood probability of entity e being the link target of mention m:[6]

$$P(e|m) = \frac{n(m,e)}{\sum_{e' \in \mathcal{E}} n(m,e')} \ . \tag{5.3}$$

Commonness, while typically estimated using Wikipedia (see, e.g., [22, 61]), is not bound to that. It can be based on any entity-annotated text that is large enough to generate meaningful statistics. Using Wikipedia is convenient as the links are of high quality and can be extracted easily from the wiki markup, but a machine-annotated corpus may also be used for the same purpose (see Sect. 5.9.2). We also note that commonness may be pre-computed and conveniently stored in the entity surface form dictionary along with the corresponding entity; see Fig. 5.4.

---

[6]The attentive reader may notice the similarity to link probability in Eq. (5.2). The difference is that link probability is the likelihood of a given mention being linked to *any* entity, while commonness is the likelihood of a given mention referring to a *particular* entity.

It has also been shown that commonness follows a power law distribution with a long tail of extremely unlikely aliases [61]. Thus one can safely discard entities at the tail end of the distribution (0.001 is a sensible threshold).

5.6 Disambiguation

The last step, which is the heart and soul of the entity linking process, is disambiguation: selecting a single entity, or none, from the set of candidate entities identified for each mention. The simplest solution to resolving ambiguity is to resort to the "most common sense," i.e., select the entity that is most commonly referred to by that mention. This is exactly what the commonness measure, which was discussed in the previous section, captures; see Eq. (5.3). Despite being a naïve solution, it "is a very reliable indicator of the correct disambiguation" [68]. Relying solely on commonness can yield correct answers in many cases and represents a solid baseline [43]. For accurate entity disambiguation, nevertheless, we need to incorporate additional clues.

> Modern disambiguation approaches consider three types of evidence: *prior importance* of entities and mentions, *contextual similarity* between the text surrounding the mention and the candidate entity, and *coherence* among all entity linking decisions in the document.

We start off in Sect. 5.6.1 by presenting a set of features for capturing the above three types of evidence. Next, in Sect. 5.6.2, we discuss specific disambiguation approaches that combine this evidence in some way (e.g., using supervised learning or graph-based approaches). The selection of the single best entity for each mention may optionally be followed by a subsequent pruning step: rejecting low confidence or semantically meaningless annotations. We discuss pruning in Sect. 5.6.3.

5.6.1 Features

We discuss features by dividing them into three main groups:

- *Prior importance features* may rely on the entity alone, $f(e)$, or the mention and the entity in combination, $f(e,m)$. In either case, the score is estimated based on prior importance without taking the mention's context into account.
- *Contextual features* are guided by the intuition that the context surrounding an ambiguous entity mention provides valuable additional information for disambiguating it. These features could be written as $f(e,m;d)$, emphasizing that the context is based on the input document. Since we process one document at a time, we will omit d for notational convenience, and simply write $f(e,m)$.

Table 5.3 Features for entity disambiguation

Group	Feature	Description
Prior importance (context-independent)		
	$P(\text{keyphrase}|m)$	Keyphraseness (likelihood of m being linked)
	$P(\text{link}|m)$	Link probability (likelihood of m being linked)
	$P(e|m)$	Commonness (the probability of e being the link target of m)
	$P_{link}(e)$	Fraction of links in the knowledge repository pointing to e
	$P_{pageviews}(e)$	Fraction of (Wikipedia) page views e receives
Contextual		
	$sim_F(m,e)$	Similarity between the context of a mention d_m and the entity's description d_e according to some similarity function F (e.g., cosine, Jaccard, dot product, KL divergence, etc.)
Entity-relatedness		
	$WLM(e,e')$	Wikipedia link-based measure, a.k.a. relatedness
	$PMI(e,e')$	Pointwise mutual information
	$Jaccard(e,e')$	Jaccard similarity
	$\chi^2(e,e')$	χ^2 statistic
	$P(e'|e)$	Conditional probability

- *Entity-relatedness features* aim at measuring the degree of semantic relatedness between a pair of entities, $f(e,e')$. The ultimate goal is to measure the *coherence* of entity annotations in a document; as we shall see later, this boils down to pairwise entity relatedness.

We discuss these feature groups in turn, highlighting some of the most effective features within each. Table 5.3 provides an overview. The reader will note the large number of features, which reflects the broad diversity of factors that need to be taken into account for effective disambiguation. Unfortunately, there is no systematic and comprehensive feature comparison available. The decision on what features to use (or design) should take into account the characteristics of the particular dataset and examine the trade-off between effectiveness and efficiency.

5.6.1.1 Prior Importance Features

The first group of features consider a single mention m and/or entity e, where e is one of the candidate annotations for that mention, $e \in \mathcal{E}_m$. Neither the text nor other mentions in the document are taken into account, hence the context-independence. We have already introduced keyphraseness (Eq. (5.1)), link probability (Eq. (5.2)), and commonness (Eq. (5.3)), which all belong to this category. These are all related to the popularity of a mention or the popularity of a particular entity given a mention.

To measure the popularity of the entity itself, we present two simple estimates. The first feature is *link prior*, defined as the fraction of all links in the knowledge repository that are incoming links to the given entity [68]:

$$P_{link}(e) = \frac{|\mathcal{L}_e|}{\sum_{e' \in \mathcal{E}} |\mathcal{L}_{e'}|} \,,$$

where $|\mathcal{L}_e|$ denotes the total number of incoming links entity e has. In the case of Wikipedia, $\mathcal{L}_e$ is the number of all articles that link to the entity's Wikipedia page. In the case of a knowledge base, where entities are represented as SPO triples, it is the number of triples where e stands as object.

Entity popularity may also be estimated based on traffic volume, e.g., by utilizing the Wikipedia page view statistics of the entity's page [29]:

$$P_{pageviews}(e) = \frac{pageviews(e)}{\sum_{e' \in \mathcal{E}} pageviews(e')} \,,$$

where $pageviews(e)$ denotes the total number of page views (measured over a certain time period).

When mention detection is performed using NER as opposed to a dictionary-based approach, the match between the mention and the candidate entity's known surface forms should also be considered. Common name-based similarity features include, among others, whether (1) the mention matches exactly the entity name, (2) the mention starts or ends with the entity name, (3) the mention is contained in the entity name or vice versa, and (4) string similarity between the mention and the entity name (e.g., edit distance) [72]. Additionally, the type of the mention, as detected by the NER (i.e., PER, ORG, LOC, etc.), may be compared against the type of the entity in the knowledge repository [14].

5.6.1.2 Contextual Features

One of the simplest and earliest techniques is to compare the surrounding context of a mention with the textual representation (entity description) of the given candidate entity [2, 12]. The context of a mention, denoted as d_m, can be a window of text around the mention, such as the sentence or paragraph containing the mention, or even the entire document. The textual representation of the entity, denoted as d_e, is based on the entity's description in the knowledge repository. As disambiguation is most commonly performed against Wikipedia, it could be, e.g., the whole Wikipedia entity page [2], the first description paragraph of the Wikipedia page [49], or the top-k terms with the highest TF-IDF score from the entity's Wikipedia page [68].[7]

---

[7]Entity descriptions may also be assembled from a document collection, cf. Sect. 3.2.1. However, those approaches assume that some documents have already been annotated with entities.

Both the mention's context and the entity are commonly represented as bag-of-words. Let $sim_F(m, e)$ denote the contextual similarity between the mention and the entity, using some similarity function F. There is a range of options for the function F, with *cosine similarity* being the most commonly used, see, e.g., [2, 49, 57, 68]:

$$sim_{cos}(m, e) = \frac{\mathbf{d}_m \cdot \mathbf{d}_e}{\| \mathbf{d}_m \| \| \mathbf{d}_e \|},$$

where $\mathbf{d}_m$ and $\mathbf{d}_e$ are the term vectors corresponding to the mention's and entity's representations. Other options for the similarity function F include (but are not limited to): dot product [49], Kullback–Leibler divergence [37], or Jaccard similarity (between word sets) [49].

The representation of context does not have to be limited to bag-of-words. It is straightforward to extend the notion of term vectors to *concept vectors*, to better capture the semantics of the context. Concepts to embed as term vectors could include, among others, named entities (identified using NER) [14], Wikipedia categories [12], anchor text [49], or keyphrases [37].

Additional possibilities to compute context similarity include topic modeling [67, 84] and augmenting the entity's representation using an external corpus [52].

5.6.1.3 Entity-Relatedness Features

In addition to the textual context around a mention, other entities that co-occur in the document can also serve as clues for disambiguation. It can reasonably be assumed that a document focuses on one or at most a few topics. Consequently, the entities mentioned in a document should be topically related to each other. This topical coherence is captured by developing some measure of *relatedness* between a pair of entities. The pairwise entity relatedness scores are then utilized by the disambiguation algorithm to optimize coherence over the set of candidate entities in the document. Notice that we have already touched upon this idea briefly earlier, in Sect. 5.6.1.2, when considering named entities as context. The key difference is that there named entities were treated as string tokens while here we consider the actual entities (given by their identifiers) that are candidates for a particular mention.

Milne and Witten [60] formalize the notion of semantic relatedness for entity linking by introducing the *Wikipedia link-based measure* (WLM), which in later works is often referred to simply as *relatedness*. Modeled after the normalized Google distance measure [10], a close relationship is assumed between two entities if there is a large overlap between the entities linking to them:

$$WLM(e, e') = 1 - \frac{\log(\max(|\mathcal{L}_e|, |\mathcal{L}_{e'}|)) - \log(|\mathcal{L}_e \cap \mathcal{L}_{e'}|)}{\log(|\mathcal{E}|) - \log(\min(|\mathcal{L}_e|, |\mathcal{L}_{e'}|))}, \tag{5.4}$$

where $\mathcal{L}_e$ is the set of entities that link to e and $|\mathcal{E}|$ is the total number of entities. If either of the entities has no links or the two entities have no common links, the score

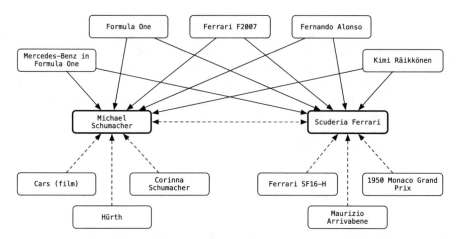

Fig. 5.5 Obtaining the Wikipedia link-based measure between MICHAEL SCHUMACHER and SCUDERIA FERRARI from incoming Wikipedia links (only a selection of links is shown). Solid arrows represent shared links

is set to zero. Figure 5.5 provides an illustration. While relatedness has originally been proposed for incoming Wikipedia links, it may also be considered for outgoing links [68] or for the union of incoming and outgoing links [5]. Also notice that we can equivalently work with relationships in a knowledge base.[8]

Milne and Witten's relatedness measure is the most widely used one and is regarded as the state of the art (see, e.g., [22, 34, 37, 49, 61, 68]), but there are other options, including the Jaccard similarity [30], pointwise mutual information (PMI) [68], or the χ^2 statistic [5]. Notice that all these are symmetric, i.e., $f(e, e') = f(e', e)$ for a particular relatedness function f.

Ceccarelli et al. [5] argue that "a relatedness function should not be symmetric." For example, the relatedness of the UNITED STATES given NEIL ARMSTRONG is intuitively larger than the relatedness of NEIL ARMSTRONG given the UNITED STATES. One effective asymmetric feature they introduce is the conditional probability of an entity given another entity:

$$P(e'|e) = \frac{|\mathcal{L}_{e'} \cap \mathcal{L}_e|}{|\mathcal{L}_e|} .$$

There is obviously a large number of ways one could define relatedness. As we shall see later (in Sect. 5.6.2), having a single relatedness function is preferred to keep the disambiguation process simple (or at least not to make it more complicated). Ceccarelli et al. [5] show that various relatedness measures (a total of 27 in their experiments) can effectively be combined into a single relatedness score using a machine learning approach.

---

[8]There is a link from e_1 to e_2 if there exists an SPO triple where e_1 appears as subject and e_2 appears as object (the predicate is not considered).

All the features we have presented here so far are based on links. The main reason for favoring link-based features over content-based ones is that the former are cheaper to compute. We need to keep in mind, however, that for entities that do not have many links associated with them (e.g., long-tail entities or entities that have been only recently added to the knowledge repository), these techniques do not work very well. In those cases, one can estimate the semantic relatedness between a pair of entities based on (1) the similarity of the contexts in which they occur (e.g., using keyphrases [36] or n-grams [73]) or (2) the assigned types (e.g., by considering their distance in the type hierarchy [73]).

5.6.2 Approaches

Formally, the disambiguation task is to find the assignment of entities to mentions in a given document: $\Gamma : \mathcal{M}_d \rightarrow \mathcal{E} \bigcup \{\emptyset\}$, where $\emptyset$ denotes the NIL entity assignment. We shall now present various methods and algorithms for establishing this mapping.

> Effective disambiguation needs to combine *local compatibility* (which includes prior importance and contextual similarity) and *coherence* with the other entity linking decisions in the document.

The overall objective function thus can be written as:

$$\Gamma^* = \arg\max_{\Gamma} \left(\sum_{(m,e)\in\Gamma} \phi(m,e) + \psi(\Gamma) \right),$$

where $\phi(m,e)$ denotes the local compatibility between the mention and the assigned entity, $\psi(\Gamma)$ is the coherence function for all entity annotations in the document, and Γ is a solution (set of mention-entity pairs). This optimization problem is shown to be NP-hard [37, 49, 68, 73], therefore approaches need to resort to approximation algorithms and heuristics.

We distinguish between two main disambiguation strategies, based on whether they consider mentions (1) *individually*, one mention at a time, or (2) *collectively*, all mentions in the document jointly.

Individual disambiguation approaches most commonly cast the task of entity disambiguation as a ranking problem. Each mention is annotated with the highest scoring entity (or as NIL, if the highest score falls below a given threshold):

$$\Gamma^*_{local}(m) = \arg\max_{e\in\mathcal{E}_m} score(e;m) . \tag{5.5}$$

As discussed earlier, this ranking may be based on a prior popularity (i.e., commonness) alone: $score(e;m) = P(e|m)$. For effective disambiguation, however, it is key

Table 5.4 Entity disambiguation approaches

Approach	Context	Entity interdependence
Most common sense	None	None
Individual local disambiguation	Text	None
Individual global disambiguation	Text and entities	Pairwise
Collective disambiguation	Text and entities	Collective

to consider the context of the mention. Learning-to-rank approaches are well suited for combining multiple signals and have indeed been the most popular choice for this task, see, e.g., [2, 14, 68, 73, 84, 85]. It is important to point out that the fact that mentions are disambiguated individually does not imply that these disambiguation decisions are independent of each other. The interdependence between entity linking decisions may be ignored or may be incorporated (in a pairwise fashion). We refer to these two variants as *local* and *global* approaches, respectively.

Instead of considering each mention individually, once, one might attempt to jointly disambiguate all mentions in the text. Collective disambiguation typically involves an inference process where entity assignments are iteratively updated until some target criterion is met. Table 5.4 provides an overview of approaches.[9]

A final note before we enter into the discussion of specific methods. It is generally assumed (following the *one sense per discourse* assumption [26]) that all the instances of a mention refer to the same entity within the document. If that assumption is lifted, one might employ an iterative algorithm that shrinks the disambiguation context from document to paragraph or even to the sentence level, if necessary [12].

5.6.2.1 Individual Local Disambiguation

Early entity linking approaches [2, 58] focused on *local compatibility* based on contextual features, such as the similarity between the document and the entity's description. Statistics extracted from large-scale entity-annotated data (e.g., Wikipedia), i.e., prior importance, can also be incorporated in the local compatibility score. That is, $score(e; m) = \phi(e, m)$. The local compatibility score can be written in the form of a simple linear combination of features:

$$\phi(e, m) = \sum_i \lambda_i f_i(e, m), \tag{5.6}$$

---

[9]Certain approaches from the literature are not immediately straightforward to categorize. We are guided by the following simple rule: It is individual disambiguation if a candidate entity is assigned a score once, and that score does not change. In the case of collective disambiguation, the initially assigned score changes over the course of multiple successive iterations.

where $f_i(e,m)$ can be either a context-independent or a context-dependent feature (see Sects. 5.6.1.1 and 5.6.1.2), and λ_i is the corresponding feature weight. Note that other entity assignments in the document are not taken into consideration.

The idea is to learn the "optimal" combination of features (which is not limited to being a linear combination) from training data. Working within a learning-to-rank framework, each entity-mention pair becomes an instance, described by a feature vector. In the training dataset, the target label is set to 1 for the correct entity and 0 for all other candidate entities.

5.6.2.2 Individual Global Disambiguation

Entity linking can be improved by considering what other entities are mentioned in the document, an idea that was first proposed by Cucerzan [12]. The underlying assumption is that "entities are correlated and consistent with the main topic of the document" [27]. Cucerzan [12] attempts to find an assignment of entities to mentions such that it maximizes the similarity between each entity in the assignment and all possible disambiguations of all other mentions in the document. This can be incorporated as a feature function $f(e,m;\tilde{d})$, where $\tilde{d}$ is a high-dimensional extended document vector that contains all candidate entities for all other mentions in the document.[10] The function then measures the similarity as a scalar product between the entity and the extended document given a particular representation (e.g., topic words or IDs). A disadvantage of this approach is that the extended document vector contains noisy data, as it includes all the incorrect disambiguations as well.

Another disambiguation strategy, proposed by Milne and Witten [61], is to first identify a set of unambiguous mentions. These are then used as context to disambiguate the other mentions in the document. The two main features used for disambiguation are commonness (Eq. (5.3)) and relatedness (Eq. (5.4)). The disadvantage of this approach is the assumption that there exist unambiguous mentions (which, in practice, translates to documents needing to be sufficiently long).

The general idea behind global approaches is to optimize the *coherence* of the disambiguations (entity linking decisions). A true global optimization would be NP-hard, however a good approximation can be computed efficiently by considering pairwise interdependencies for each mention independently. For this reason, the pairwise entity relatedness scores (which we have introduced in Sect. 5.6.1.3) need to be aggregated into a single number. This number will tell us how coherent the given candidate entity is with the rest of the entities in the document. We discuss two specific realizations of this idea.

Ratinov et al. [68] first perform local disambiguation and use the predictions of that system (i.e., the top ranked entity for each mention) in a second, global

---

[10]Following Cucerzan [12], we use the distinctive notation $\tilde{d}$ to "emphasize that this vector contains information that was not present in the original document."

disambiguation round. Let $\mathcal{E}^\star$ denote the set of linked entities identified by the local disambiguator. The coherence of entity e with all other linked entities in the document is given by:

$$\psi_j(e,\mathcal{E}^\star) = \sum_{\substack{e' \in \mathcal{E}^\star \\ e' \neq e}} g_j(e,e'),\tag{5.7}$$

where g_j is a particular pairwise entity relatedness function. We may now extend our scoring function with a second component consisting of global features:

$$score(e;m) = \underbrace{\sum_i \lambda_i f_i(e,m)}_{\phi(e,m)} + \sum_j \Big(\lambda_j \underbrace{\sum_{\substack{e' \in \mathcal{E}^\star \\ e' \neq e}} g_j(e,e')}_{\psi_j(e,\mathcal{E}^\star)} \Big).$$

The λ_i and λ_j coefficients are trained using supervised learning.

An alternative approach is given by Ferragina and Scaiella [22], capitalizing on the fact that commonness and Milne and Witten's relatedness are the two most important features. Their system, called TAGME, introduces a voting mechanism, illustrated in Fig. 5.6, that allows for the combination of these two features, without involving supervised learning. Similarly to Cucerzan [12], a score for a given mention-entity pair is determined by a "collective agreement" between the entity and all possible disambiguations of all other mentions in the document, but in TAGME this is achieved computationally much more efficiently (specifically, time complexity is linear in the number of mentions [21]). Formally, given the set of all mentions in the document $\mathcal{M}_d$, the score of a candidate entity e for a particular mention m is defined as:

$$score(e;m) = \sum_{\substack{m' \in \mathcal{M}_d \\ m' \neq m}} vote(m',e).\tag{5.8}$$

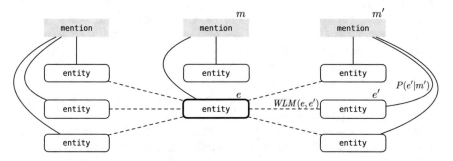

Fig. 5.6 TAGME's voting mechanism. Solid lines connect mentions with the respective candidate entities. A given candidate entity (indicated with the thick border) receives votes from all candidate entities of all mentions in the text (dashed lines)

The *vote* function estimates the agreement between the given entity e and all candidate entities of mention m'. It is computed as the average relatedness between each possible disambiguation e' of m', weighted by its commonness score:

$$vote(m', e) = \frac{\sum_{e' \in \mathcal{E}_{m'}} WLM(e, e') P(e'|m')}{|\mathcal{E}_{m'}|}.$$

Simply returning the entity with the highest score, as defined by Eq. (5.8), is insufficient to obtain an accurate disambiguation, it needs to be combined with other features. For example, this score could be plugged into Eq. (5.6) as a feature function f_i. Another possibility is proposed in [22] in the form of a simple but robust heuristic. Only the highest scoring entities are considered for a given mention, then the one with the highest commonness score among those is selected:

$$\Gamma(m) = \underset{e \in \mathcal{E}_m}{\arg\max} \{P(e|m) : e \in \text{top}_\epsilon [score(e; m)]\}. \tag{5.9}$$

That is, the *score* defined in Eq. (5.8) merely acts as a filter. According to Eq. (5.9), only entities in the top ϵ percent of the scores are retained (with ϵ set to 30% in [21]). Out of the remaining entities, the most common sense of the mention will be finally selected.

Individual disambiguation approaches are inherently limited to incorporating interdependencies between entities in a pairwise fashion. This still enforces some degree of coherence among the linked entities, while remaining computationally efficient. Next, we will look at how to model and exploit interdependencies globally.

5.6.2.3 Collective Disambiguation

The main difference when moving from individual to collective disambiguation is how the maximization of coherence between all entity linking decisions in the document is attempted. As we have already pointed out, this optimization is NP-hard. Kulkarni et al. [49] were the first to undertake direct optimization by turning it into a binary integer linear program, and then relaxing it to a linear program (LP). Coherence is measured as the sum of pairwise relatedness between all pairs of linked entities in the document. They show that LP relaxations often give optimal integral solutions. Kulkarni et al. [49] also present a direct greedy hill-climbing approach as an alternative to linear programming, which is comparable both in speed and accuracy to linear programming relaxation.

More recent approaches use a graph structure for collective disambiguation, an idea that was proposed by two independent groups, at about the same time [34, 37]. Mention–entity and entity–entity relations in a document can naturally be represented as a weighted (undirected) graph (termed *referent graph* in [34]). The node set contains all mentions and all candidate entities corresponding to those mentions. There are two types of edges;

- *Mention–entity edges* capture the local compatibility between the mention and the entity. Edge weights $w(m, e)$ could be measured using a combination of context-independent and context-dependent features, as expressed in Eq. (5.6).
- *Entity–entity edges* represent the semantic relatedness between a pair of entities. Edge weights, $w(e, e')$, are set based on Milne and Witten's relatedness (cf. Eq. (5.4)) [34, 37], but other entity-relatedness functions may also be used.

This graph representation is illustrated in Fig. 5.7. While there is no additional type of evidence compared to what was considered before (namely, local compatibility and pairwise entity relatedness), this representation allows for various graph algorithms to be applied. We note that the graph construction might involve additional heuristics (e.g., "robustness tests" in [37]); we omit these in our discussion.

Hoffart et al. [37] pose the problem of entity disambiguation as that of finding a dense subgraph that contains "all mention nodes and exactly one mention-entity edge for each mention." They propose a greedy algorithm, shown in Algorithm 5.1, that starts from the full graph and iteratively removes the entity node with the lowest weighted degree (along with all its incident edges), provided that each mention node remains connected to at least one entity. The *weighted degree* of an entity node, $wd(e)$ is defined as the sum of the weights of its incident edges. The density of the graph is measured as the *minimum weighted degree* among its entity nodes. From the graphs that are produced in each iteration, the one with the highest density is kept as the solution. This ensures that weak links are captured and the solution is not dominated by a few prominent entities with very high weighted degree.

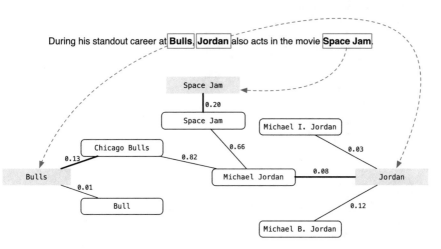

Fig. 5.7 Graph representation for collective disambiguation. Mention nodes are shaded, entity nodes are rounded rectangles. Note that the dashed arrows are not part of the graph. Thick lines indicate the correct mention-entity assignments. Example is taken from [34]

A number of heuristics are applied to ensure that the algorithm is robust. In a pre-processing phase, entities that are "too distant" from the mention nodes are removed. For each entity, the distance from the set of mention nodes is computed as a sum of the squared shortest path distances. Then, only the $k \times |\mathcal{M}_d|$ closest entities are kept ($\mathcal{E}_c$), where $|\mathcal{M}_d|$ is the number of mentions in the document, and k is set to 5 based on experiments. This smaller graph is used as the input to the greedy algorithm. At the end of the iterations, the solution graph may still contain mentions that are connected to more than one entity. The final solution, which maximizes the sum of edge weights, is selected in a post-processing phase. If the graph is sufficiently small, it is feasible to exhaustively consider all possible mention-entity pairs. Otherwise, a faster local (hill-climbing) search algorithm may be used.

Han et al. [34] employ a different graph-based algorithm, *random walk with restarts* [79], for collective disambiguation. "A random walk with restart is a stochastic process to traverse a graph, resulting in a probability distribution over the vertices corresponding to the likelihood those vertices are visited" [31].

Algorithm 5.1: Graph-based entity disambiguation [37]

Input: weighted graph G of mentions and entities
Output: result graph with one edge per mention

```
   /* pre-processing phase                                           */
 1 foreach entity node e do
 2 |    diste ← sum of (weighted) shortest paths to each mention
 3 end
 4 keep entities Ec with lowest diste, drop the others

   /* main loop                                                      */
 5 objective ← mine∈Ec wd(e)/|Ec|
 6 while G has non-taboo entity do
        /* entity is taboo if last candidate for any mention          */
 7 |    e ← non-taboo entity with lowest wd(e)
 8 |    Ec ← Ec \ e
 9 |    remove e with all its incident edges from G
10 |    mwd ← mine∈Ec wd(e) / |Ec|
11 |    if mwd > objective then
12 |    |    solution ← G
13 |    |    objective ← mwd
14 |    end
15 end

   /* post-processing phase                                          */
16 if feasible then
17 |    process solution by enumerating all possible mention-entity pairs
18 else
19 |    process solution by local search
20 end
```

Let $\mathbf{v}$ be a starting vector (initial evidence), holding the prior importance associated with each mention node:

$$\mathbf{v}(m) = \frac{TFIDF(m)}{\sum_{m' \in \mathcal{M}_d} TFIDF(m')} ,$$

where $TFIDF(m)$ is the TF-IDF score of mention m.[11] For entity nodes $\mathbf{v}(e) = 0$. Notice that this formulation assumes a directed graph, with edges going from mentions to entities, but not the other way around.

Given the initial evidence, it is propagated through the two types of edges in the graph. We write $\mathbf{T}$ to denote the evidence propagation matrix. The evidence propagation ratio from a mention to its candidate entities is defined as:

$$\mathbf{T}(m \to e) = \frac{w(m \to e)}{\sum_{e' \in \mathcal{E}_m} w(m \to e')} ,$$

and between entities is defined as:

$$\mathbf{T}(e \to e') = \frac{w(e \to e')}{\sum_{e'' \in \mathcal{E}_d} w(e \to e'')} .$$

Let $\mathbf{r}^i$ be a vector holding the probability distribution over nodes at iteration i, corresponding to the likelihood that those nodes are visited. Initially, it is set to be the starting (initial evidence) vector: $\mathbf{r}^0 = \mathbf{v}$. Then, the probability distribution is updated iteratively until convergence:

$$\mathbf{r}^{i+1} = (1 - \alpha) \, \mathbf{r}^i \, \mathbf{T} + \alpha \, \mathbf{v} ,$$

where α is the restart probability (set to 0.1 in [34]).

Once the random walk process has converged to a stationary distribution $\mathbf{r}$, the referent entity for mention m is determined according to:

$$\Gamma(m) = \arg\max_{e \in \mathcal{E}_m} \phi(m, e) \, \mathbf{r}(e) ,$$

where $\phi(m, e)$ is the local compatibility between the mention and the entity, according to Eq. (5.6).

In conclusion, collective disambiguation approaches tend to perform better than individual ones, and they work especially well "when a text contains mentions of a sufficiently large number of entities within a thematically homogeneous context" [37]. On the other hand, the space of possible entity assignments grows combinatorially, which takes a toll on efficiency, in particular for long documents.

---

[11]The TF component is the normalized frequency of the mention in the document, while the IDF part can be computed using Wikipedia or the Google N-gram dataset.

5.6.3 Pruning

Candidate annotations produced by the disambiguation phase can possibly be pruned to discard meaningless or low-confidence annotations. The simplest possible solution is to control this by a confidence threshold; if $score(e;m) < \tau$ then back off from annotating mention m. The threshold τ can be learned from training data.

More advanced ways to pruning are also conceivable; we highlight three of those here. Milne and Witten [60] employ a machine learned classifier to retain only entities that are "relevant enough" to be linked in the sense of what a human editor would consider annotation-worthy (for instance, in Wikipedia, only the first occurrence of an entity is linked). The set of features includes link probability, relatedness, disambiguation confidence, and location and spread of mentions. Ratinov et al. [68] approach pruning as an optimization problem: They decide, for each mention, whether switching the top-ranked disambiguation to NIL would improve the objective function. Finally, Ferragina and Scaiella [22] define the coherence of entity e with all other candidate entity annotations in the text as:

$$coherence(e, \mathcal{E}^{\star}) = \frac{1}{|\mathcal{E}^{\star}| - 1} \sum_{\substack{e' \in \mathcal{E}^{\star} \\ e' \neq e}} WLM(e, e') \, ,$$

where $\mathcal{E}^{\star}$ is the set of linked entities. For each entity, this coherence score is combined with link probability (either as a simple average or as a linear combination) into a pruning score ρ, which is then checked against the pruning threshold.

5.7 Entity Linking Systems

Table 5.5 presents a selection of prominent entity linking systems that have been made publicly available. Their brief summaries follow below.

Table 5.5 Overview of publicly available entity linking systems

System	Reference KR	Online demo	Web API	Source code
AIDA[a]	YAGO2	Yes	Yes	Yes (Java)
DBpedia Spotlight[b]	DBpedia	Yes	Yes	Yes (Java)
Illinois Wikifier[c]	Wikipedia	No	No	Yes (Java)
TAGME[d]	Wikipedia	Yes	Yes	Yes (Java)
Wikipedia Miner[e]	Wikipedia	No	No	Yes (Java)

[a]http://www.mpi-inf.mpg.de/yago-naga/aida/
[b]http://spotlight.dbpedia.org/
[c]http://cogcomp.cs.illinois.edu/page/download_view/Wikifier
[d]https://tagme.d4science.org/tagme/
[e]https://github.com/dnmilne/wikipediaminer

- *AIDA* [37] performs collective disambiguation using a graph-based approach, which we detailed in Sect. 5.6.2.3. Annotations are done against the YAGO2 knowledge base.
- *DBpedia Spotlight* [57] implements a rather straightforward local disambiguation approach; vector space representations of entities are compared against the paragraphs of their mentions using the cosine similarity. Instead of using standard IDF, Mendes et al. [57] introduce the inverse candidate frequency (ICF) weight and employ TF-ICF term weighting. They annotate with DBpedia entities, which can be restricted to certain types or even to a custom entity set defined by a SPARQL query.
- *Illinois Wikifier* [68], a.k.a. GLOW, implements both local and (individual) global disambiguation; their global disambiguation approach is discussed in Sect. 5.6.2.2. In version 2 of their system, Cheng and Roth [9] focus on eliminating mistakes that are "obvious" (to humans) by better understanding the relational structure of the text (e.g., resolving coreference).
- *TAGME* [22] is one of the most popular entity linking systems. It has been designed specifically for efficient annotation of short texts, but it is shown to deliver competitive results on long texts as well. TAGME's one-mention-at-a-time global disambiguation approach is detailed in Sect. 5.6.2.2. The authors have also published an extended report [21] with more algorithmic details and experiments. We further refer to [35] for additional notes on reproducibility.
- *Wikipedia Miner* [61] is a seminal entity linking system that was first to combine commonness and relatedness for (local) disambiguation. It was also the first system with an open-sourced implementation and with wikification provided as a web service (at the time of writing, it is no longer available). See [62] for more technical details and experimental results.

The above selection concentrates on systems that are accompanied by a scholarly publication detailing the underlying methods and approaches. There is a large number of annotation services offered by commercial parties, including but not limited to: AlchemyAPI,[12] AYLIEN Text Analysis API,[13] Google Cloud Natural Language API,[14] Microsoft Entity Linking service,[15] Open Calais,[16] and Rosette Entity Linking API.[17]

---

[12]http://www.alchemyapi.com/.

[13]http://aylien.com/.

[14]https://cloud.google.com/natural-language/.

[15]https://www.microsoft.com/cognitive-services/en-us/entity-linking-intelligence-service.

[16]http://www.opencalais.com.

[17]https://www.rosette.com/function/entity-linking/.

5.8   Evaluation

In this section, we introduce evaluation measures and test collections.

5.8.1   Evaluation Measures

The overall (end-to-end) performance of an entity linking system is evaluated by comparing the system-generated annotations against a human-annotated gold standard. The measures are set-based: *precision*, *recall*, and *F-measure*. Precision is computed as the fraction of correctly linked entities that have been annotated by the system, while recall is the fraction of correctly linked entities that should be annotated. Since these measures are typically computed over a collection of documents, they can be either *micro-averaged* (aggregated across mentions) or *macro-averaged* (aggregated across documents).

Let us formalize these notions. We write $\mathcal{A}_d$ to denote the annotations generated by the entity linking system and $\hat{\mathcal{A}}_d$ to denote the reference (ground truth) annotations for a single document d. Further, let $\mathcal{A}_{\mathcal{D}}$ include all annotations for a set $\mathcal{D}$ of documents: $\mathcal{A}_{\mathcal{D}} = \cup_{d \in \mathcal{D}} \mathcal{A}_d$. Analogously, $\hat{\mathcal{A}}_{\mathcal{D}}$ is the collection of reference annotations for $\mathcal{D}$. Micro-averaged precision and recall are then defined as:

$$P_{mic} = \frac{|\mathcal{A}_{\mathcal{D}} \cap \hat{\mathcal{A}}_{\mathcal{D}}|}{|\mathcal{A}_{\mathcal{D}}|} , \quad R_{mic} = \frac{|\mathcal{A}_{\mathcal{D}} \cap \hat{\mathcal{A}}_{\mathcal{D}}|}{|\hat{\mathcal{A}}_{\mathcal{D}}|} ,$$

where $|\mathcal{A}_{\mathcal{D}} \cap \hat{\mathcal{A}}_{\mathcal{D}}|$ denotes the number of matching annotations between the systems and the gold standard (to be defined more precisely later).

Macro-averaged precision and recall are computed as follows:

$$P_{mac} = \frac{1}{|\mathcal{D}|} \sum_{d \in \mathcal{D}} \frac{|\mathcal{A}_d \cap \hat{\mathcal{A}}_d|}{|\mathcal{A}_d|} , \quad R_{mac} = \frac{1}{|\mathcal{D}|} \sum_{d \in \mathcal{D}} \frac{|\mathcal{A}_d \cap \hat{\mathcal{A}}_d|}{|\hat{\mathcal{A}}_d|} .$$

The *F-measure* is computed from the overall precision (P) and recall (R):

$$F1 = \frac{2\,P\,R}{P + R} . \tag{5.10}$$

When entity mentions are also given as input to the entity linking system, *accuracy* is used to assess system performance. Accuracy is defined as the number of correctly linked entity mentions divided by the total number of entity mentions. Thus, in this case, accuracy = precision = recall = F1.

When comparing annotations, the linked entities must match, but we may decide to be lenient with respect to their mentions, i.e., the mention offsets. Let $a = (e, m_i, m_t)$ be an annotation generated by the system and $\hat{a} = (\hat{e}, \hat{m}_i, \hat{m}_t)$ be the corresponding reference annotation. We define an indicator function for *perfect match* (PM) as follows:

$$match_{PM}(a,\hat{a}) = \begin{cases} 1, & e = \hat{e}, \, m_i = \hat{m}_i, \, m_i = \hat{m}_i \\ 0, & \text{otherwise}. \end{cases}$$

Alternatively, we can lessen the requirements for mentions such that it is sufficient for them to overlap. For example, "the Madison Square" and "Madison Square Garden" would be accepted as a match as long as they link to the same entity. The indicator function for *relaxed match* (RM) is defined as:

$$match_{RM}(a,\hat{a}) = \begin{cases} 1, & e = \hat{e}, \, [m_i, m_t] \text{ overlaps with } [\hat{m}_i, \hat{m}_t] \\ 0, & \text{otherwise}. \end{cases}$$

Using either flavor of the match function, the number of matching annotations is computed as:

$$|\mathcal{A} \cap \hat{\mathcal{A}}| = \sum_{a \in \mathcal{A}, \hat{a} \in \hat{\mathcal{A}}} match(a,\hat{a}) .$$

5.8.2 Test Collections

Early work used Wikipedia both as the reference KR and as the ground truth for annotations [12, 58, 61]. Using Wikipedia articles as input documents, the task is to "recover" links that were created by Wikipedia contributors. Over the years, the focus has shifted toward entity linking "in the wild," using news articles or web pages as input. This subsection presents the various test collections that have been used in entity linking evaluation. We discuss resources developed by individual researchers and those devised at world-wide evaluation campaigns separately. The main test collections and their key characteristics are summarized in Table 5.6.

5.8.2.1 Individual Researchers

Cucerzan [12] annotated 20 news articles from MSNBC with a total of 755 linkable entity mentions, out of which 113 are NIL (i.e., there is no corresponding Wikipedia article). Milne and Witten [61] used a subset of 50 documents from the AQUAINT text corpus (a collection of newswire stories). Following Wikipedia's style, only the first mention of each entity is linked and only the most important entities are retained. Unlike others, they annotated not only proper nouns but concepts as well. Kulkarni et al. [49] collected and annotated over hundred popular web pages from

Table 5.6 Entity linking and wikification test collections

Name	Reference KR	Document type(s)	#Docs	Annotations type	#Mentions	All[a]	NIL[b]
Individual researchers							
MSNBC [12]	Wikipedia	News	20	Entities	755	Yes	Yes
AQUAINT [61]	Wikipedia	News	50	Entities and concepts	727	No	No
IITB [49]	Wikipedia	Web pages	107	Entities	17,200	Yes	Yes
ACE2004 [68]	Wikipedia	News	57	Entities	306	Yes	Yes
CoNLL-YAGO [37]	YAGO2	News	1393	Entities	34,956	Yes	Yes
Evaluation campaigns							
TAC EL 2009 [74]	Wikipedia	News	3904	Entities	3904	No	Yes
TAC EL 2010 [45]	Wikipedia	News and web	2240	Entities	2240	No	Yes
TAC EL 2011 [51]	Wikipedia	News and web	2250	Entities	2250	No	Yes
TAC EL 2012 [20]	Wikipedia	News and web	2229	Entities	2229	No	Yes
TAC EL 2013 [18]	Wikipedia	News and web	2190	Entities	2190	No	Yes
TAC ELD 2014 [19]	Wikipedia	News and web	138	Entities	5598	Yes	Yes
TAC ELD 2015 [17]	Freebase	News and web	167	Entities	15,581	Yes	Yes
ERD Challenge [3]	Freebase	Web pages	200	Entities	Unknown	Yes	No

[a]Whether all entity mentions are annotated in the documents
[b]Whether out-of-KR entities are annotated as NIL

a handful of domains (sport, entertainment, science and technology, and health). Annotators were instructed to be as exhaustive as possible; this resulted in a total of 17,200 entity mentions with 40% of them annotated as NIL. Ratinov et al. [68] took a subset of the ACE 2004 Coreference dataset as a starting point and annotated mentions (specifically, "the first nominal mention of each co-reference chain" [68]) using crowdsourcing. Hoffart et al. [37] created a dataset based on the CoNLL 2003 Named Entity Recognition task. They annotated 1393 Reuters newswire articles with entities from YAGO2. The collection is split into train, test-A, and test-B partitions. Notably, the original dataset has since been extended to include the corresponding Wikipedia and Freebase entity identifiers as well. Guo and Barbosa [31] released a newer version of the MSNBC, AQUAINT, and ACE2004 datasets, with the annotations aligned to the 2013 June version of Wikipedia.[18]

5.8.2.2   INEX Link-the-Wiki

The Link-the-Wiki track ran at INEX between 2007 and 2010 [40–42, 80] with the objective of evaluating link discovery methods. The assumed user scenario is that of creating a new article in Wikipedia; a link discovery system can then automatically suggest both outgoing and incoming links for that article. Note that link detection

---

[18]http://www.cs.ualberta.ca/~denilson/data/deos14_ualberta_experiments.tgz.

here is approached as a recommendation task, and—unlike in traditional entity linking—it is assumed that a human editor will process the results. Evaluation was performed by selecting an existing Wikipedia article and eradicating all links to and from that page ("orphaning" it), thereby simulating that this is the new document that is being added. The recommended links were assessed manually through a purpose-built interface (with the links originally present in Wikipedia also added to the pool of assessed results). The task is addressed both at the document level (i.e., document-to-document links) and at the element level (i.e., for each prospective anchor, ranking "best entry points" within target documents). System performance is measured using standard IR measures (e.g., MAP). The 2009 and 2010 editions of the track also experimented with a different encyclopedia (Te Ara). That collection, albeit much smaller in size, makes the linking task markedly more complex than using Wikipedia for the reasons that (1) it does not include hyperlinks at all and (2) articles do not represent entities. Since the INEX Link-the-Wiki setup is quite different from our interpretation of the entity linking task, we do not include it in Table 5.6. For further details, we refer to [39].

5.8.2.3 TAC Entity Linking

Entity linking has been running since 2009 at the Knowledge Base Population (KBP) track of the Text Analysis Conference (TAC) [44–47, 55]. Since the track's inception, the task setup has undergone several changes. We start by presenting the initial setting (from 2009) and then discuss briefly how it evolved since.

The Entity Linking (EL) task at TAC KBP is to determine for a given mention string, originating from a particular document, which KB entity is being referred to, or if the entity is not present in the reference KB (NIL). Thus, the focus is on evaluating a single mention per document (referred to as *query*), which is identified in advance, rather than systematically annotating all mentions. The mentions are selected manually by "cherry-picking" those that they are sufficiently confusable, i.e., they have either zero or several KB matches. Additionally, entities with numerous nicknames and shortened or misspelled name variants are also targeted. Entities are either of type person (PER), organization (ORG), or geopolitical entities (GPE). The reference KB is derived from Wikipedia and is further restricted to entities having an infobox. From 2010, web data was also included in addition to newswire documents. An optional entity linking task was also conducted, where systems can only utilize the attributes present in the KB and may not consult the associated Wikipedia page (thereby simulating a setting where a salient and novel entity appears that does not yet have a Wikipedia page). The 2011 edition saw the introduction of two new elements: (1) clustering together NIL mentions (referring to the same out-of-KB entity) and (2) cross-lingual entity linking ("link a given entity from a Chinese document to an English KB" [44]). The cross-lingual version was extended with Spanish in 2012.

In 2014, the task was broadened to end-to-end entity linking and also got re-branded as Entity Discovery and Linking (EDL). Participating systems were required to automatically identify (and classify) all entity mentions, link each entity mention to the KB, and cluster all NIL mentions [46]. In 2015, the EDL task was expanded from monolingual to trilingual coverage (English, Chinese, and Spanish), extended with two additional entity types (location and facility), and the knowledge repository changed from Wikipedia to a curated subset of Freebase.

Monolingual EDL corresponds to our notion of the entity linking task, except for the two additional subtasks it addresses: (1) classifying entity mentions according to entity types and (2) clustering NIL mentions. These are not intrinsic to entity linking but are important for knowledge base population (which is the ultimate goal of TAC KBP). Evaluation for these subtasks is isolated from entity linking performance. Another key difference in EDL, compared to conventional entity linking, is that EDL focuses only on a handful of entity types.

In Table 5.6 we list the number of test queries (i.e., mentions) for the monolingual entity linking task; additionally, a number of training queries were also made available both as part of the evaluation campaign and in follow-up work [14].

5.8.2.4 Entity Recognition and Disambiguation Challenge

The Entity Recognition and Disambiguation (ERD) Challenge was organized in 2014 by representatives of major web search engine companies, in an effort to make entity linking evaluation more realistic [3]. There are two main differences in contrast to TAC KBP. First, entity linking systems are evaluated in an end-to-end fashion, without providing mention segmentations. Second, each participating team was required to set up a web service for their entity linking system, such that processing times can be measured.

In the "long text" track, the documents to be annotated were pages crawled from the Web (see Sect. 7.3.4.4 for the "short text" track). The reference knowledge repository was Freebase, with entities restricted to specific types and to those having an associated English Wikipedia page. Only proper noun entities are annotated. A set of 100 documents was made available for development and a disjoint set of additional 100 documents was used for testing. Half of the documents were sampled from general web pages, the other half were news articles from msn. com. Evaluation was performed by sending an "evaluation request" to the server hosting the challenge. The evaluation server then sent a set of documents to the participating team's web service for annotation. The returned results were evaluated, with evaluation scores posted on the challenge's leaderboard. Online evaluation took place over a period of time and was divided into train and test phases.

This type of live, online evaluation has its advantages and disadvantages. Asking participants to provide their entity linking system as a service ensures an absolutely fair comparison since (1) the process is completely automated with no possibility of human intervention and (2) annotations for test documents are not released (eliminating the risks of overfitting to a particular test collection). The main drawback is that evaluation is subject to the availability of the evaluation service. Further, as only overall evaluation scores are made available, a detailed success/failure analysis of the generated annotations is not possible. At the time of writing, the evaluation service is no longer available.[19]

5.8.3 Component-Based Evaluation

The pipeline architecture (cf. Fig. 5.2) makes the evaluation of entity linking systems especially challenging. The main research focus often lies in the disambiguation component, which is the heart of the entity linking process and lends itself to creative algorithmic solutions. However, disambiguation effectiveness is largely influenced by the preceding steps. The fact that improvements are observed on the end-to-end task does not necessarily mean that one disambiguation approach is better than another; it might be a result of more effective mention detection, candidate entity ranking, etc. In general, a fair comparison between two alternative approaches for a given component of an entity linking system can only be made if they share all other elements of the processing pipeline.

The first systematic investigation in this direction was performed by Hachey et al. [32], who implemented and compared three systems: two of the early seminal entity linking systems [2, 12] and the top performing system at TAC 2009 EL [82]. A surprising finding of their study is that much of the variation between the studied systems originates from candidate ranking and not from disambiguation.

Ceccarelli et al. [4] introduced Dexter,[20] an open source framework for entity linking, "where spotting, disambiguation and ranking are well separated and easy to isolate in order to study their performance" [4]. Dexter implements TAGME [22], Wikipedia Miner [61], and the collective linking approach of Han et al. [34].

Cornolti et al. [11] developed and made publicly available the BAT-Framework[21] for comparing publicly available entity annotation systems, namely: AIDA [37], Illinois Wikifier [68], TAGME [22], Wikipedia Miner [61], and DBpedia Spotlight [57]. These systems are evaluated on a number of test collections corresponding to different document genres (news, web pages, and tweets). Linking is done against Wikipedia. Building on top of the BAT-Framework, Usbeck et al. [81] introduced GERBIL,[22] an open-source web-based platform for comparing entity annotation systems. GERBIL extends the BAT-Framework by being able to link to any knowledge repository, not only to Wikipedia. It also includes additional evaluation measures (e.g., for dealing with NIL annotations). Any annotation service can easily be benchmarked by providing a URL to a REST interface that conforms to a given protocol specification. Finally, GERBIL provides persistent URLs for experimental settings, thereby allowing for reproducibility and archival of experimental results.

---

[19]http://web-ngram.research.microsoft.com/erd2014/.

[20]http://dexter.isti.cnr.it.

[21]http://acube.di.unipi.it/bat-framework/.

[22]http://gerbil.aksw.org.

5.9   Resources

Section 5.7 has introduced entity linking systems that are made publicly available as open source and/or are exposed as a web service (cf. Table 5.5). With these, anyone can annotate documents with entities from a given catalog. We have also discussed benchmarking platforms and test collections in Sect. 5.8. These are essential for those that wish to develop and evaluate their own entity linking system and compare it against existing solutions. This section presents additional resources that may prove useful when building/improving an entity linking system. It could, however, also be the case that one's interest lies not in the entity linking process itself, but rather in the resulting annotations. In Sect. 5.9.2, we present a large-scale web crawl that has been annotated with entities; this resource can be particularly of use for those that are merely "users" of entity annotations and wish to utilize them in downstream processing for some other task.

5.9.1   A Cross-Lingual Dictionary for English Wikipedia Concepts

Recall that a key source of entity surface forms is anchor texts from intra-Wikipedia links (cf. Sect. 5.4.1). The same idea could be extended to inter-Wikipedia links, by considering non-Wikipedia web pages that link to Wikipedia articles. The resource constructed by Spitkovsky and Chang [76] (Google) does exactly this. In addition, they also collect links that point to non-English versions of a given English Wikipedia article. (Notice that the mappings are to all Wikipedia articles, thus there is no distinction made between concepts and entities.) The end result is a cross-lingual surface form dictionary, with names of concepts and entities on one side and Wikipedia articles on the other. The dictionary also contains statistical information, including raw counts and mapping probabilities (i.e., commonness scores). This resource is of great value for the reason that it would be difficult to reconstruct without having access to a comprehensive web crawl. See Table 5.7 for an excerpt.

5.9.2   Freebase Annotations of the ClueWeb Corpora

ClueWeb09 and ClueWeb12 are large-scale web crawls that we discussed earlier in this book (see Sect. 2.1.1). Researchers from Google annotated the English-language web pages from these corpora with entities from the Freebase knowledge

Table 5.7 Excerpt from the dictionary entries matching the surface form (*s*) "Hank Williams" from the Cross-Lingual Dictionary for English Wikipedia Concepts [76]

Entity (*e*)	$P(e|s)$
Hank_Williams	0.990125
Your_Cheatin'_Heart	0.006615
Hank_Williams,_Jr.	0.001629
I	0.000479
Stars_&_Hank_Forever:_The_American_Composers_Series	0.000287
I'm_So_Lonesome_I_Could_Cry	0.000191
I_Saw_the_Light_(Hank_Williams_song)	0.000191
Drifting_Cowboys	0.000095
Half_as_Much	0.000095
Hank_Williams_(Clickradio_CEO)	0.000095

Table 5.8 Excerpt from the Freebase Annotations of the ClueWeb Corpora (FACC)

Mention	Byte offsets	Entity (*e*)	$P(e|m,d)$	$P(e|d)$
PDF	21089, 21092	/m/0600q	0.997636	0.000066
FDA	21303, 21306	/m/032mx	0.999825	0.000571
Food and Drug Administration	21312, 21340	/m/032mx	0.999825	0.000571

base [25], and made these annotations publicly available.[23,24] The system that was used for generating the annotations is proprietary, and as such there is no information disclosed about the underlying algorithm and techniques. It is known, however, that the annotations strove for high precision (which, by necessity, is at the expense of recall) and are of generally high quality. Table 5.8 shows a small excerpt with the annotations created for one of the ClueWeb12 web pages. It can be seen from the table that in addition to the mention (given both as a text span and as byte offsets in the file) and the linked entity, there are two types of confidence scores. The first one ($P(e|m,d)$) is the posterior of an entity given both the mention and the context, while the second one ($P(e|d)$) is the posterior that ignores the mention string and only considers the context of the mention.

5.10   Summary

This chapter has dealt with the task of entity linking: annotating an input text with entity identifiers from a reference knowledge repository. The canonical entity linking approach consists of a pipeline of three components. The first component,

---

[23]http://lemurproject.org/clueweb09/.

[24]http://lemurproject.org/clueweb12/.

mention detection, is responsible for identifying text spans that may refer to an entity. This is commonly performed using an extensive dictionary of entity surface forms provided in the reference knowledge repository (and possibly augmented with additional name variants from external sources). The second component, *candidate selection*, restricts the set of candidate entities for each mention, by eliminating those that are unlikely to be good link targets for that mention (even though one of their surface forms matches the mention). The third component, *disambiguation*, selects a single entity (or none) from the set of candidate entities identified for each mention. For effective disambiguation, one needs to consider the *local compatibility* between the linked entity and its context as well as the *coherence* between the linked entity and all other entities linked in the document. Two main families of approaches have been delineated, based on whether they perform disambiguation for each mention individually in a single pass or for all entity mentions collectively, using some iterative process. The former is more efficient (an order of magnitude faster), while the latter is more accurate (up to 25% higher F1-score, depending on the particular dataset).

A direct comparison of entity linking systems, based on the reported evaluation scores, is often problematic, due to differences in task definition and evaluation methodology. Further, it is typically difficult to untangle how much each pipeline component has contributed to the observed differences. There are standardization efforts addressing these issues, such as GERBIL [81], by providing an experimental platform for evaluation and diagnostics on reference datasets. According to the results in [81], the best systems reach, depending on the dataset, an F1-score of 0.9.

We have stated that the task of entity linking is one part of the bridge between unstructured and unstructured data. So, how does entity linking enable the task of knowledge base population? Once a document has been found to mention a given entity, that document may be checked to possibly discover new facts with which the knowledge base entry of that entity may be updated. The practical details of this approach will be discussed in the next chapter. Entity annotations can also be utilized to improve document retrieval, as we shall see in Chap. 8.

5.11   Further Reading

Nadeau and Sekine [64] survey the first 15 years of named entity recognition, from 1991 to 2006. An excellent recent survey about entity linking by Shen et al. [72] covers much of the same material as this chapter, with some further pointers. Entity linking is still a very active area of research, with new approaches springing up. Due to space considerations, we did not include approaches based on topic modeling (i.e., LDA-inspired models), see, e.g., [33, 38, 48, 67]. Most recently, semantic embeddings and neural models are gaining popularity in this domain too [24, 28, 77, 87]. Instead of relying on fully automatic techniques, Demartini et al. [13] incorporate human intelligence in the entity linking process, by dynamically generating micro-tasks on an online crowdsourcing platform.

Both entity linking and word sense disambiguation address the lexical ambiguity of language; we have discussed the similarities and differences between the two tasks in Sect. 5.1.2. Moro et al. [63] bring the two tasks to a common ground and present a unified graph-based approach to entity linking and WSD. Motivated by the interdependencies of entity annotation tasks, Durrett and Klein [15] develop a joint model for coreference resolution, named entity recognition, and entity linking.

Finally, in this chapter we have concentrated entirely on a monolingual setting. Cross-lingual entity linking is currently being investigated at the TAC Knowledge Base Population track; for further details, we refer to the TAC proceedings.

References

1. Aho, A.V., Corasick, M.J.: Efficient string matching: An aid to bibliographic search. Commun. ACM **18**(6), 333–340 (1975). doi: 10.1145/360825.360855
2. Bunescu, R., Paşca, M.: Using encyclopedic knowledge for named entity disambiguation. In: Proceedings of the 11th Conference of the European Chapter of the Association for Computational Linguistics, EACL '06, pp. 9–16. Association for Computational Linguistics (2006)
3. Carmel, D., Chang, M.W., Gabrilovich, E., Hsu, B.J.P., Wang, K.: ERD'14: Entity recognition and disambiguation challenge. SIGIR Forum **48**(2), 63–77 (2014). doi: 10.1145/2701583.2701591
4. Ceccarelli, D., Lucchese, C., Orlando, S., Perego, R., Trani, S.: Dexter: An open source framework for entity linking. In: Proceedings of the Sixth International Workshop on Exploiting Semantic Annotations in Information Retrieval, ESAIR '13, pp. 17–20. ACM (2013a). doi: 10.1145/2513204.2513212
5. Ceccarelli, D., Lucchese, C., Orlando, S., Perego, R., Trani, S.: Learning relatedness measures for entity linking. In: Proceedings of the 22nd ACM International Conference on Conference on Information & Knowledge Management, CIKM '13, pp. 139–148. ACM (2013b). doi: 10.1145/2505515.2505711
6. Chakrabarti, K., Chaudhuri, S., Cheng, T., Xin, D.: A framework for robust discovery of entity synonyms. In: Proceedings of the 18th ACM SIGKDD International Conference on Knowledge Discovery and Data Mining, KDD '12, pp. 1384–1392. ACM (2012). doi: 10.1145/2339530.2339743
7. Chaudhuri, S., Ganti, V., Xin, D.: Exploiting web search to generate synonyms for entities. In: Proceedings of the 18th International Conference on World Wide Web, WWW '09, pp. 151–160. ACM (2009). doi: 10.1145/1526709.1526731
8. Cheng, T., Lauw, H.W., Paparizos, S.: Entity synonyms for structured web search. IEEE Transactions on Knowledge and Data Engineering **24**(10), 1862–1875 (2012). doi: 10.1109/TKDE.2011.168
9. Cheng, X., Roth, D.: Relational inference for wikification. In: Proceedings of the 2013 Conference on Empirical Methods in Natural Language Processing, pp. 1787–1796. Association for Computational Linguistics (2013)
10. Cilibrasi, R.L., Vitanyi, P.M.B.: The Google similarity distance. IEEE Transactions on Knowledge and Data Engineering **19**(3), 370–383 (2007). doi: 10.1109/TKDE.2007.48
11. Cornolti, M., Ferragina, P., Ciaramita, M.: A framework for benchmarking entity-annotation systems. In: Proceedings of the 22nd International Conference on World Wide Web, WWW '13, pp. 249–260. ACM (2013). doi: 10.1145/2488388.2488411

12. Cucerzan, S.: Large-scale named entity disambiguation based on Wikipedia data. In: Proceedings of the 2007 Joint Conference on Empirical Methods in Natural Language Processing and Computational Natural Language Learning, EMNLP-CoNLL '07, pp. 708–716. Association for Computational Linguistics (2007)

13. Demartini, G., Difallah, D.E., Cudré-Mauroux, P.: ZenCrowd: Leveraging probabilistic reasoning and crowdsourcing techniques for large-scale entity linking. In: Proceedings of the 21st International Conference on World Wide Web, WWW '12, pp. 469–478. ACM (2012). doi: 10.1145/2187836.2187900

14. Dredze, M., McNamee, P., Rao, D., Gerber, A., Finin, T.: Entity disambiguation for knowledge base population. In: Proceedings of the 23rd International Conference on Computational Linguistics, COLING '10, pp. 277–285. Association for Computational Linguistics (2010)

15. Durrett, G., Klein, D.: A joint model for entity analysis: Coreference, typing, and linking. In: Transactions of the Association for Computational Linguistics, vol. 2, pp. 477–490 (2014)

16. Eckhardt, A., Hreško, J., Procházka, J., Smrf, O.: Entity linking based on the co-occurrence graph and entity probability. In: Proceedings of the First International Workshop on Entity Recognition and Disambiguation, ERD '14, pp. 37–44. ACM (2014). doi: 10.1145/2633211.2634349

17. Ellis, J., Getman, J., Fore, D., Kuster, N., Song, Z., Bies, A., Strassel, S.M.: Overview of linguistic resources for the TAC KBP 2015 evaluations: Methodologies and results. In: Proceedings of the 2015 Text Analysis Conference, TAC '15. NIST (2015)

18. Ellis, J., Getman, J., Mott, J., Li, X., Griffitt, K., Strassel, S.M., Wright, J.: Linguistic resources for 2013 knowledge base population evaluations. In: Proceedings of the 2013 Text Analysis Conference, TAC '13. NIST (2013)

19. Ellis, J., Getman, J., Strassel, S.M.: Overview of linguistic resources for the TAC KBP 2014 evaluations: Planning, execution, and results. In: Proceedings of the 2014 Text Analysis Conference, TAC '14. NIST (2014)

20. Ellis, J., Li, X., Griffitt, K., Strassel, S.M., Wright, J.: Linguistic resources for 2012 knowledge base population evaluations. In: Proceedings of the 2012 Text Analysis Conference, TAC '12. NIST (2012)

21. Ferragina, P., Scaiella, U.: Fast and accurate annotation of short texts with Wikipedia pages. CoRR **abs/1006.3** (2010a)

22. Ferragina, P., Scaiella, U.: TAGME: On-the-fly annotation of short text fragments (by Wikipedia entities). In: Proceedings of the 19th ACM International Conference on Information and Knowledge Management, CIKM '10, pp. 1625–1628. ACM (2010b). doi: 10.1145/1871437.1871689

23. Finkel, J.R., Grenager, T., Manning, C.: Incorporating non-local information into information extraction systems by Gibbs sampling. In: Proceedings of the 43rd Annual Meeting on Association for Computational Linguistics, ACL '05, pp. 363–370. Association for Computational Linguistics (2005). doi: 10.3115/1219840.1219885

24. Francis-Landau, M., Durrett, G., Klein, D.: Capturing semantic similarity for entity linking with convolutional neural networks. In: Proceedings of the North American Association for Computational Linguistics, NAACL '16. Association for Computational Linguistics (2016)

25. Gabrilovich, E., Ringgaard, M., Subramanya, A.: FACC1: Freebase annotation of Clueweb corpora, version 1. Tech. rep., Google, Inc. (2013)

26. Gale, W.A., Church, K.W., Yarowsky, D.: One sense per discourse. In: Proceedings of the Workshop on Speech and Natural Language, HLT '91, pp. 233–237. Association for Computational Linguistics (1992). doi: 10.3115/1075527.1075579

27. Ganea, O.E., Ganea, M., Lucchi, A., Eickhoff, C., Hofmann, T.: Probabilistic bag-of-hyperlinks model for entity linking. In: Proceedings of the 25th International Conference on World Wide Web, WWW '16, pp. 927–938. International World Wide Web Conferences Steering Committee (2016). doi: 10.1145/2872427.2882988

28. Ganea, O.E., Hofmann, T.: Deep joint entity disambiguation with local neural attention. In: Proceedings of the 2017 Conference on Empirical Methods in Natural Language Processing, pp. 2619–2629. Association for Computational Linguistics (2017). doi: 10.18653/v1/D17-1277

29. Gattani, A., Lamba, D.S., Garera, N., Tiwari, M., Chai, X., Das, S., Subramaniam, S., Rajaraman, A., Harinarayan, V., Doan, A.: Entity extraction, linking, classification, and tagging for social media: A Wikipedia-based approach. Proceedings of the VLDB Endowment **6**(11), 1126–1137 (2013). doi: 10.14778/2536222.2536237

30. Guo, S., Chang, M.W., Kiciman, E.: To link or not to link? - A study on end-to-end tweet entity linking. In: Proceedings of the 2013 Conference of the North American Chapter of the Association for Computational Linguistics: Human Language Technologies, pp. 1020–1030. Association for Computational Linguistics (2013)

31. Guo, Z., Barbosa, D.: Robust entity linking via random walks. In: Proceedings of the 23rd ACM International Conference on Conference on Information and Knowledge Management, CIKM '14, pp. 499–508. ACM (2014). doi: 10.1145/2661829.2661887

32. Hachey, B., Radford, W., Nothman, J., Honnibal, M., Curran, J.R.: Evaluating entity linking with Wikipedia. Artif. Intell. **194**, 130–150 (2013). doi: 10.1016/j.artint.2012.04.005

33. Han, X., Sun, L.: An entity-topic model for entity linking. In: Proceedings of the 2012 Joint Conference on Empirical Methods in Natural Language Processing and Computational Natural Language Learning, EMNLP-CoNLL '12, pp. 105–115. Association for Computational Linguistics (2012)

34. Han, X., Sun, L., Zhao, J.: Collective entity linking in web text: A graph-based method. In: Proceedings of the 34th International ACM SIGIR Conference on Research and Development in Information Retrieval, SIGIR '11, pp. 765–774. ACM (2011). doi: 10.1145/2009916.2010019

35. Hasibi, F., Balog, K., Bratsberg, S.E.: On the reproducibility of the TAGME entity linking system. In: Proceedings of the 38th European conference on Advances in Information Retrieval, ECIR '16, pp. 436–449. Springer (2016). doi: 10.1007/978-3-319-30671-1_32

36. Hoffart, J., Seufert, S., Nguyen, D.B., Theobald, M., Weikum, G.: KORE: Keyphrase overlap relatedness for entity disambiguation. In: Proceedings of the 21st ACM International Conference on Information and Knowledge Management, CIKM '12, pp. 545–554. ACM (2012). doi: 10.1145/2396761.2396832

37. Hoffart, J., Yosef, M.A., Bordino, I., Fürstenau, H., Pinkal, M., Spaniol, M., Taneva, B., Thater, S., Weikum, G.: Robust disambiguation of named entities in text. In: Proceedings of the Conference on Empirical Methods in Natural Language Processing, EMNLP '11, pp. 782–792. Association for Computational Linguistics (2011)

38. Houlsby, N., Ciaramita, M.: A scalable Gibbs sampler for probabilistic entity linking. In: Advances in Information Retrieval, *Lecture Notes in Computer Science*, vol. 8416, pp. 335–346. Springer (2014). doi: 10.1007/978-3-319-06028-6_28

39. Huang, W.C.D.: Evaluation framework for focused link discovery. Ph.D. thesis, Queensland University of Technology (2011)

40. Huang, W.C.D., Geva, S., Trotman, A.: Overview of the INEX 2008 Link the Wiki track. In: Geva, S., Kamps, J., Trotman, A. (eds.) Advances in Focused Retrieval, *Lecture Notes in Computer Science*, vol. 5631, pp. 314–325. Springer (2009). doi: 10.1007/978-3-642-03761-0_32

41. Huang, W.C.D., Geva, S., Trotman, A.: Overview of the INEX 2009 Link the Wiki track. In: Geva, S., Kamps, J., Trotman, A. (eds.) Focused Retrieval and Evaluation, *Lecture Notes in Computer Science*, vol. 6203, pp. 312–323. Springer (2010). doi: 10.1007/978-3-642-14556-8_31

42. Huang, W.C.D., Xu, Y., Trotman, A., Geva, S.: Overview of INEX 2007 Link the Wiki track. In: Fuhr, N., Kamps, J., Lalmas, M., Trotman, A. (eds.) Focused Access to XML Documents, *Lecture Notes in Computer Science*, vol. 4862, pp. 373–387. Springer (2008). doi: 10.1007/978-3-540-85902-4_32

43. Ji, H., Grishman, R.: Knowledge base population: Successful approaches and challenges. In: Proceedings of the 49th Annual Meeting of the Association for Computational Linguistics: Human Language Technologies - Volume 1, HLT '11, pp. 1148–1158. Association for Computational Linguistics (2011)

44. Ji, H., Grishman, R., Dang, H.T.: Overview of the TAC 2011 Knowledge Base Population track. In: Proceedings of the 2010 Text Analysis Conference, TAC '11. NIST (2011)

45. Ji, H., Grishman, R., Dang, H.T., Griffitt, K., Ellis, J.: Overview of the TAC 2010 Knowledge Base Population track. In: Proceedings of the 2010 Text Analysis Conference, TAC '10. NIST (2010)
46. Ji, H., Nothman, J., Hachey, B.: Overview of TAC-KBP2014 Entity discovery and linking tasks. In: Proceedings of the 2014 Text Analysis Conference, TAC '14. NIST (2014)
47. Ji, H., Nothman, J., Hachey, B., Florian, R.: Overview of TAC-KBP2015 Tri-lingual entity discovery and linking. In: Proceedings of the 2015 Text Analysis Conference, TAC '15. NIST (2015)
48. Kataria, S.S., Kumar, K.S., Rastogi, R.R., Sen, P., Sengamedu, S.H.: Entity disambiguation with hierarchical topic models. In: Proceedings of the 17th ACM SIGKDD International Conference on Knowledge Discovery and Data Mining, KDD '11, pp. 1037–1045. ACM (2011). doi: 10.1145/2020408.2020574
49. Kulkarni, S., Singh, A., Ramakrishnan, G., Chakrabarti, S.: Collective annotation of Wikipedia entities in web text. In: Proceedings of the 15th ACM SIGKDD International Conference on Knowledge Discovery and Data Mining, KDD '09, pp. 457–466. ACM (2009). doi: 10.1145/1557019.1557073
50. Lample, G., Ballesteros, M., Subramanian, S., Kawakami, K., Dyer, C.: Neural architectures for named entity recognition. In: Proceedings of the 2016 Conference of the North American Chapter of the Association for Computational Linguistics: Human Language Technologies, pp. 260–270. Association for Computational Linguistics (2016). doi: 10.18653/v1/N16-1030
51. Li, X., Ellis, J., Griffit, K., Strassel, S., Parker, R., Wright, J.: Linguistic resources for 2011 knowledge base population evaluation. In: Proceedings of the 2011 Text Analysis Conference, TAC '11. NIST (2011)
52. Li, Y., Wang, C., Han, F., Han, J., Roth, D., Yan, X.: Mining evidences for named entity disambiguation. In: Proceedings of the 19th ACM SIGKDD International Conference on Knowledge Discovery and Data Mining, KDD '13, pp. 1070–1078. ACM (2013). doi: 10.1145/2487575.2487681
53. Ling, X., Weld, D.S.: Fine-grained entity recognition. In: In Proceedings of the 26th AAAI Conference on Artificial Intelligence, AAAI '12 (2012)
54. Ma, X., Hovy, E.: End-to-end sequence labeling via bi-directional LSTM-CNNs-CRF. In: Proceedings of the 54th Annual Meeting of the Association for Computational Linguistics (Volume 1: Long Papers), pp. 1064–1074. Association for Computational Linguistics (2016). doi: 10.18653/v1/P16-1101
55. McNamee, P., Dang, H.T.: Overview of the TAC 2009 Knowledge Base Population track. In: Proceedings of the 2009 Text Analysis Conference, TAC '09. NIST (2009)
56. Medelyan, O., Witten, I.H., Milne, D.: Topic indexing with Wikipedia. In: Bunescu, R., Gabrilovich, E., Mihalcea, R. (eds.) Proceedings of AAAI Workshop on Wikipedia and Artificial Intelligence: An Evolving Synergy, vol. 1, pp. 19–24. AAAI, AAAI Press (2008)
57. Mendes, P.N., Jakob, M., García-Silva, A., Bizer, C.: Dbpedia Spotlight: Shedding light on the web of documents. In: Proceedings of the 7th International Conference on Semantic Systems, I-Semantics '11, pp. 1–8 (2011)
58. Mihalcea, R., Csomai, A.: Wikify! - Linking documents to encyclopedic knowledge. In: Proceedings of the Sixteenth ACM Conference on Conference on Information and Knowledge Management, CIKM '07, pp. 233–242. ACM (2007). doi: 10.1145/1321440.1321475
59. Miller, G.A.: WordNet: A lexical database for English. Communications of the ACM **38**(11), 39–41 (1995). doi: 10.1145/219717.219748
60. Milne, D., Witten, I.H.: An effective, low-cost measure of semantic relatedness obtained from Wikipedia links. In: Proceeding of AAAI Workshop on Wikipedia and Artificial Intelligence: An Evolving Synergy, pp. 25–30. AAAI Press (2008a)
61. Milne, D., Witten, I.H.: Learning to link with Wikipedia. In: Proceedings of the 17th ACM Conference on Information and Knowledge Management, CIKM '08, pp. 509–518 (2008b)
62. Milne, D., Witten, I.H.: An open-source toolkit for mining Wikipedia. Artificial Intelligence **194**, 222–239 (2013)

63. Moro, A., Raganato, A., Navigli, R.: Entity linking meets word sense disambiguation: A unified approach. Transactions of the Association for Computational Linguistics **2**, 231–244 (2014)
64. Nadeau, D., Sekine, S.: A survey of named entity recognition and classification. Lingvisticae Investigationes **30**(1), 3–26 (2007). doi: 10.1075/li.30.1.03nad
65. Navigli, R.: Word sense disambiguation: A survey. ACM Comput. Surv. **41**(2), 10:1–10:69 (2009). doi: 10.1145/1459352.1459355
66. Ng, V.: Supervised noun phrase coreference research: The first fifteen years. In: Proceedings of the 48th Annual Meeting of the Association for Computational Linguistics, ACL '10, pp. 1396–1411. Association for Computational Linguistics (2010)
67. Pilz, A., Paaß, G.: From names to entities using thematic context distance. In: Proceedings of the 20th ACM International Conference on Information and Knowledge Management, CIKM '11, pp. 857–866. ACM (2011). doi: 10.1145/2063576.2063700
68. Ratinov, L., Roth, D., Downey, D., Anderson, M.: Local and global algorithms for disambiguation to Wikipedia. In: Proceedings of the 49th Annual Meeting of the Association for Computational Linguistics: Human Language Technologies - Volume 1, HLT '11, pp. 1375–1384. Association for Computational Linguistics (2011)
69. Sekine, S.: Extended named entity ontology with attribute information. In: Proceedings of the Sixth International Language Resources and Evaluation, LREC '08. ELRA (2008)
70. Sekine, S., Nobata, C.: Definition, dictionaries and tagger for extended named entity hierarchy. In: Proceedings of the Fourth International Conference on Language Resources and Evaluation, LREC '04. ELRA (2004)
71. Sekine, S., Sudo, K., Nobata, C.: Extended named entity hierarchy. In: Third International Conference on Language Resources and Evaluation, LREC '02. ELRA (2002)
72. Shen, W., Wang, J., Han, J.: Entity linking with a knowledge base: Issues, techniques, and solutions. IEEE Trans. Knowl. Data Eng. **27**(2), 443–460 (2015). doi: 10.1109/TKDE.2014.2327028
73. Shen, W., Wang, J., Luo, P., Wang, M.: LIEGE: link entities in web lists with knowledge base. In: Proceedings of the 18th ACM SIGKDD International Conference on Knowledge Discovery and Data Mining, KDD '12, pp. 1424–1432. ACM (2012). doi: 10.1145/2339530.2339753
74. Simpson, H., Strassel, S., Parker, R., McNamee, P.: Wikipedia and the web of confusable entities: Experience from entity linking query creation for TAC 2009 Knowledge base population. In: Proceedings of the Seventh International Conference on Language Resources and Evaluation, LREC '10. ELRA (2010)
75. Singh, S., Subramanya, A., Pereira, F., McCallum, A.: Large-scale cross-document coreference using distributed inference and hierarchical models. In: Proceedings of the 49th Annual Meeting of the Association for Computational Linguistics: Human Language Technologies - Volume 1, HLT '11, pp. 793–803. Association for Computational Linguistics (2011)
76. Spitkovsky, V.I., Chang, A.X.: A cross-lingual dictionary for English Wikipedia concepts. In: Proceedings of the Eight International Conference on Language Resources and Evaluation, LREC '12. ELRA (2012)
77. Sun, Y., Lin, L., Tang, D., Yang, N., Ji, Z., Wang, X.: Modeling mention, context and entity with neural networks for entity disambiguation. In: Proceedings of the 24th International Conference on Artificial Intelligence, IJCAI '15, pp. 1333–1339. AAAI Press (2015)
78. Sundheim, B.M.: Overview of results of the MUC-6 evaluation. In: Message Understanding Conference, MUC-6, pp. 13–31 (1995). doi: 10.3115/1072399.1072402
79. Tong, H., Faloutsos, C., Pan, J.Y.: Fast random walk with restart and its applications. In: Proceedings of the Sixth International Conference on Data Mining, ICDM '06, pp. 613–622. IEEE Computer Society (2006). doi: 10.1109/ICDM.2006.70
80. Trotman, A., Alexander, D., Geva, S.: Overview of the INEX 2010 Link the Wiki track. In: Geva, S., Kamps, J., Schenkel, R., Trotman, A. (eds.) Comparative Evaluation of Focused Retrieval, *Lecture Notes in Computer Science*, vol. 6932, pp. 241–249. Springer (2011). doi: 10.1007/978-3-642-23577-1_22

81. Usbeck, R., Röder, M., Ngonga Ngomo, A.C., Baron, C., Both, A., Brümmer, M., Ceccarelli, D., Cornolti, M., Cherix, D., Eickmann, B., Ferragina, P., Lemke, C., Moro, A., Navigli, R., Piccinno, F., Rizzo, G., Sack, H., Speck, R., Troncy, R., Waitelonis, J., Wesemann, L.: GERBIL – general entity annotation benchmark framework. In: Proceedings of the 24th International World Wide Web Conference, WWW '15. International World Wide Web Conferences Steering Committee (2015). doi: 10.1145/2736277.2741626

82. Varma, V., Reddy, V.B., Kovelamudi, S., Bysani, P., Gsk, S., Kumar, N.K., B, K.R., Kumar, K., Maganti, N.: IIIT Hyderabad at TAC 2009. In: Proceedings of the 2009 Text Analysis Conference, TAC '09. NIST (2009)

83. Yosef, M.A., Bauer, S., Hoffart, J., Spaniol, M., Weikum, G.: HYENA: Hierarchical type classification for entity names. In: Proceedings of the 24th International Conference on Computational Linguistics, COLING '12, pp. 1361–1370. Association for Computational Linguistics (2012)

84. Zhang, W., Sim, Y.C., Su, J., Tan, C.L.: Entity linking with effective acronym expansion, instance selection and topic modeling. In: Proceedings of the Twenty-Second International Joint Conference on Artificial Intelligence - Volume Volume Three, IJCAI'11, pp. 1909–1914. AAAI Press (2011). doi: 10.5591/978-1-57735-516-8/IJCAI11-319

85. Zheng, Z., Li, F., Huang, M., Zhu, X.: Learning to link entities with knowledge base. In: Human Language Technologies: The 2010 Annual Conference of the North American Chapter of the Association for Computational Linguistics, HLT '10, pp. 483–491. Association for Computational Linguistics (2010)

86. Zhou, G., Su, J.: Named entity recognition using an HMM-based chunk tagger. In: Proceedings of the 40th Annual Meeting on Association for Computational Linguistics, ACL '02, pp. 473–480. Association for Computational Linguistics (2002). doi: 10.3115/1073083.1073163

87. Zwicklbauer, S., Seifert, C., Granitzer, M.: Robust and collective entity disambiguation through semantic embeddings. In: Proceedings of the 39th International ACM SIGIR Conference on Research and Development in Information Retrieval, SIGIR '16, pp. 425–434. ACM (2016). doi: 10.1145/2911451.2911535

Chapter 6
Populating Knowledge Bases

A knowledge base (KB) contains information about entities and their properties (types, attributes, and relationships). In the case of large KBs, the number of entities is in the millions and the number of facts is in the billions. Commonly, this information is represented in the form of (sets of) subject-predicate-object (SPO) triples, according to the RDF data model (cf. Sect. 2.3). KBs are utilized in a broad variety of information access tasks, including entity retrieval (Chaps. 3 and 4), entity linking (Chap. 5), and semantic search (Chaps. 7–9). Two main challenges associated with knowledge bases are that (1) they are inherently incomplete (and will always remain so, despite any effort), and (2) they need constant updating over time as new facts and discoveries may turn the content outdated, inaccurate, or incomplete.

Knowledge base population (KBP) refers to the task of discovering new facts about entities from a large text corpus, and augmenting a KB with these facts. KBP is a broad problem area, with solutions ranging from fully automated systems to setups with a human content editor in the loop, who is in charge of any changes made to the KB. Our interest in this chapter will be on the latter type of systems, which "merely" provide assistance with the labor-intensive manual process.

Specifically, we will focus on a streaming setting, with the goal to discover and extract new information about entities as it becomes available. This information can then be used to augment an existing KB. This flavor of KBP has been termed *knowledge base acceleration* (KBA) [27]. KBA systems "seek to help humans expand knowledge bases [...] by automatically recommending edits based on incoming content streams" [7].

There is a practical real-world motivation behind this particular problem formulation. Many large knowledge repositories are maintained by a small workforce of content editors. Due to the scarcity of human resources, "most entity profiles lag far behind current events" [26]. For example, Frank et al. [27] show that the median time elapsed between the publication dates of news articles that are cited in Wikipedia articles of living people and the dates of the corresponding edits to

© The Author(s) 2018   189
K. Balog, *Entity-Oriented Search*, The Information Retrieval Series 39,
https://doi.org/10.1007/978-3-319-93935-3_6

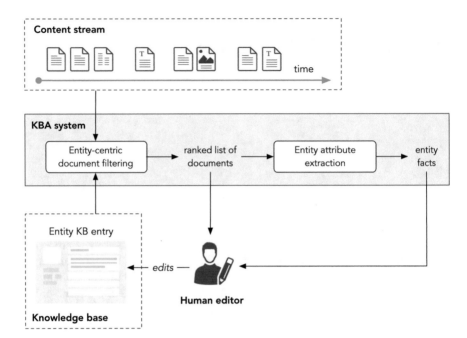

Fig. 6.1 Overview of knowledge base acceleration

Wikipedia is over a year. Thus, we wish to help editors stay on top of changes by automatically identifying content (news articles, blog posts, etc.) that may imply modifications to the KB entries of a certain set of entities of interest (i.e., entities that a given editor is responsible for).

Accelerating the knowledge base entry of a given target entity entails two main steps. The first is about identifying relevant documents that contain new facts about that entity. In a streaming setting, the processing of new documents is performed in batches, as they become available. (We note that the same techniques may be applied in a static, i.e., non-streaming, setting as well.) We address this subtask in Sect. 6.2. The second step concerns the extraction of facts from those documents. This is restricted to a predefined set of entity properties, thus may be seen as the problem of filling slots in the entity's knowledge base entry. We focus on this subtask in Sect. 6.3. Depending on the degree of automation, the human editor may use only the first step and process the documents manually that were filtered by the KBA system. Alternatively, she may operate with the extracted facts and—after reviewing them—apply the recommended changes to the KB by a click of a mouse. Figure 6.1 illustrates the process.

Before discussing the two main components of KBA systems, we shall begin in Sect. 6.1 with giving a brief overview of the broader problem area of extracting structured information from unstructured data.

6.1 Harvesting Knowledge from Text

The broad goal of information extraction (IE) is to automatically extract structured (factual) information from unstructured or semi-structured sources. While IE has its roots in natural language processing, the topic of extracting structured data now engages many different research communities, including those of information retrieval and databases. There is a wide spectrum of potential input sources, including web pages, news articles, social media and user-generated content (blogs, tweets, reviews), corporate and medical reports, web tables, and search queries, just to name a few. The spectrum of methods and techniques is similarly broad, ranging from hand-written regular expressions to probabilistic graphical models. Finally, the type of structure extracted extends from atomic units, such as entities, types, attributes, and relationships to higher-order structures such as lists, tables, and ontologies. For a general overview on IE, the reader is referred to [38, 68].

The area of large-scale knowledge acquisition, i.e., extracting information related to entities from large document corpora with the aim to build or extend a knowledge base, has become a major research avenue over the past decade [14, 87]. There are many dimensions along which approaches may be categorized. We highlight two particular dimensions.

Closed vs. Open Information Extraction Traditional information extraction is *closed*, in a sense that we wish to populate an existing knowledge base with additional facts about entities, where all entities, types, and relationships already have canonicalized (unique) identifiers assigned to them in the KB. That is, we are not aiming to discover new entities, entity types, or relationship types, but only instantiations of those in the form of new facts. This paradigm is also referred to as *ontology-based IE* [71]. *Open* information extraction, on the other hand, aims to discover new entities and new relationship types and extract all instances of those. In other words, there is no pre-specified vocabulary, the elements of facts are represented as textual phrases, which—at this stage—are not linked to a knowledge base. For example, given the sentence "Born in Honolulu, Hawaii, Obama is a US Citizen," a closed IE system would represent the contained information (using the entities and predicates from the DBpedia knowledge base) as:[1]

`<dbr:Barack_Obama>`	`<dbo:nationality>`	`<dbr:United_States>`
`<dbr:Barack_Obama>`	`<dbo:birthPlace>`	`<dbr:Honolulu>`

---

[1]We know (can infer) from the knowledge base that Honolulu is the state capital of Hawaii, therefore it is not needed to add a third triple asserting the relationship between Obama and Hawaii.

Instead, an open IE system might extract the following two triples:

```
(Obama; is; US citizen)
(Obama; born in; Honolulu, Hawaii)
```

Non-targeted vs. Targeted Extraction *Non-targeted* (or unfocused) extraction processes a data corpus (e.g., a collection of documents) in batch mode and extracts "whatever facts it can find" [88]. In West et al. [88], it is referred to as the "push" model. The Never-Ending Language Learner (NELL) project [49], which performs continuous machine reading of the Web, provides one specific example of non-targeted extraction. *Targeted* (or focused) extraction, on the other hand, refers to extracting facts for specific entities (a.k.a. "slot filling"). In [88], it is referred to as the "pull" paradigm. For example, West et al. [88] leverage question answering techniques, and aim to learn the best set of queries to ask, to find missing entity attributes.

The methods we will be discussing in this chapter fall under closed and targeted information extraction. Specifically, our interest will be in filtering an incoming stream of documents for information pertinent to a pre-defined set of entities, and extracting values of specific properties of entities from those documents in order to extend an existing knowledge base. Below, we briefly look at a number of other tasks that fall within the general problem area of harvesting entity-related information from text. These are loosely organized around entity types (Sect. 6.1.1), attributes (Sect. 6.1.2), and relationships (Sect. 6.1.3). Note that we are not aiming for an in-depth or exhaustive treatment of these topics here. Our goal is merely to give a high-level overview of research in this area.

6.1.1   Class-Instance Acquisition

There is a large body of research on semantic class learning (a.k.a. *hyponym extraction*), which focuses on "is-A" (or "instanceOf") relations between instances and classes (hypernyms). In our case, instances are entities, while classes refer to entity types (e.g., "country") or semantic categories (e.g., "countries with nuclear power plants"). We briefly review related work from two directions: (1) obtaining additional instances (entities) that belong to a given semantic class (entity type) and (2) obtaining (additional) semantic classes (entity types) for a given instance (entity).

6.1.1.1   Obtaining Instances of Semantic Classes

Concept expansion is the problem of expanding a seed set of instances that belong to the same semantic class with additional instances. *Set expansion* is a closely related task, where the underlying semantic class is not defined explicitly via a textual label, only implicitly through the seed set of examples. The inherent ambiguity of the seeds makes set expansion a considerably harder task than concept expansion. Both concept and set expansion are commonly cast as a ranking task: Given a set of seed instances and (optionally) a class label, return a ranked list of instances that belong to the same target class.

One particular application of class-instance acquisition techniques is to extend tail types or categories of knowledge bases. We note that this is similar to the task of *similar entity search* (using the class label as the keyword query), which we have discussed in Sect. 4.5. However, since the overall goal here is KB expansion, the ranking is performed over a different data source than the KB, such as web documents [59] (often accessed via general-purpose web search engines [85, 86]), HTML tables [82], HTML lists [33], search queries [57], or a combination of multiple sources [61]. We note that many of the concept expansion approaches concentrate on noun phrases as instances, which are not necessarily entities (e.g., "red," "green," and "blue" are instances of the concept "colors"). Moreover, even when specifically entities are targeted, they may be identified only via their surface forms (noun phrases), and not as unique identifiers. In that case, an additional linking step is required to anchor them in a knowledge base [52]. When entities are coming from a knowledge base, then the class-instance acquisition process may incorporate additional semantic constraints, by considering entity properties in the KB [77].

Regarding the size of the seed set, Pantel et al. [59] show that "only few seeds (10–20) yield best performance and that adding more seeds beyond this does not on average affect performance in a positive or negative way." Vyas et al. [81] investigate the impact that the composition of seed sets has on the quality of set expansions and show that seed set composition can significantly affect expansion performance. They further show that, in many cases, the seed sets produced by an average (i.e., not expert) editor are worse than randomly chosen seed sets. Finally, they propose algorithms that can improve the quality of set expansion by removing certain entities from the seed set. Inevitably, the lists generated by set expansion systems will contain errors. Vyas and Pantel [80] propose semi-supervised techniques, which require minimal annotation effort, to help editors clean the resulting expanded sets.

6.1.1.2   Obtaining Semantic Classes of Instances

Reversing the concept expansion task, we now take a particular instance as input and seek to obtain semantic classes for that instance. In the context of knowledge bases, this problem is known as *entity typing*: Given an entity as input, automatically assign (additional) types to that entity from a type catalog. This task is addressed in two flavors: when the entity already exists in the KB and when it does not.

We start with the task of assigning types to entities that are already in the knowledge base. It is shown in [60] that, while seemingly a straightforward approach, traditional reasoning cannot be used to tackle this problem, due to the incompleteness and noisy nature of large-scale KBs. Gangemi et al. [30] extract types for DBpedia entities, based on the corresponding natural language definitions of those entities in Wikipedia (i.e., Wikipedia abstracts). Their approach relies on a number of heuristics and makes heavy use of Wikipedia's markup conventions. A more general approach is to frame the type prediction task as a hierarchical multi-label classification problem ("hierarchical because we assume the types to be structured in a hierarchy, and it is a multilabel problem because instances are allowed to have more than one type" [46]). Paulheim and Bizer [60] exploit links between instances in the knowledge base as indicators for types, without resorting to features specific to any particular KB. Melo et al. [46] present a top-down prediction approach and employ local classifiers and local feature selection per node. "The local training sets are created including the instances belonging to the target class as positive examples and the instances belonging to its sibling classes as negative examples" [46]. This method is shown to improve scalability without sacrificing performance.

The other flavor of entity typing is when the entity is not present in the KB. This is closely related to the task of identifying *emerging* or *tail* entities, i.e., entities that are not (yet) prominent enough to be included in a human-curated knowledge base. Often, a tail entity is informally defined as not being present in Wikipedia [43, 50, 52]. Fine-grained typing of entities has been long established as a subtask in named entity recognition [25, 69] (cf. Sect. 5.1.1). These early works, however, are limited to flat sets of several dozen types. It has been only more recently that entity typing is performed against type systems of large-scale knowledge bases, i.e., hierarchical type taxonomies with hundreds of types [43, 52, 92]. Lin et al. [43] introduce the *unlinkable noun phrase* problem: Given a noun phrase that cannot be linked to a knowledge base, determine whether it is an entity, and, if so, return its fine-grained semantic types. To decide whether the noun phrase is an entity, they train a classifier with features primarily derived from mentions of the noun phrase over time in a timestamped corpus (specifically, the Google Books Ngram Corpus [47]). In stage two, the semantic type of a noun phrase is predicted by (1) finding the textual relations it occurs with, (2) finding linked entities that share the same textual relations, and (3) propagating types from those linked entities. HYENA [92] employs a multi-label classification approach, using a rich set of mention, part-of-speech, and gazetteer features. The PEARL system [52] considers typed relational patterns (learned by PATTY [53]) and the type signatures of patterns the new entity occurs with (e.g., "⟨singer⟩ released ⟨album⟩" or "⟨person⟩ nominated for ⟨award⟩"). This is similar in spirit to Lin et al. [43], but the patterns in [43] are not typed. Additionally, PEARL attempts to resolve inconsistencies among the predicted types using integer linear programming. Mohapatra et al. [50] point out the synergy between named entity linking and entity typing: "knowing the types of unfamiliar entities helps disambiguate mentions, and words in mention contexts help assign types to entities." Mohapatra et al. [50] present a bootstrapping system for solving the two tasks jointly.

6.1.2   Class-Attribute Acquisition

Another area of automatic knowledge acquisition concerns the discovery of relevant attributes (and relationship types) of classes. For example, for the class "country," the set of attributes includes "capital city," "population," "currency," etc. The task is commonly formulated as a ranking problem: Given an input class label, return a ranked list of attributes. Candidate attributes are generally obtained by employing an *instance-driven* extraction strategy, i.e., by "inspecting the attributes of individual instances from that class" [58], thereby enjoying access to substantially more input data. We will look at attribute extraction for specific instances (entities) in Sect. 6.3. Alternatively, attributes may be extracted by using only the class label, without the set of instances, see, e.g., [78, 79].

A broad variety of data sources have been utilized, including web documents [78], web tables [12, 91], web search query logs [56, 58], and Wikipedia category labels [54]. Methods for attribute extraction range from simple extraction patterns [4] and term frequency statistics [58, 78] to probabilistic topic modeling techniques [64].

6.1.3   Relation Extraction

Relation extraction refers to the task of extracting relationships between entities from a data collection (usually, from documents). Relationships are generally defined in the form of tuples, however, most relation extraction systems focus on binary relationships between named entities; these take the form of subject-predicate-object triples (e.g., "X marriedTo Y" or "X locatedIn Y"). Early work has focused on pre-defined relationships between pairs of entities. Traditional relation extraction benchmarking campaigns include the Message Understanding Conference (MUC-7, *template relations* task [15]) and the Automatic Content Extraction program (ACE, *relation detection and characterization* task [18]). Relation extraction is naturally approached by learning a binary classifier, using feature-based approaches [39, 96] or kernel methods [11, 95]. These *supervised* models can achieve high accuracy, but they require lots of hand-labeled training data. Moreover, they do not generalize to different (new) relationships.

To overcome these limitations, *semi-supervised* approaches, and in particular *boostrapping*, have been proposed and gained attention. The idea behind bootstrapping is to learn new extraction patterns by using previous instances extracted by the system as training data. More specifically, the general bootstrapping algorithm, for a given relationship, is as follows:

1. Take as input a small number of seed tuples (i.e., pairs of entities) that engage in the required relationship.
2. Find occurrences of seed tuples in documents and extract contextual patterns.
3. Identify the best (top-k) patterns and add those to the pattern set.

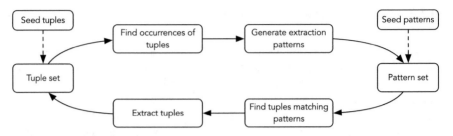

Fig. 6.2 Bootstrapping for relation extraction

4. Find new instances of the relationship using the pattern set.
5. Add those instances to the tuple set and repeat from step 2.

The output is a set of tuples along with a set of patterns. Alternatively, the process
may also be initialized with a small set of golden seed patterns (extraction rules),
instead of the seed tuples [22] (and then start the bootstrapping process with step
4). See Fig. 6.2 for an illustration. Early examples of bootstrapping systems are
DIPRE [10], Snowball [2], and KnowItAll [22]. More recent approaches include
SOFIE [71] and NELL [14] . All these systems learn extractors for a predefined
set of relationship types. While they clearly need less labeled data than supervised
approaches, extending them to new relationships still requires seed data.

Distant supervision (also called *weak supervision*) uses a knowledge base
to provide a "training set of relations and entity pairs that participate in those
relations" [48]. Wu and Weld [89] applied this idea to learn relationships from
Wikipedia, by matching attributes from infoboxes to corresponding sentences
in the article. Others have used Freebase facts to train relational extractors on
Wikipedia [48] or on a news corpus [65]. According to the *distant supervision
assumption*, if two entities engage in a certain relationship, then any sentence
containing those two entities might express that relationship [48]. Or, in a stronger
form, "all sentences that mention these two entities express that relation" [65]. It
has been argued that this assumption is too strong, and various refinements have
been proposed [65, 76]. Riedel et al. [65] employ the following relaxation: "if
two entities participate in a relation, at least one sentence that mentions these
two entities might express that relation" [65], thereby casting the problem as a
form of multi-instance learning. Surdeanu et al. [76] assume that a mention of
two entities together expresses exactly one relationship, but the two entities might
be related with different predicates across different mentions. This corresponds
to a multi-instance multi-label learning setting. Distant supervision is shown
to be scalable [35]. However, the training data is noisy and does not contain
explicit negative examples. We further note that training data generation requires
high-quality entity linking. State-of-the-art systems, such as Google's Knowledge
Vault [19] and DeepDive [70], employ distant supervision and probabilistically
combine results from multiple extractors.

Open information extraction is an alternative paradigm, pioneered by the Text-Runner system [8], which employs *self-supervision*; training examples are generated heuristically using a small set of extraction patterns. The idea was taken up and developed further in a number of influential systems, including WOE [90], ReVerb [24], R2A2 [23], OLLIE [44], and ClausIE [16]. Open IE methods are domain-independent and very scalable but also highly susceptible to noise. Note that unlike previous approaches, they do not use canonical names for relationships, i.e., are "schema-less." To be able to use the extracted relationships for KB expansion, the natural language relation phrases need to be aligned with (i.e., mapped to) existing knowledge base predicates; see, e.g., [13].

All the above methods are based on the idea of two named entities appearing in the same sentence. Some systems operate on the level of entity mentions (identified by a named entity recognizer), while others ground entities in a knowledge base (by performing entity linking). For the purpose of KBP, the latter one is more convenient. Finally, while most systems operate on unstructured text, there are also other sources that provide relational information, e.g., web tables [12, 42].

6.2 Entity-Centric Document Filtering

Document filtering is the task of identifying documents from a stream, from which relevant information can be extracted. This problem dates back to the Topic Detection and Tracking (TDT) track of the Text Retrieval Conference (TREC), which focused on finding and following events in broadcast news stories [3]. In this section, we are focusing on an entity-centric variant of the document filtering task. Unlike traditional filter topics, which are defined by a set of keyword queries, entities are described by (semi-)structured entries in a knowledge repository.

> **Definition 6.1** *Entity-centric document filtering* is the task of analyzing a time-ordered stream of documents and assigning a score to each document based on how relevant it is to a given target entity.

Consider the following scenario. A human content editor is responsible for the maintenance of the entries of one or multiple entities in a knowledge repository, e.g., Wikipedia. The filtering system monitors a stream of documents (news articles, blog posts, tweets, etc.) and identifies those that contain novel information pertinent to the target entities. Each document is assigned a (relevance) score; whenever the editor checks the system, which may be several times a day or once a week, the system presents a relevance-ordered list of documents that have been discovered since the last time the system was checked. The editor reviews some or all of these documents; the stopping criterion may be a score threshold or her time available.

Upon examining a document, the editor may decide to make changes to the KR entry of a given target entity. That document will then serve as provenance, i.e., the change made in the knowledge repository can be tracked back to this document as the original source. For example, in Wikipedia, changes are substantiated by adding a citation; the corresponding source is listed at the bottom of the article under the "References" section.

An abstraction of the above scenario was studied at the Knowledge Base Acceleration track at TREC 2012–2014, termed as *cumulative citation recommendation*: Filter a time-ordered corpus for documents that are highly relevant to a predefined set of entities [27]. One particularly tricky aspect of this task—and many other IR problems for that matter—is how to define relevance. At TREC KBA, it was initially based on the notion of "citation worthiness," i.e., whether a human editor would cite that document in the Wikipedia article of the target entity. This was later refined by requiring the presence of new information that would change an already up-to-date KR entry, rendering that document "vital." See Sect. 6.2.5.2 for details. We note that the need for this type of filtering also arises in other domains besides KBA, for instance, in social media, for business intelligence, crisis management, or celebrity tracking [97].

One of the main challenges involved in the entity-centric filtering task is how to distinguish between documents that are only tangentially relevant and those that are vitally relevant to the entity. Commonly, this distinction is captured by training supervised learning models, leveraging a rich set of signals as features, to replicate human judgments.

The attentive reader might have noticed that we are talking about knowledge repositories here, and not about knowledge bases. This is on purpose. For the document filtering task, we are unconcerned whether knowledge is in semi-structured or in structured format, since it is the human editor that needs to make manual updates to the entity's (KR or KB) entry.

6.2.1 Overview

Formally, the entity-centric filtering task is defined as follows. Let $\mathcal{D}$ be a time-ordered document stream and e be a particular target entity. The entity-centric filtering system is to assign a numerical score, $score(d; e)$, to each document $d \in \mathcal{D}$.

Figure 6.3 shows the canonical architecture for this task. For each document in the stream, the processing starts with an entity detection step: Does the document mention the target entity? (That is, for a single target entity. In the case of multiple

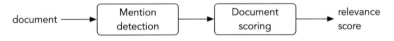

Fig. 6.3 Entity-centric document filtering architecture

Table 6.1 Notation used in Sect. 6.2

Symbol	Meaning
a	Entity aspect ($a \in \mathcal{A}$)
$\mathcal{A}$	Set of entity aspects
d	Document ($d \in \mathcal{D}$)
$\mathcal{D}$	Collection (stream) of documents
e	Entity ($e \in \mathcal{E}$)
h	Time (measured in hours)
$\mathcal{L}_e$	Set of entities e links to
T_d	Publication time of document d
$\mathcal{T}_e$	Type of entity e

target entities, this step has to be performed for each entity.) If the answer is yes, then, as a second step, the document needs to be scored do determine its relevance ("citation-worthiness") with respect to the target entity; otherwise, it is no longer of interest (i.e., gets zero assigned as score).

The mention detection component is discussed in Sect. 6.2.2. For document scoring, we present both unsupervised and supervised approaches in Sect. 6.2.3. State-of-the-art systems employ supervised learning based on a rich array of features; we present a selection of effective features in Sect. 6.2.4. Table 6.1 summarizes the notation used in this section.

6.2.2 Mention Detection

This step is responsible for determining whether a particular document contains a mention of a given target entity. Since this operation needs to be performed for all pairs of streaming documents and target entities, efficiency is paramount here. We wish to obtain high recall, so as not to miss any document that is potentially of interest. At the same time, we would also like to keep the number of false positives low. Therefore, much as it was done for entity linking in Sect. 5.4, mention detection relies on known surface forms of the target entity. We let $\mathcal{S}_e$ denote the set of known surface forms for entity e, where $s \in \mathcal{S}_e$ is a particular surface form. By definition, each entity has at least one canonical name, therefore $\mathcal{S}_e \neq \emptyset$. With $\mathcal{S}_e$ at hand, determining whether the document mentions the entity is conceptually as simple as:

$$mentions(d, e) = \begin{cases} 1, & \exists s \in \mathcal{S}_e : contains(d, s) \\ 0, & \text{otherwise}, \end{cases}$$

where $contains(d, s)$ denotes a case-insensitive string matching function that searches for string s in d.

The only remaining issue concerns the construction of the set of surface forms. For entities that already have an entry in the knowledge repository, this is readily available. For example, Balog et al. [7] extract surface forms (aliases) from DBpedia. For certain entity types, name variants may be generated automatically by devising simple heuristics, e.g., using only the last names of people [7] or dropping the type labels of business entities ("Inc.," "Co.," "Ltd.," etc.). Finally, considering the KBA problem setting, it is reasonable to assume that a human knowledge engineer would be willing to manually expand the set of surface forms. This was also done at TREC KBA 2013, where the oracle baseline system relied on a hand-crafted list of entity surface forms [26].

6.2.3   Document Scoring

Next, we need to estimate $score(d; e)$ for document d, which, as we know, potentially contains the target entity e. For convenience, we shall assume that this score is between 0 and 1.[2] Further, we assume that, for each target entity, a set of documents and corresponding (manually assigned) relevance labels are made available as training data. We distinguish between non-relevant (R0) and relevant documents, where relevance has two levels. Following the latest TREC KBA terminology (cf. Sect. 6.2.5.2), we shall refer to the lower relevance level (R1) as *useful* and the higher relevance level (R2) as *vital*.

It is worth pointing out that none of the document scoring techniques we will discuss attempt to explicitly disambiguate the mentioned entities (i.e., no entity linking is performed). The reason is that entity disambiguation requires the presence of either a textual description or relationship information about the given entity (cf. Sect. 5.6). In the KBA context, where target entities are often long-tail entities, this information is not necessarily available (yet) in the knowledge repository.

6.2.3.1   Mention-Based Scoring

A simple baseline approach, which was employed at TREC KBA, is to assign a score "based on the number of matches of tokens in the name" [28]. That is, the longest observed mention of e in the document is considered, normalized by the longest known surface form of e, thereby producing a score in $(0, 1]$. Formally:

$$score(d; e) = \frac{\max(\{l_s : s \in \mathcal{S}_e, contains(d, s)\})}{\max(\{l_{s'} : s' \in \mathcal{S}_e\})},$$

where l_s denotes the character length of surface form s.

---

[2]At TREC KBA, the score needs to be in the range $(0, 1000]$, but mapping to that scale is straightforward.

Other simple mention-based scoring formulas may also be imagined, e.g., based on the total number of occurrences of the entity in the document.

6.2.3.2 Boolean Queries

Efron et al. [21] perform filtering by developing highly accurate Boolean queries, called *sufficient queries*, for each entity. A sufficient query is defined as a "Boolean query with enough breadth and nuance to identify documents relevant to an entity e without further analysis or estimation." Essentially, the mention detection and scoring steps are performed jointly, by using a single Boolean query per target entity. Specifically, a sufficient query consists of two parts: (1) a *constraint clause*, which is the surface form of the entity, and (2) a *refinement clause*, which is a set of zero or more n-grams (bigrams in [21]), selected from the set of relevant training documents. An example query, expressed using the Indri query language, is

```
#band(#1(phyllis lambert) #syn(#1(canadian architect) #1(public
art))),
```

where #1 matches as an exact phrase, #band is a binary AND operator, and #syn enumerates the elements of the refinement clause. That is, the document must contain the name of the entity (constraint clause) and any of the n-grams that are listed in the refinement clause.

The same idea may also be expressed in terms of our framework (cf. Fig. 6.3), by viewing the constraint clause as the mention detection step, and the refinement clause as a Boolean scoring mechanism that assigns either 0 or 1 as score.

6.2.3.3 Supervised Learning

The predominant approach to entity-centric filtering is to employ supervised learning. The notion of citation-worthiness "is not a precise definition that can easily be captured algorithmically" [7]. It is a combination of multiple factors that make a document useful/vital. The idea, therefore, is to focus on capturing and extracting as many of these contributing factors as possible, as features. Then, we let a machine learning algorithm figure out how to best combine these signals based on a manually labeled set of training documents. We shall discuss specific features in Sect. 6.2.4. For now, our concern is the supervised learning part, i.e., what type of model we want to train and how.

The entity-centric filtering task may be approached both as a classification and as a ranking problem. In the former case, a binary decision is made, where the classifier's confidence may be translated into a relevance score. In the latter case, relevance is directly predicted, and thresholding is left to the end user, i.e., she decides when to stop in the ranked list. We look at these two possibilities in more detail below. In terms of performance, ranking-based approaches were found to perform better in [6]. Later studies, however, have shown that it varies whether classification or ranking is more effective, depending on the (sub-)set of features used and on the relevance criteria that are targeted [31, 83].

Classification Balog et al. [7] propose two classification methods, referred to as *2-step* and *3-step classification* approaches. The first step, in both cases, corresponds to the mention detection phase, which may also be seen as a binary classification decision (whether the document contains the target entity or not). This is then followed by one or two additional binary classification steps. The 2-step approach makes a single classification decision to decide whether the document is vital or not. The 3-step approach first tries to separate non-relevant documents from relevant ones (R0 vs. R1 or R2), and in a subsequent step distinguish between useful and vital documents (R1 vs. R2). See Fig. 6.4 for an illustration. Notably, the same set of features are used in all classification steps. The difference lies in how documents get labeled as positive/negative instances during training. In 2-step classification, non-relevant documents constitute the negative class, while vital documents are the positive class; useful documents are not used so as not to "soften the distinction between the two classes that we are trying to separate" [7]. In 3-step classification, the "Relevant?" decision uses non-relevant and relevant documents as negative and positive instances, respectively, while the "Vital?" decision uses useful and vital documents as negatives and positives, respectively.

According to the experiments reported in [7], these two methods deliver very similar performance, making the simpler 2-step approach the recommended choice. We note that this stance may need to be revisited depending on the amount of training data available.

2-step classification

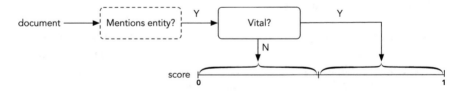

3-step classification

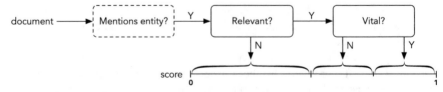

Fig. 6.4 Multi-step classification approaches proposed in [7]. Each box represents a binary classifier. The dashed box corresponds to the mention detection step

Learning-to-Rank Filtering may also be approached as a learning-to-rank problem: Estimate a numerical score for a document-entity pair. We refer back to Sect. 3.3.3 for a crash course on learning-to-rank. It is simpler than classification-based approaches in the sense that there is no additional mapping involved, i.e., the target score directly corresponds to the relevance level. Balog and Ramampiaro [6] compared pointwise, pairwise, and listwise learning-to-rank approaches and found that the pointwise approach yielded the best performance.

Global vs. Entity-Specific Models Another issue is whether to train a single global model, which is used for all target entities, or a separate entity-specific model for each entity. Commonly, the former option is selected [6, 7, 63, 83], as there is limited training data available for any individual entity. Also, the global model generalizes to previously unseen entities. Wang et al. [84] argue that "these models ignore the distinctions between different entities and learn a set of fixed model parameters for all entities, which leads [to] unsatisfactory performance when dealing with a diverse entity set." Therefore, they propose an entity-type-dependent discriminative mixture model, which involves an intermediate latent layer to model each entity's distribution across entity types.

6.2.4 Features

We present a selection of features from [7, 63], organized into the following four main groups:

- *Document features* estimate the "citation worthiness" of the document, independent of the target entity.
- *Entity features* capture characteristics of the target entity, based on its entry in the knowledge repository.
- *Document-entity features* express the relation between a particular document and the target entity.
- *Temporal features* model events happening around the target entity.

We discuss each feature group in turn. Table 6.2 provides a summary. We note that these features may be used both with binary classification and learning-to-rank approaches (cf. Sect. 6.2.3.3).

6.2.4.1 Document Features

Simple document features include various measurements of length, the document's source, and its language [7]. Reinanda et al. [63] consider several intrinsic characteristics of a document to help identify vital documents. In particular: informativeness, entity-saliency, and timeliness.

Table 6.2 Overview of features for the entity-centric document filtering task

Group	Feature	Description	Src.	Val.
Document features				
	l_{f_d}	Length of doc. field f (title, body, anchors)	S	N
	$len(d)$	Length of d in number of chunks or sentences	S	N
	$src(d)$	Document source (news, social, linking, etc.)	S	C
	$lang(d)$	Whether the document's language is English	S	B
	$aspectSim(d,a)$	Similarity between d and aspect a	S, KR	N
	$numEntities(d)$	Number of unique entities mentioned in d	S	N
	$numMentions(d)$	Total number of entity mentions in d	S	N
	$timeMatchY(d)$	Number of temporal expressions matching the document's creation year	S	N
	$timeMatchYM(d)$	Number of temporal expressions matching the document's creation year and month	S	N
	$timeMatchYMD(d)$	Number of temporal expressions matching the document's creation year, month, and day	S	N
Entity features				
	$\mathcal{T}_e$	Type of the entity (PER, ORG, etc.)	KR	C
	l_e	Length of the entity's description	KR	N
	$|\mathcal{L}_e|$	Number of related entities	KR	N
Document-entity features				
	$n(f_d,e)$	Number of mentions of e in document field f	S	N
	$firstPos(d,e)$	Term position of the first mention of e in d	S	N
	$firstPosNorm(d,e)$	$firstPos(d,e)$ normalized by the document length	S	N
	$lastPos(d,e)$	Term position of the last mention of e in d	S	N
	$lastPosNorm(d,e)$	$lastPos(d,e)$ normalized by the document length	S	N
	$spread(d,e)$	Spread (distance between first and last mentions)	S	N
	$spreadNorm(d,e)$	$spread(d,e)$ normalized by the document length	S	N
	$numSentences(d,e)$	Number of sentences mentioning e	S	N
	$mentionFrac(d,e)$	Mentions of e divided by all entity mentions in d	S	N
	$sim_{Jac}(d,e)$	Jaccard similarity between d and e	S, KR	N
	$sim_{cos}(d,e)$	Cosine similarity between d and e	S, KR	N
	$sim_{KL}(d,e)$	KL divergence between d and e	S, KR	N
	$numRelated(f_d,e)$	Number of unique related entities mentioned in document field f	S	N
Temporal features				
	$sv(e)$	Average hourly stream volume	S	N
	$sv_h(e)$	Stream volume over the past h hours	S	N
	$\Delta sv_h(e)$	Change in stream volume over the past h hours	S	N
	$isBurstSv_h(e)$	Whether there is a burst in stream volume	S	B
	$wpv(e)$	Average hourly Wikipedia page views	U	N

(continued)

Table 6.2 (continued)

Group	Feature	Description	Src.	Val.
	$wpv_h(e)$	Wikipedia page views volume in the past h hours	U	N
	$\Delta wpv_h(e)$	Change in Wikipedia page views volume	U	N
	$isBurstWpv_h(e)$	Whether there is a burst in Wikipedia page views	U	B
	$burstValue(d,e)$	Document's burst value, based on Wikipedia page views	U	N

Source can be stream (S), knowledge repository (KR), or usage data (U). Value can be numerical (N), categorical (C), or Boolean (B)

The intuition behind *informativeness* features is that a document that is rich in facts is more likely to be vital. One way to measure this is to consider the aspects of entities that are mentioned in the document, where *aspects* are defined as "key pieces of information with respect to an entity" [63]. The set of aspects $\mathcal{A}$ is constructed by taking the top-k ($k = 50$ in [63]) most common section headings (with stopwords removed) of Wikipedia articles belonging to a set of categories of interest (here, *Person* or *Location*). Then, a textual representation of each entity aspect $a \in \mathcal{A}$ is constructed by aggregating the contents of those sections. Finally, the cosine similarity between bag-of-words representations of document d and aspect a are computed:

$$aspectSim(d,a) = \cos(\mathbf{d}, \mathbf{a}) .$$

Reinanda et al. [63] also consider relations from both open and closed information extraction systems, but those are reported to perform worse than Wikipedia aspects.

Entity-saliency refers to the notion of how prominent the target entity is in the document, by considering other entities mentioned in the document. Here, we only consider two simple features of this kind that do not depend on the specific target entity. One is the number of unique entities mentioned in the document ($numEntities(d)$), the other is the total number of entity mentions in d ($numMentions(d)$).

The *timeliness* of a document captures "how timely a piece of information mentioned in the document is" [63]. It is measured by comparing the document's creation/publication time, T_d (which is obtained from document metadata), with temporal expressions present in the document. Specifically, the following equation counts the occurrences of the year of the document's creation time appearing in d:

$$timeMatchY(d) = count(year(T_d), d) .$$

Similarly, (year and month) and (year, month, and date) expressions of T_d are also counted:

$$timeMatchYM(d) = count(yearmonth(T_d), d) ,$$

$$timeMatchYMD(d) = count(yearmonthday(T_d), d) .$$

6.2.4.2 Entity Features

These features capture our knowledge about a given target entity e, including (1) the type of the entity $(\mathcal{T}_e)^3$; (2) the length of the entity's description in the knowledge repository (l_e); and (3) the number of related entities $|\mathcal{L}_e|$, where $\mathcal{L}_e$ is defined as the set of entities that e links to [7]. Alternative interpretations of $\mathcal{L}_e$ are also possible.

6.2.4.3 Document-Entity Features

Document-entity features are meant to express how vital a given document d is for the target entity e. We assume a fielded document representation, which would typically include title, body, and anchors fields. Several features are developed in [7] to characterize the occurrences of the target entity in the document, including the total number of mentions in each document field, the first and last position of mentions in the document's body, and the spread between the first and last mentions. To measure entity-salience, Reinanda et al. [63] calculate the number of sentences mentioning the target entity and the fraction of the target entity's mentions with respect to all entity mentions in the document. All these features may be computed for both full and partial entity name matches (i.e., requiring the detection of one of the surface forms for a full match and considering only the last names of people for a partial match) [7].

A second group of features looks at the similarity between the term-based representations of the target entity (i.e., the entity's description in the knowledge repository, e.g., the Wikipedia article corresponding to the entity) and the document, employing various similarity measures (cosine, Jaccard, or KL divergence) [7]. Finally, we may also take into consideration what other entities, related to e, are mentioned in the document [7]. Formally:

$$numRelated(f_d, e) = \sum_{e' \in \mathcal{L}_e} \mathbb{1}(n(f_d, e')) ,$$

where $\mathcal{L}_e$ is the set of related entities, and $\mathbb{1}(n(f_d, e'))$ is 1 if entity e' is mentioned in field f of document d, otherwise 0.

6.2.4.4 Temporal Features

Temporal features attempt "to capture if something is happening around the target entity at a given point in time" [7]. The volume of mentions of the entity in the streaming corpus offers one such temporal signal. Another—independent—signal may be obtained from an usage log that indicates how often people searched for

---

^{3}Unlike in other parts of the book, type here is not a set but is assumed to take a single value (PER, ORG, LOC, etc.).

or looked at the given entity. For example, Balog et al. [7] leverage Wikipedia page view statistics for that purpose. In both cases, the past h hours are observed in a sliding windows manner ($h = \{1, 2, 3, 6, 12, 24\}$), to detect changes or bursts. Formally, let $sv(e)$ denote the *stream volume* of entity e, i.e., the average number of documents mentioning e per hour. This is computed over the training period and serves as a measure of the general popularity of the entity. Further, $sv_h(e)$ is the number of documents mentioning e over the past h hours. The change relative to the normal volume, over the past h hours, is expressed as:

$$\Delta sv_h(e) = \frac{sv_h(e)}{h \times sv(e)} \, .$$

All these "raw" quantities, i.e., $sv(e)$, $sv_h(e)$, and $\Delta sv_h(e)$, are used as features. In addition, Balog et al. [7] detect bursts by checking if $\Delta sv_h(e)$ is above a given threshold τ, which they set to 2:

$$isBurstSv_h(e) = \begin{cases} 1, & \Delta sv_h(e) \geq \tau \\ 0, & \text{otherwise} . \end{cases}$$

For Wikipedia page views, the definitions follow analogously, using the page view count instead of the number of documents mentioning the entity. Alternatively, Wang et al. [83] define the *burst value* for each document-entity pair as follows:

$$burstValue(d, e) = \frac{N \times wpvd(e, T_d)}{\sum_{i=1}^{N} wpvd(e, i)} , \tag{6.1}$$

where N is the total number of days covered by the stream corpus, $wpvd(e, i)$ is the number of page views of the Wikipedia page of entity e during the ith day of the streaming corpus, and T_d refers to the date when document d was published ($T_d \in [1..N]$). We note that this formulation considers information "from the future" (i.e., beyond the given document's publication date); this may be avoided by replacing N with T_d in Eq. (6.1).

6.2.5 Evaluation

We discuss the evaluation of entity-centric filtering systems as it was carried out at the TREC 2012–2014 Knowledge Base Acceleration track.

6.2.5.1 Test Collections

For each edition of the TREC KBA track, a streaming test collection was developed. Content from the 2012 corpus is included in the 2013 corpus. Similarly, the 2013 corpus is subsumed by the 2014 one. The document collection is composed of

Table 6.3 TREC KBA test collections for entity-centric document filtering

Name	Time period	Size	#Docs	#Target entities
Stream Corpus 2012 [27][a]	Oct 2011–Apr 2012	1.9TB	462.7M	29
Stream Corpus 2013 [26][b]	Oct 2011–Feb 2013	6.5TB	1B	141
Stream Corpus 2014 [28][c]	Oct 2011–Apr 2013	10.9TB	1.2B	109

Size is for compressed data
[a] http://trec-kba.org/kba-stream-corpus-2012.shtml
[b] http://trec-kba.org/kba-stream-corpus-2013.shtml
[c] http://trec-kba.org/kba-stream-corpus-2014.shtml

three main sources: news (global public news wires), social (blogs and forums), and linking (content from an URL-shortening service). The 2013 and 2014 versions include additional substreams of web content. The streaming corpus is divided into hourly batches, allowing entities to evolve over time.

The 2014 corpus adds to the previous two versions not only in terms of duration but also in terms of NLP tagging information, by deploying BBN's Serif NLP system [9]. The official corpus (listed in Table 6.3) is tagged with the recognized named entities. An extended version (16.1 TB in size) also contains within-document coreference resolution and dependency parsing annotations. In both cases, annotations are available only for the (likely) English documents. Furthermore, Google has released Freebase annotations for the TREC KBA 2014 Stream Corpus (FAKBA1).[4] A total of 9.4 billion entity mentions are recognized and disambiguated in 394M documents. The format is identical to that of the ClueWeb Freebase annotations, cf. Sect. 5.9.2.

The set of target entities for 2012 consists of 27 people and 2 organizations from Wikipedia. The 2013 set focuses on 14 inter-related communities of entities, a total of 98 people, 19 organizations, and 24 facilities from Wikipedia and Twitter. The 2014 entities are primarily long-tail entities that lacked Wikipedia entries; 86 people, 16 organizations, and 7 facilities were hand-picked from within a given geographic region (between Seattle and Vancouver).

6.2.5.2 Annotations

The annotation guidelines evolved over the years. Initially, human annotators were given the following instructions: "Use the Wikipedia article to identify (disambiguate) the entity, and then imagine *forgetting* all info in the Wikipedia article and asking whether the text provides any information about the entity" [27].

---

[4]http://trec-kba.org/data/fakba1/.

Documents were annotated along two dimensions: (1) whether it mentions the target entity explicitly and (2) whether it is relevant. The annotation matrix is shown in Fig. 6.5, where rows correspond to mentions and columns denote the level of relevance. The relevance levels were defined as follows:

- *Garbage*: not relevant, e.g., spam.
- *Neutral*: not relevant, i.e., no information can be learned about the target entity.
- *Relevant*: relates indirectly to the target entity, e.g., discusses topics or events that likely impact the entity.
- *Central*: relates directly to the target entity, i.e., the entity is a central figure in the mentioned topics or events. The document would be cited in the Wikipedia article of the entity.

For the 2013 edition, the instructions given to the assessors were revised, in order to make a better distinction between the two highest relevance levels. The "central" judgment was replaced by "vital," reserving this rating to documents that would trigger changes to an already up-to-date KR entry. The second highest rating level was also renamed, from "relevant" to "useful," to include documents that are "citation-worthy as background information that might be used when writing an initial dossier but do not present timely or 'fresh' changes to the entity" [26]. That is:

- *Useful*: contains citation-worthy background information that should be included in the entity's KR entry.
- *Vital*: presents timely information that would imply changes to an already up-to-date KR entry.

According to Frank et al. [26], these changes "removed a large area of subjectivity in the notion of 'citation-worthiness'" (as observed by the increased inter-annotator agreement). However, deciding what to include in the entity's KR entry and what up-to-dateness means still involves subjective judgment. For less noteworthy entities there seems to be increased subjectivity, hence lower inter-annotator agreement [28].

Theoretically speaking, a document may be relevant, even if it does not mention the target directly (e.g., through relations to other entities mentioned in the

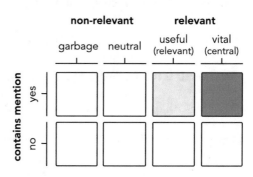

Fig. 6.5 Document annotation matrix from the TREC KBA track. The goal is to identify vital documents (top right corner). The labels in parentheses correspond to the 2012 terminology

document). In practice, relevance without an explicit mention of the target entity rarely happens. In 2014, the annotations were simplified by dropping the "mention" dimension; the "useful" and "vital" ratings imply that the entity was mentioned in the document.

Crucially, for all three editions of the TREC KBA track, the judgments were performed *pre-hoc*, i.e., there might be relevant documents (true positives) that were not seen by annotators. The recall of annotations is estimated to be over 90% [27]. Documents that are not primarily English were discarded. Furthermore, there is no novelty requirement, i.e., all documents that report the same event within a given timeframe are deemed citation-worthy. That timeframe is decided subjectively by the assessor; generally, less than a week and more than an hour. Annotations for an early portion of the stream are provided as training data.

6.2.5.3   Evaluation Methodology

Systems are required to process the stream corpus in chronological order, in hourly batches, and assign a confidence score in (0,1000] to each citation-worthy document. At any given point in time, systems may only access information about entities from the past. Evaluation uses set-based measures: F1-measure and scaled utility. See Eq. (5.10) for the definition of the F1-measure. *Scaled utility* is a measure from general information filtering that measures a system's ability to accept relevant and reject non-relevant documents from a stream [66]. Formally,

$$SU = \frac{\max(T11NU, MinNU) - MinNU}{1 - MinNU}, \tag{6.2}$$

where T11U is the linear utility, which gives a credit of 2 for a relevant document retrieved (i.e., true positives) and a debit of 1 for a non-relevant document retrieved (i.e., false positives):

$$T11U = 2 \times TP - FP.$$

A normalized version is computed using:

$$T11NU = \frac{T11U}{MaxU},$$

where MaxU is the maximum possible utility:

$$MaxU = 2 \times (TP + FN).$$

Finally, MinNU in Eq. (6.2) is a tunable parameter, set to -0.5.

entity ID	document ID	score
Aharon_Barak	1328055120-f6462409e60d2748a0adef82fe68b86d	1000
Aharon_Barak	1328057880-79cdee3c9218ec77f6580183cb16e045	500
Aharon_Barak	1328057280-80fb850c089caa381a796c34e23d9af8	500
Aharon_Barak	1328056560-450983d117c5a7903a3a27c959cc682a	480
Aharon_Barak	1328056560-450983d117c5a7903a3a27c959cc682a	450
Aharon_Barak	1328056260-684e2f8fc90de6ef949946f5061a91e0	430
Aharon_Barak	1328056560-be417475cca57b6557a7d5db0bbc6959	428
Aharon_Barak	1328057520-4e92eb721bfbfdfa0b1d9476b1ecb009	428
Aharon_Barak	1328058660-807e4aaeca58000f6889c31c24712247	380
Aharon_Barak	1328060040-7a8c209ad36bbb9c946348996f8c616b	380
Aharon_Barak	1328063280-1ac4b6f3a58004d1596d6e42c4746e21	375
Aharon_Barak	1328064660-1a0167925256b32d715c1a3a2ee0730c	315
Aharon_Barak	1328062980-7324a71469556bcd1f3904ba090ab685	263

(Positive: rows 1–8; Negative: rows 9–13; τ cutoff line between score 428 and 380)

Fig. 6.6 Scoring of documents for a given target entity, using a cutoff value of 400. Figure is based on [6]

To be able to apply the above set-based measures, the output of the filtering system needs to be divided into positive and negative sets (which are then compared against the positive/negative classes defined in the ground truth). It is done by employing a confidence threshold τ. Documents with a confidence score $\geq \tau$ are treated as positive instances, while all remaining documents are negatives. This idea is illustrated in Fig. 6.6. At TREC KBA, a scoring tool sweeps the confidence cutoff between 0 and 1000 in steps of 50. For each entity, the highest performing cutoff value is chosen, then F1/SU scores are averaged over the set of all target entities (i.e., macro-averaging is used). Alternatively, a single cutoff value may be applied for all entities [6]. The strictness of evaluation is considered on two levels: (1) only central/vital documents are accepted as positive, and (2) both relevant/useful and central/vital documents are treated as positives.

6.2.5.4 Evaluation Methodology Revisited

The official TREC KBA evaluation methodology is time-agnostic. Even though systems process the stream corpus in hourly batches, evaluation considers all returned documents (above the threshold) as a single set, thereby ignoring the streaming nature of the task. Intuitively, not all *time batches* (e.g., hourly periods) are equally important; e.g., some might be high intensity burst periods, while others hardly contain any relevant documents. Additionally, the cutoff threshold τ is a free parameter; it remains an open issue how to set it in a way that it enables a fair comparison between systems. Furthermore, the actual ranking of documents within a given batch (which are above the cutoff threshold) is not taken into account. Dietz et al. [17] present a time-aware evaluation framework, which is composed of three main elements: (1) slicing the entire evaluation period into time batches, (2) evaluating the performance of each batch using standard rank-based measures, (3) aggregating the batch evaluation results into a single system-level score.

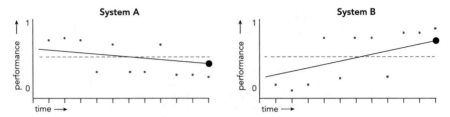

Fig. 6.7 Time-aware evaluation of document filtering systems proposed by Kenter et al. [40]. Systems A and B have the same average performance (dashed line), however, A degrades over time while B improves. The solid straight line is the fitted trend line. Systems are compared in terms of "estimated end-point performance" (the large dots)

Kenter et al. [40] argue that a crucial aspect of the document filtering task, which sets it apart from other document classification problems, is how the performance of a system changes over time. They propose to capture if a system improves, degrades, or remains the same over time, by employing trend analysis. The idea is to measure performance at regular time intervals, using any existing evaluation measure (either set-based or rank-based). Then, by fitting a straight line to these data points, one can estimate the expected performance of the system at the end of the evaluation period. This "estimated end-point performance" can serve as the basis of comparison across systems; see Fig. 6.7.

6.3 Slot Filling

Once documents that potentially contain new information about a given target entity have been identified, the next step is to examine the contents of those documents for new facts. The extracted facts can then be used to populate the knowledge base entry of the target entity. (Notice the shift from a knowledge repository to a knowledge base, as we shall now operate on the level of specific facts.) We look at one particular variant of this task, which assumes that the set of predicates ("slots") that are of interest is provided.

Definition 6.2 *Slot filling* is the task of extracting the corresponding values for a pre-defined set $\mathcal{P}$ of predicates, for a given target entity e, from a previously identified set of documents $\mathcal{D}_e$.

Stating the problem in terms of SPO triples, we are to find triples (e, p, o), where e is the target entity, $p \in \mathcal{P}$ is one of the target predicates (slots), and the object value o is to be extracted from $\mathcal{D}_e$. The resulting triples (facts) can then be used to populate the

Table 6.4 Entity attributes targeted at the TAC 2009 KBP slot filling task [45]

Entity type	Attributes
Person	alternate names; age; date and place of birth; date, place, and cause of death;
	national or ethnic origin; places of residence; spouses, children, parents, siblings, and other familial relationships; schools attended; job titles; employers; organizational memberships; religion; criminal charges
Organization	alternate names; political or religious affiliation; top members/employees;
	number of employees; members; member of; subsidiaries; parents; founder; date founded;
	date dissolved; location of headquarters; shareholders; website
Geo-political entity	alternate names; capital; subsidiary organizations; top employees;
	active political parties; when established; population; currency

KB. Note that, depending on the object's value, a distinction can be made between attributes (objects with a literal value) and relationships (where the object is another entity). In this section, we shall assume that objects are strings (literals), hence we are always targeting attributes. Consequently, we will refer to the predicates for which values are to be found as *attributes* and to the corresponding object values as *slot values*. These values may subsequently be linked within the knowledge base. We also note that relationship extraction is typically separately addressed through a dedicated set of techniques, which we have already surveyed in Sect. 6.1.3.

As an illustration, Table 6.4 lists the attributes that were targeted by the slot filling task of the Knowledge Base Population (KBP) track at the 2009 Text Analysis Conference (TAC). According to the categorization scheme presented in Sect. 6.1, slot filling is a closed and targeted information extraction task. We are targeting specific entities and are interested in a closed set of attributes.

6.3.1 Approaches

There are two main groups of approaches to slot filling: (1) *pattern-based* methods, which extract and generalize lexical and syntactic patterns (semi-)automatically [41, 72, 93], and (2) *supervised classification* methods, which consider entities and candidate slot values as instances and train a classifier through distant supervision [1, 67, 73]. Here, we will focus exclusively on the latter group, as it is more relevant to our entity-oriented perspective.

Training data for slot filling consists of documents that are explicitly marked up with slot values. The challenge is that such training datasets are restricted in size. Thus, "traditional supervised learning, based directly on the training data, would provide limited coverage" [36]. On the other hand, the facts that are already in the knowledge base can be exploited to produce training examples [48].

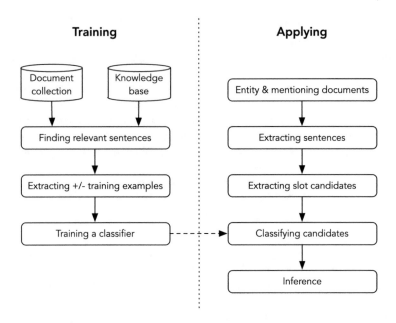

Fig. 6.8 Slot filling architecture using distant supervision

Figure 6.8 shows the architecture of a typical slot filling system that involves distant supervision [73, 75]. The steps of the training phase, for a given attribute $p \in \mathcal{P}$, are the following.

1. Finding sentences in the document collection that mention entities with already existing values for attribute p in the KB. The slot value candidates are identified in these sentences using an *extended named entity recognizer*. This extended recognizer should be able to detect text spans that are possible values of the targeted attribute and label those with a type (e.g., temporal expressions, job titles, religion, etc.).

2. All sentences that contain an entity and its corresponding slot value from the KB, are positive training examples. All other sentences, which contain a candidate slot value with the matching type (e.g., a temporal expression) but not with the correct slot value from the KB, are taken as negative training examples.

3. A classifier is trained on entity and slot value candidate pairs. It can be a global multi-class classifier for all attributes or multiple binary classifiers for each individual attribute. Commonly, three main groups of features are used: (i) information about the entity and the slot value candidate (e.g., terms and the type of the slot value candidate as labeled by the extended NER); (ii) surface features (e.g., terms surrounding the entity and the slot value candidate); (iii) syntactic features (e.g., dependency path between the entity and the slot value candidate) [75].

Applying the learned model involves the following steps:

1. Retrieving sentences that mention the target entity.
2. Extracting candidate slot values using the extended named entity recognizer.
3. Classifying the target entity and candidate slot value pairs according to the learned model.
4. A final inference step decides what slot values to return as output. For example, certain attributes can take only a single value (e.g., age or city of birth) while others can take a list of values (e.g., parents or alternate names). If multiple slot values are extracted for the same attribute, then those decisions need to be merged. One can return the slot value with the highest overall score or the one with the highest combined score. Additionally, a confidence threshold may be applied; slot values below that threshold are discarded.

6.3.2 Evaluation

The slot filling task has been addressed at the Web People Search (WePS) evaluation campaign, targeting only persons [5], and at the Knowledge Base Population track of the Text Analysis Conference (TAC KBP), targeting persons, organizations, and geo-political entities [37, 45, 74]. For details of the test collections, evaluation methodology, and evaluation metrics, we refer to the overview papers of the respective campaigns [5, 74].

Slot filling remains to be an extremely challenging task. According to the 2014 TAC KBP evaluation results, the best performing team achieved an F1-score of 0.36, which is approximately at 52% of human annotator performance [74].

6.4 Summary

This chapter has addressed the topic of populating knowledge bases with new facts about entities. We have approached this task using a two-step pipeline. The first component, *entity-centric document filtering*, aims to identify documents, from an incoming stream, that contain information that would trigger an update to a given entity's entry in the knowledge repository. The second component, *slot filling*, focuses on extracting new facts about an entity from a given input document. More specifically, for a pre-defined set of predicates, this component aims to find the values corresponding to those predicates.

When it comes to making actual updates to a knowledge base, one could argue that manual curation by humans is, and will remain, indispensable. Otherwise, systems could enter into a vicious cycle of feeding themselves false information that they themselves would be generating. The techniques presented in this chapter, however, represent crucial tools to support knowledge engineers and to help them perform their work effectively.

Knowledge bases are (some of) the core building blocks of next generation search engines. As we shall see in Part III, they can be utilized, among other tasks, to help understand users' information needs, improve the ranking of documents, present rich search results, aid users in formulating their queries, and recommend related content.

6.5 Further Reading

The automatic updating and expansion of knowledge bases still presents a number of challenges. A knowledge base can never be truly complete and up-to-date. In reality, some information is changing so rapidly that it would be difficult (or hardly meaningful) to materialize it in a knowledge base. Examples include movie sales, stock prices, or chart positions of songs. Preda et al. [62] refer to it as *active knowledge*, and propose to gather and integrate data from web services on the fly, transparently to the user, as if it was already contained in the knowledge base. In a recent study, Galárraga et al. [29] make the first steps toward predicting the completeness of certain properties of objects in a KB.

Cold start knowledge base population refers to the ambitious goal of constructing a KB from raw input data (e.g., a document collection). Cold start KBP encompasses a number of subtasks, including the discovery of *emerging entities* [32, 34], the clustering of entity mentions, relation extraction, and slot filling. Knowledge can also be inferred from what is already in the knowledge graph, without resorting to external sources, referred to as the problem of *link prediction*. For example, if we know that "playerA playsFor teamX" and "playerB teamMates playerA," then "playerB playsFor teamX" may be inferred. We refer to Nickel et al. [55] for an overview of approaches.

Another active area is concerned with the *quality* of data. Nakashole and Mitchell [51] compute believability, i.e., the likelihood that a given statement is true. Dong et al. [20] evaluate the trustworthiness of sources by the correctness of their factual information. They show that trustworthiness is almost orthogonal to the popularity of the source. For example, many websites with high PageRank scores are "gossip sites," and generally are considered less reliable. Conversely, some less popular websites contain very accurate information. Zaveri et al. [94] present a survey of data quality issues for Linked Data.

References

1. Adel, H., Roth, B., Schütze, H.: Comparing convolutional neural networks to traditional models for slot filling. In: Proceedings of the 2016 Conference of the North American Chapter of the Association for Computational Linguistics: Human Language Technologies, NAACL '16, pp. 828–838. Association for Computational Linguistics (2016)

2. Agichtein, E., Gravano, L.: Snowball: Extracting relations from large plain-text collections. In: Proceedings of the Fifth ACM Conference on Digital Libraries, DL '00, pp. 85–94. ACM (2000). doi: 10.1145/336597.336644
3. Allan, J.: Topic detection and tracking: Event-based information organization. Kluwer Academic Publishers (2002)
4. Almuhareb, A., Massimo, P.: Finding attributes in the web using a parser. In: Proceedings of the Corpus Linguistics Conference (2005)
5. Artiles, J., Borthwick, A., Gonzalo, J., Sekine, S., Amigó, E.: Weps-3 evaluation campaign: Overview of the web people search clustering and attribute extraction tasks. In: CLEF 2010 LABs and Workshops, Notebook Papers, 22–23 September 2010, Padua, Italy (2010)
6. Balog, K., Ramampiaro, H.: Cumulative citation recommendation: Classification vs. ranking. In: Proceedings of the 36th international ACM SIGIR conference on Research and development in information retrieval, SIGIR '13, pp. 941–944 (2013). doi: 10.1145/2484028.2484151
7. Balog, K., Ramampiaro, H., Takhirov, N., Nørvåg, K.: Multi-step classification approaches to cumulative citation recommendation. In: Proceedings of the 10th Conference on Open Research Areas in Information Retrieval, OAIR '13, pp. 121–128 (2013)
8. Banko, M., Cafarella, M.J., Soderland, S., Broadhead, M., Etzioni, O.: Open information extraction from the web. In: Proceedings of the 20th International Joint Conference on Artificial Intelligence, IJCAI'07, pp. 2670–2676. Morgan Kaufmann Publishers Inc. (2007)
9. Boschee, E., Weischedel, R., Zamanian, A.: Automatic information extraction. In: Proceedings of the International Conference on Intelligence Analysis (2005)
10. Brin, S.: Extracting patterns and relations from the world wide web. In: Selected Papers from the International Workshop on The World Wide Web and Databases, WebDB '98, pp. 172–183. Springer (1999)
11. Bunescu, R.C., Mooney, R.J.: A shortest path dependency kernel for relation extraction. In: Proceedings of the Conference on Human Language Technology and Empirical Methods in Natural Language Processing, HLT '05, pp. 724–731. Association for Computational Linguistics (2005). doi: 10.3115/1220575.1220666
12. Cafarella, M.J., Halevy, A., Wang, D.Z., Wu, E., Zhang, Y.: Webtables: Exploring the power of tables on the web. Proc. VLDB Endow. 1(1), 538–549 (2008). doi: 10.14778/1453856.1453916
13. Cai, Q., Yates, A.: Large-scale semantic parsing via schema matching and lexicon extension. In: Proceedings of the 51st Annual Meeting of the Association for Computational Linguistics, ACL '13, pp. 423–433. Association for Computational Linguistics (2013)
14. Carlson, A., Betteridge, J., Kisiel, B., Settles, B., Jr., E.R.H., Mitchell, T.M.: Toward an architecture for never-ending language learning. In: Proceedings of the Twenty-Fourth Conference on Artificial Intelligence (AAAI 2010) (2010)
15. Chinchor, N.A.: Overview of MUC-7/MET-2. In: Proceedings of the 7th Message Understanding Conference, MUC-7 (1998)
16. Del Corro, L., Gemulla, R.: Clausie: Clause-based open information extraction. In: Proceedings of the 22nd International Conference on World Wide Web, WWW '13, pp. 355–366. ACM (2013). doi: 10.1145/2488388.2488420
17. Dietz, L., Dalton, J., Balog, K.: Time-aware evaluation of cumulative citation recommendation systems. In: SIGIR 2013 Workshop on Time-aware Information Access (TAIA2013) (2013)
18. Doddington, G., Mitchell, A., Przybocki, M., Ramshaw, L., Strassel, S., Weischedel, R.: The automatic content extraction (ACE) program – tasks, data, and evaluation. In: Proceedings of the Fourth International Conference on Language Resources and Evaluation, LREC '04. ELRA (2004)
19. Dong, X., Gabrilovich, E., Heitz, G., Horn, W., Lao, N., Murphy, K., Strohmann, T., Sun, S., Zhang, W.: Knowledge Vault: A web-scale approach to probabilistic knowledge fusion. In: Proceedings of the 20th ACM SIGKDD International Conference on Knowledge Discovery and Data Mining, KDD '14, pp. 601–610. ACM (2014). doi: 10.1145/2623330.2623623
20. Dong, X.L., Gabrilovich, E., Murphy, K., Dang, V., Horn, W., Lugaresi, C., Sun, S., Zhang, W.: Knowledge-based trust: Estimating the trustworthiness of web sources. Proc. VLDB Endow. 8(9), 938–949 (2015). doi: 10.14778/2777598.2777603

21. Efron, M., Willis, C., Sherman, G.: Learning sufficient queries for entity filtering. In: Proceedings of the 37th International ACM SIGIR Conference on Research and Development in Information Retrieval, SIGIR '14, pp. 1091–1094. ACM (2014). doi: 10.1145/2600428.2609517
22. Etzioni, O., Cafarella, M., Downey, D., Kok, S., Popescu, A.M., Shaked, T., Soderland, S., Weld, D.S., Yates, A.: Web-scale information extraction in knowitall: (preliminary results). In: Proceedings of the 13th International Conference on World Wide Web, WWW '04, pp. 100–110. ACM (2004). doi: 10.1145/988672.988687
23. Etzioni, O., Fader, A., Christensen, J., Soderland, S., Mausam, M.: Open information extraction: The second generation. In: Proceedings of the Twenty-Second International Joint Conference on Artificial Intelligence - Volume One, IJCAI'11, pp. 3–10. AAAI Press (2011). doi: 10.5591/978-1-57735-516-8/IJCAI11-012
24. Fader, A., Soderland, S., Etzioni, O.: Identifying relations for open information extraction. In: Proceedings of the Conference on Empirical Methods in Natural Language Processing, EMNLP '11, pp. 1535–1545. Association for Computational Linguistics (2011)
25. Fleischman, M., Hovy, E.: Fine grained classification of named entities. In: Proceedings of the 19th International Conference on Computational Linguistics - Volume 1, COLING '02, pp. 1–7. Association for Computational Linguistics (2002). doi: 10.3115/1072228.1072358
26. Frank, J.R., Bauer, S.J., Kleiman-Weiner, M., Roberts, D.A., Tripuraneni, N., Zhang, C., Ré, C., Voorhees, E.M., Soboroff, I.: Evaluating stream filtering for entity profile updates for TREC 2013. In: Proceedings of The Twenty-Second Text REtrieval Conference, TREC '13 (2013)
27. Frank, J.R., Kleiman-Weiner, M., Roberts, D.A., Niu, F., Zhang, C., Ré, C., Soboroff, I.: Building an entity-centric stream filtering test collection for TREC 2012. In: The Twenty-First Text REtrieval Conference Proceedings, TREC '12 (2012)
28. Frank, J.R., Kleiman-Weiner, M., Roberts, D.A., Voorhees, E.M., Soboroff, I.: Evaluating stream filtering for entity profile updates in TREC 2012, 2013, and 2014. In: Proceedings of The Twenty-Third Text REtrieval Conference, TREC '14 (2014)
29. Galárraga, L., Razniewski, S., Amarilli, A., Suchanek, F.M.: Predicting completeness in knowledge bases. In: Proceedings of the Tenth ACM International Conference on Web Search and Data Mining, WSDM '17, pp. 375–383. ACM (2017). doi: 10.1145/3018661.3018739
30. Gangemi, A., Nuzzolese, A.G., Presutti, V., Draicchio, F., Musetti, A., Ciancarini, P.: Automatic typing of DBpedia entities. In: Proceedings of the 11th International Conference on The Semantic Web, ISWC '12, pp. 65–81. Springer (2012)
31. Gebremeskel, G.G., He, J., de Vries, A.P., Lin, J.J.: Cumulative citation recommendation: A feature-aware comparison of approaches. In: 25th International Workshop on Database and Expert Systems Applications, DEXA '14, pp. 193–197 (2014)
32. Graus, D., Odijk, D., de Rijke, M.: The birth of collective memories: Analyzing emerging entities in text streams. arXiv preprint arXiv:1701.04039 (2017)
33. He, Y., Xin, D.: SEISA: Set expansion by iterative similarity aggregation. In: Proceedings of the 20th International Conference on World Wide Web, WWW '11. ACM (2011). doi: 10.1145/1963405.1963467
34. Hoffart, J., Altun, Y., Weikum, G.: Discovering emerging entities with ambiguous names. In: Proceedings of the 23rd International Conference on World Wide Web, WWW '14, pp. 385–396. ACM (2014). doi: 10.1145/2566486.2568003
35. Hoffmann, R., Zhang, C., Weld, D.S.: Learning 5000 relational extractors. In: Proceedings of the 48th Annual Meeting of the Association for Computational Linguistics, ACL '10, pp. 286–295. Association for Computational Linguistics (2010)
36. Ji, H., Grishman, R.: Knowledge base population: Successful approaches and challenges. In: Proceedings of the 49th Annual Meeting of the Association for Computational Linguistics: Human Language Technologies - Volume 1, HLT '11, pp. 1148–1158. Association for Computational Linguistics (2011)
37. Ji, H., Grishman, R., Dang, H.T.: Overview of the TAC 2011 Knowledge Base Population track. In: Proceedings of the 2010 Text Analysis Conference, TAC '11. NIST (2011)
38. Jiang, J.: Information Extraction from Text, pp. 11–41. Springer (2012). doi: 10.1007/978-1-4614-3223-4_2

39. Kambhatla, N.: Combining lexical, syntactic, and semantic features with maximum entropy models for extracting relations. In: Proceedings of the ACL 2004 on Interactive Poster and Demonstration Sessions, ACLdemo '04. Association for Computational Linguistics (2004). doi: 10.3115/1219044.1219066

40. Kenter, T., Balog, K., de Rijke, M.: Evaluating document filtering systems over time. Inf. Process. Manage. **51**(6), 791–808 (2015). doi: 10.1016/j.ipm.2015.03.005

41. Li, Y., Chen, S., Zhou, Z., Yin, J., Luo, H., Hong, L., Xu, W., Chen, G., Guo, J.: PRIS at TAC2012 KBP track. In: Text Analysis Conference, TAC '12. NIST (2012)

42. Limaye, G., Sarawagi, S., Chakrabarti, S.: Annotating and searching web tables using entities, types and relationships. Proc. VLDB Endow. **3**(1–2), 1338–1347 (2010). doi: 10.14778/1920841.1921005

43. Lin, T., Mausam, Etzioni, O.: No noun phrase left behind: Detecting and typing unlinkable entities. In: Proceedings of the 2012 Joint Conference on Empirical Methods in Natural Language Processing and Computational Natural Language Learning, EMNLP-CoNLL '12, pp. 893–903. Association for Computational Linguistics (2012)

44. Mausam, Schmitz, M., Bart, R., Soderland, S., Etzioni, O.: Open language learning for information extraction. In: Proceedings of the 2012 Joint Conference on Empirical Methods in Natural Language Processing and Computational Natural Language Learning, EMNLP-CoNLL '12, pp. 523–534. Association for Computational Linguistics (2012)

45. McNamee, P., Dang, H.T., Simpson, H., Schone, P., Strassel, S.: An evaluation of technologies for knowledge base population. In: Proceedings of the International Conference on Language Resources and Evaluation, LREC '10 (2010)

46. Melo, A., Paulheim, H., Völker, J.: Type prediction in RDF knowledge bases using hierarchical multilabel classification. In: Proceedings of the 6th International Conference on Web Intelligence, Mining and Semantics, WIMS '16, pp. 14:1–14:10. ACM (2016). doi: 10.1145/2912845.2912861

47. Michel, J.B., Shen, Y.K., Aiden, A.P., Veres, A., Gray, M.K., Team, T.G.B., Pickett, J.P., Holberg, D., Clancy, D., Norvig, P., Orwant, J., Pinker, S., Nowak, M.A., Aiden, E.L.: Quantitative analysis of culture using millions of digitized books. Science (2010)

48. Mintz, M., Bills, S., Snow, R., Jurafsky, D.: Distant supervision for relation extraction without labeled data. In: Proceedings of the Joint Conference of the 47th Annual Meeting of the ACL and the 4th International Joint Conference on Natural Language Processing of the AFNLP: Volume 2 - Volume 2, ACL '09, pp. 1003–1011. Association for Computational Linguistics (2009)

49. Mitchell, T., Cohen, W., Hruschka, E., Talukdar, P., Betteridge, J., Carlson, A., Dalvi, B., Gardner, M., Kisiel, B., Krishnamurthy, J., Lao, N., Mazaitis, K., Mohamed, T., Nakashole, N., Platanios, E., Ritter, A., Samadi, M., Settles, B., Wang, R., Wijaya, D., Gupta, A., Chen, X., Saparov, A., Greaves, M., Welling, J.: Never-ending learning. In: Proceedings of the Twenty-Ninth AAAI Conference on Artificial Intelligence (AAAI-15) (2015)

50. Mohapatra, H., Jain, S., Chakrabarti, S.: Joint bootstrapping of corpus annotations and entity types. In: Proceedings of the 2013 Conference on Empirical Methods in Natural Language Processing, EMNLP '13, pp. 436–446. Association for Computational Linguistics (2013)

51. Nakashole, N., Mitchell, T.M.: Language-aware truth assessment of fact candidates. In: Proceedings of the 52nd Annual Meeting of the Association for Computational Linguistics, ACL '14, pp. 1009–1019 (2014)

52. Nakashole, N., Tylenda, T., Weikum, G.: Fine-grained semantic typing of emerging entities. In: 51st Annual Meeting of the Association for Computational Linguistics, ACL '13, pp. 1488–1497. ACL (2013)

53. Nakashole, N., Weikum, G., Suchanek, F.: PATTY: a taxonomy of relational patterns with semantic types. In: Proceedings of the 2012 Joint Conference on Empirical Methods in Natural Language Processing and Computational Natural Language Learning, EMNLP-CoNLL '12, pp. 1135–1145. Association for Computational Linguistics (2012)

54. Nastase, V., Strube, M.: Decoding Wikipedia categories for knowledge acquisition. In: Proceedings of the 23rd National Conference on Artificial Intelligence - Volume 2, AAAI'08, pp. 1219–1224. AAAI Press (2008)
55. Nickel, M., Murphy, K., Tresp, V., Gabrilovich, E.: A review of relational machine learning for knowledge graphs. Proceedings of the IEEE **104**(1), 11–33 (2016). doi: 10.1109/JPROC.2015.2483592
56. Paşca, M.: Organizing and searching the World Wide Web of facts – step two: Harnessing the wisdom of the crowds. In: Proceedings of the 16th International Conference on World Wide Web, WWW '07, pp. 101–110. ACM (2007a). doi: 10.1145/1242572.1242587
57. Paşca, M.: Weakly-supervised discovery of named entities using web search queries. In: Proceedings of the 16th ACM conference on Conference on information and knowledge management, CIKM '07, pp. 683–690. ACM (2007b). doi: 10.1145/1321440.1321536
58. Paşca, M., Van Durme, B.: What you seek is what you get: Extraction of class attributes from query logs. In: Proceedings of the 20th International Joint Conference on Artificial Intelligence, IJCAI'07, pp. 2832–2837. Morgan Kaufmann Publishers Inc. (2007)
59. Pantel, P., Crestan, E., Borkovsky, A., Popescu, A.M., Vyas, V.: Web-scale distributional similarity and entity set expansion. In: Proceedings of the 2009 Conference on Empirical Methods in Natural Language Processing: Volume 2 - Volume 2, EMNLP '09, pp. 938–947. Association for Computational Linguistics (2009)
60. Paulheim, H., Bizer, C.: Type inference on noisy RDF data. In: Proceedings of the 12th International Semantic Web Conference - Part I, ISWC '13, pp. 510–525. Springer (2013). doi: 10.1007/978-3-642-41335-3_32
61. Pennacchiotti, M., Pantel, P.: Entity extraction via ensemble semantics. In: Proceedings of the 2009 Conference on Empirical Methods in Natural Language Processing: Volume 1 - Volume 1, EMNLP '09, pp. 238–247. Association for Computational Linguistics (2009)
62. Preda, N., Kasneci, G., Suchanek, F.M., Neumann, T., Yuan, W., Weikum, G.: Active knowledge: Dynamically enriching RDF knowledge bases by web services. In: Proceedings of the 2010 ACM SIGMOD International Conference on Management of Data, SIGMOD '10, pp. 399–410. ACM (2010). doi: 10.1145/1807167.1807212
63. Reinanda, R., Meij, E., de Rijke, M.: Document filtering for long-tail entities. In: Proceedings of the 25th ACM International on Conference on Information and Knowledge Management, CIKM '16, pp. 771–780. ACM (2016). doi: 10.1145/2983323.2983728
64. Reisinger, J., Paşca, M.: Latent variable models of concept-attribute attachment. In: Proceedings of the Joint Conference of the 47th Annual Meeting of the ACL and the 4th International Joint Conference on Natural Language Processing of the AFNLP: Volume 2 - Volume 2, ACL '09, pp. 620–628. Association for Computational Linguistics (2009)
65. Riedel, S., Yao, L., McCallum, A.: Modeling relations and their mentions without labeled text. In: Proceedings of the 2010 European Conference on Machine Learning and Knowledge Discovery in Databases: Part III, ECML PKDD'10, pp. 148–163. Springer (2010)
66. Robertson, S., Soboroff, I.: The TREC 2002 Filtering track report. In: Proceedings of the Eleventh Text Retrieval Conference, TREC '02 (2002)
67. Roth, B., Barth, T., Wiegand, M., Singh, M., Klakow, D.: Effective slot filling based on shallow distant supervision methods. In: Text Analysis Conference, TAC '13. NIST (2013)
68. Sarawagi, S.: Information extraction. Foundations and Trends in Databases **1**(3), 261–377 (2007). doi: 10.1561/1900000003
69. Sekine, S., Nobata, C.: Definition, dictionaries and tagger for extended named entity hierarchy. In: Proceedings of the Fourth International Conference on Language Resources and Evaluation, LREC '04. ELRA (2004)
70. Shin, J., Wu, S., Wang, F., De Sa, C., Zhang, C., Ré, C.: Incremental knowledge base construction using deepdive. Proc. VLDB Endow. **8**(11), 1310–1321 (2015). doi: 10.14778/2809974.2809991
71. Suchanek, F.M., Sozio, M., Weikum, G.: SOFIE: a self-organizing framework for information extraction. In: Proceedings of the 18th International Conference on World Wide Web, WWW '09, pp. 631–640. ACM (2009). doi: 10.1145/1526709.1526794

72. Sun, A., Grishman, R., Xu, W., Min, B.: New York University 2011 system for KBP slot filling. In: Text Analysis Conference, TAC '11. NIST (2011)
73. Surdeanu, M., Gupta, S., Bauer, J., McClosky, D., Chang, A.X., Spitkovsky, V.I., Manning, C.D.: Stanford's distantly-supervised slot-filling system. In: Text Analysis Conference, TAC '11. NIST (2011)
74. Surdeanu, M., Ji, H.: Overview of the English Slot Filling track at the TAC2014 Knowledge Base Population evaluation. In: Text Analysis Conference, TAC '14. NIST (2014)
75. Surdeanu, M., McClosky, D., Tibshirani, J., Bauer, J., Chang, A.X., Spitkovsky, V.I., Manning, C.D.: A simple distant supervision approach for the TAC-KBP slot filling task. In: Text Analysis Conference, TAC '10. NIST (2010)
76. Surdeanu, M., Tibshirani, J., Nallapati, R., Manning, C.D.: Multi-instance multi-label learning for relation extraction. In: Proceedings of the 2012 Joint Conference on Empirical Methods in Natural Language Processing and Computational Natural Language Learning, EMNLP-CoNLL '12, pp. 455–465. Association for Computational Linguistics (2012)
77. Talukdar, P.P., Pereira, F.: Experiments in graph-based semi-supervised learning methods for class-instance acquisition. In: Proceedings of the 48th Annual Meeting of the Association for Computational Linguistics, ACL '10, pp. 1473–1481. Association for Computational Linguistics (2010)
78. Tokunaga, K., Kazama, J., Torisawa, K.: Automatic discovery of attribute words from web documents. In: Proceedings of the Second International Joint Conference on Natural Language Processing, IJCNLP '05, pp. 106–118 (2005). doi: 10.1007/11562214_10
79. Van Durme, B., Qian, T., Schubert, L.: Class-driven attribute extraction. In: Proceedings of the 22nd International Conference on Computational Linguistics - Volume 1, COLING '08, pp. 921–928. Association for Computational Linguistics (2008)
80. Vyas, V., Pantel, P.: Semi-automatic entity set refinement. In: Proceedings of Human Language Technologies: The 2009 Annual Conference of the North American Chapter of the Association for Computational Linguistics, NAACL '09, pp. 290–298. Association for Computational Linguistics (2009)
81. Vyas, V., Pantel, P., Crestan, E.: Helping editors choose better seed sets for entity set expansion. In: Proceedings of the 18th ACM Conference on Information and Knowledge Management, CIKM '09, pp. 225–234. ACM (2009). doi: 10.1145/1645953.1645984
82. Wang, C., Chakrabarti, K., He, Y., Ganjam, K., Chen, Z., Bernstein, P.A.: Concept expansion using web tables. In: Proceedings of the 24th International Conference on World Wide Web, WWW '15, pp. 1198–1208. International World Wide Web Conferences Steering Committee (2015a). doi: 10.1145/2736277.2741644
83. Wang, J., Song, D., Liao, L., Lin, C.Y.: BIT and MSRA at TREC KBA CCR track 2013. In: Proceedings of The Twenty-Second Text REtrieval Conference, TREC '13 (2013)
84. Wang, J., Song, D., Wang, Q., Zhang, Z., Si, L., Liao, L., Lin, C.Y.: An entity class-dependent discriminative mixture model for cumulative citation recommendation. In: Proceedings of the 38th International ACM SIGIR Conference on Research and Development in Information Retrieval, SIGIR '15, pp. 635–644. ACM (2015b). doi: 10.1145/2766462.2767698
85. Wang, R.C., Cohen, W.W.: Iterative set expansion of named entities using the web. In: Proceedings of the 2008 Eighth IEEE International Conference on Data Mining, ICDM '08, pp. 1091–1096. IEEE Computer Society (2008). doi: 10.1109/ICDM.2008.145
86. Wang, R.C., Cohen, W.W.: Automatic set instance extraction using the web. In: Proceedings of the Joint Conference of the 47th Annual Meeting of the ACL and the 4th International Joint Conference on Natural Language Processing of the AFNLP: Volume 1 - Volume 1, ACL '09, pp. 441–449. Association for Computational Linguistics (2009)
87. Weikum, G., Hoffart, J., Suchanek, F.: Ten years of knowledge harvesting: Lessons and challenges. IEEE Data Eng. Bull. **39**(3), 41–50 (2016)
88. West, R., Gabrilovich, E., Murphy, K., Sun, S., Gupta, R., Lin, D.: Knowledge base completion via search-based question answering. In: Proceedings of the 23rd International Conference on World Wide Web, WWW '14, pp. 515–526. ACM (2014). doi: 10.1145/2566486.2568032

89. Wu, F., Weld, D.S.: Autonomously semantifying Wikipedia. In: Proceedings of the Sixteenth ACM Conference on Conference on Information and Knowledge Management, CIKM '07, pp. 41–50. ACM (2007). doi: 10.1145/1321440.1321449
90. Wu, F., Weld, D.S.: Open information extraction using Wikipedia. In: Proceedings of the 48th Annual Meeting of the Association for Computational Linguistics, ACL '10, pp. 118–127. Association for Computational Linguistics (2010)
91. Yakout, M., Ganjam, K., Chakrabarti, K., Chaudhuri, S.: Infogather: Entity augmentation and attribute discovery by holistic matching with web tables. In: Proceedings of the 2012 ACM SIGMOD International Conference on Management of Data, SIGMOD '12, pp. 97–108. ACM (2012). doi: 10.1145/2213836.2213848
92. Yosef, M.A., Bauer, S., Spaniol, J.H.M., Weikum, G.: HYENA: Hierarchical type classification for entity names. In: Proceedings of COLING 2012, pp. 1361–1370 (2012)
93. Yu, D., Li, H., Cassidy, T., Li, Q., Huang, H., Chen, Z., Ji, H., Zhang, Y., Roth, D.: RPI-BLENDER TAC-KBP2013 knowledge base population system. In: Text Analysis Conference, TAC '13. NIST (2013)
94. Zaveri, A., Rula, A., Maurino, A., Pietrobon, R., Lehmann, J., Auer, S.: Quality assessment for linked data: A survey. Semantic Web 7(1), 63–93 (2016). doi: 10.3233/SW-150175
95. Zhao, S., Grishman, R.: Extracting relations with integrated information using kernel methods. In: Proceedings of the 43rd Annual Meeting on Association for Computational Linguistics, ACL '05, pp. 419–426. Association for Computational Linguistics (2005). doi: 10.3115/1219840.1219892
96. Zhou, G., Su, J., Zhang, J., Zhang, M.: Exploring various knowledge in relation extraction. In: Proceedings of the 43rd Annual Meeting on Association for Computational Linguistics, ACL '05, pp. 427–434. Association for Computational Linguistics (2005). doi: 10.3115/1219840.1219893
97. Zhou, M., Chang, K.C.C.: Entity-centric document filtering: Boosting feature mapping through meta-features. In: Proceedings of the 22nd ACM International Conference on Information & Knowledge Management, CIKM '13, pp. 119–128. ACM (2013). doi: 10.1145/2505515.2505683

Part III
Semantic Search

Semantic search is not a single method or approach but rather a collection of techniques that, in combination, enable search engines to understand the concepts, meaning, and intent behind the query that the user enters into the search box and provide not only more accurate search results but also an overall improved search experience. In keeping with our orientation toward entities, we will discuss three specific themes under this broad umbrella. Chapter 7 is centered on the topic of understanding the user's information need. After all, it is quite hard to satisfy the user unless we know what she is looking for. Next, in Chap. 8, we leverage entities to improve on the classic task of ad hoc document retrieval, by going beyond lexical term matching. Finally, in Chap. 9, we look at techniques that facilitate better interaction between the user and the search system, by aiding the user in expressing her information need, presenting rich search results, and suggesting related content for further exploration. By addressing these broad themes with specific solutions based on entities, we show how the concept of entities and the associated techniques can enhance all user-facing elements of search engines.

Chapter 7
Understanding Information Needs

Understanding what the user is looking for is at the heart of delivering a quality search experience. After all, it is rather difficult to serve good results, unless we can comprehend the intent and meaning behind the user's query. *Query understanding* is the first step that takes place before the scoring of results. Its overall aim is to infer a semantically enriched representation of the information need. This involves, among others, classifying the query according to higher-level goals or intent, segmenting it into parts that belong together, interpreting the query structure, recognizing and disambiguating the mentioned entities, and determining if specific services or verticals[1] should be invoked. Such semantic analysis of queries has been a long-standing research area in information retrieval. In Sect. 7.1, we give a brief overview of IR approaches to query understanding.

In the rest of the chapter, we direct our focus of attention to representing information needs with the help of structured knowledge repositories. The catchphrase "things, not strings" was coined by Google when introducing their Knowledge Graph.[2] It aptly describes the current chapter's focus: Capturing what the query is about by automatically annotating it with entries from a knowledge repository. These semantic annotations can then be utilized in downstream processing for result ranking (see Chap. 4) and/or result presentation. Specifically, in Sect. 7.2, we seek to identify the types or categories of entities that are targeted by the query. In Sect. 7.3, we perform entity linking in queries, which is about recognizing specific entity mentions and annotating them with unique identifiers from the underlying knowledge repository. Additionally, we consider the case of unresolvable ambiguity, when queries have multiple possible interpretations.

---

[1]A *vertical* is a specific segment of online content. Some of the most common verticals include shopping, travel, job search, the automotive industry, medical information, and scholarly literature.

[2]https://googleblog.blogspot.no/2012/05/introducing-knowledge-graph-things-not.html.

K. Balog, *Entity-Oriented Search*, The Information Retrieval Series 39, https://doi.org/10.1007/978-3-319-93935-3_7

Finally, in Sect. 7.4 we automatically generate query templates that can be used to determine what vertical services to invoke (e.g., weather, travel, or jobs) as well as the parameterization (attributes) of those services.

7.1 Semantic Query Analysis

The purpose of this section is to provide an overview of the range of tasks and techniques that have been proposed for *semantic query analysis*. These methods all aim to capture the underlying intent and meaning behind the user's query. Each of these techniques addresses query understanding from a specific angle, has its own particular uses, and is often complementary to the other means of query analysis. Query understanding is a vast area, one which would probably deserve a book on its own. In this chapter, we will discuss in detail only a selection of query analysis techniques, chosen for their relevance to an entity-oriented approach. In contrast, this section is meant to help see those methods in a broader perspective.

In particular, we will look at three groups of approaches.

- *Query classification* is the task of automatically assigning a query to one or multiple pre-defined categories.
- *Query annotation* is about generating semantic markup for a query.
- *Query interpretation* aims at determining the meaning of a query as a whole, by finding out how the segmented and annotated parts of the query relate to each other.

7.1.1 Query Classification

Query classification is the problem of automatically assigning a query to one or multiple pre-defined categories, based on its intent or its topic.

7.1.1.1 Query Intent Classification

The first group of techniques aims to classify queries into categories based on the underlying user intent. Jansen and Booth [38] define user intent as "the expression of an affective, cognitive, or situational goal in an interaction with a Web search engine." When we talk about user intent, we are more concerned with how the goal is expressed and what type of resources the user desires to fulfill her information need than with the goal itself (which is something "external") [38].

In his seminal work, Broder [16] introduced a taxonomy of query intents for web search, according to the following three main categories of user goals:

- *Navigational*, where the intent is to reach a particular site ("take me to X").
- *Informational*, where the intent is to acquire information about a certain topic ("find out more about X").
- *Transactional*, where the intent is to perform some web-mediated activity (purchase, sell, download, etc.).

Broder's categorization is broadly accepted and is the most commonly used one. Rose and Levinson [63] classify queries into *informational, navigational,* and *resource* categories, with further finer-grained subcategories for informational and resource queries. Jansen et al. [39] employ a three-level classification, based on [16] and [63], and elaborate on how to operationalize each category. Lee et al. [43] classify queries as navigational or informational using past user-click behavior and anchor-link distribution as features.

There are many other ways to categorize user intent. For example, Dai et al. [21] detect whether the query has a *commercial intent* by training a binary classifier, using the content of search engine result pages (SERPs) and of the top ranked pages. Ashkan and Clarke [4] use query-based features along with the content of SERPs to classify queries along two independent dimensions: commercial/noncommercial and navigational/informational. Zhou et al. [83] predict the *vertical intent* of queries, such as image, video, recipe, news, answer, scholar, etc.; this is also related to the problem of vertical selection in aggregated search [3]. There have been studies on specific verticals, e.g., identifying queries with a *question intent*, which can be successfully answered by a community question answering vertical [75], or determining *news intent*, in order to integrate content from a news vertical into web search results [22]. Yin and Shah [81] represent generic search intents, for a given type of entity (e.g., musicians or cities), as *intent phrases* and organize them in a tree. An intent phrase is a word or phrase that frequently co-occurs with entities of the given type in queries. Phrases that represent the same intent are grouped together in the same node of the tree (e.g., "songs" and "album" for musicians); sub-concepts of that intent are represented in child nodes (e.g., "song lyrics," "discography," or "hits"). Hu et al. [35] classify search intent by mapping the query onto Wikipedia articles and categories, using random walks and explicit semantic analysis [25].

7.1.1.2 Query Topic Classification

User goals may also be captured in terms of topical categories, which may be regarded as a multiclass categorization problem. The *query topic classification* task, however, is considerably more difficult than other text classification problems, due to data sparsity, i.e., the brevity of queries.

One landmark effort that inspired research in this area was the 2005 KDD Cup competition. It presented the Internet user search query categorization challenge: Classify 800,000 web queries into 67 predefined categories, organized in a two-level hierarchy. For each query, participants may return a set of up to five categories (the order is not important). Evaluation is performed in terms of precision, recall, and F1-score. Participants were given only a very small set of 111 queries with labeled categories, necessitating creative solutions. We refer to Li et al. [48] for an overview. In a follow-up study to their winning solution, Shen et al. [68] present a method that uses as intermediate taxonomy (the Open Directory Project, ODP[3]), as a bridge connecting the target taxonomy with the queries to be classified. Another key element of their solution is that they submit both search queries and category labels to a web search engine to retrieve documents, in order to obtain a richer representation for the queries. The same idea of issuing the given query against a web search engine and then classifying the highest scoring documents is employed by Gabrilovich et al. [24] for classifying queries onto a commercial taxonomy (which is intended for web advertising). The main differences with [68] is that they (1) build the query classifier directly for the target taxonomy, which is two magnitudes larger (approx. 6000 nodes), and (2) focus on rare ("tail") queries. Instead of viewing it as a text categorization problem, Ullegaddi and Varma [76] approach query topic classification as an IR problem and attempt to learn category rankings for a query, using a set of term-based features. They also use ODP as an intermediate taxonomy and utilize documents that are assigned to each category. Again, the query is submitted against a web search engine, then the top-k highest weighted terms are extracted from the highest ranked documents to enrich its representation.

A related task is that of classifying questions in community-based question answering sites; we refer to Srba and Bielikova [70, Sect. 5.3] for an overview. In Sect. 7.2, we discuss the problem of identifying the target types of entity-oriented queries, with reference to a given type taxonomy.

7.1.2   Query Annotation

Query annotation is an umbrella term covering various techniques designed to automatically generate semantic markup for search queries, which can contribute to a better understanding of what the query is about. Query annotation includes a number of tasks, such as phrase segmentation, part-of-speech and semantic tagging, named entity recognition, abbreviation disambiguation [79], and stopword and "stop structure" detection [37]. Many of these annotation tasks have been studied by the databases and natural language processing communities as well. Most approaches focus on a particular annotation task in isolation. However, given that these annotations are often related, it is also possible to obtain them jointly by combining several independent annotations [9, 29]. Below, we lay our attention on two of the most important and widely studied query annotation tasks: *segmentation* (structural annotations) and *tagging* (linguistic and semantic annotations). We shall discuss entity annotations of queries in detail in Sect. 7.3.

---

[3]http://dmoz.org/.

7.1.2.1   Query Segmentation

Query segmentation is the task of automatically grouping the terms of the query into phrases. More precisely, a segmentation s for the query q consists of a sequence of disjunct segments $\langle s_1, \ldots, s_n \rangle$, such that each s_i is a contiguous subsequence of q and the concatenation of $s_1, \ldots, s_n$ equals q. For example, the query "*new york travel guides*" may be segmented as "*[new york] [travel guides]*."

One of the first approaches to web query segmentation is by Risvik et al. [62], who segment queries based on *connexity*, which is defined as a product of the segment's frequency in a query log and the mutual information within the segment. Pointwise mutual information, as a measure of word association, has been used in many of the later studies as well, computed either on the term level [40] or on the phrase level [36, 71], and is commonly used as a baseline. Bergsma and Wang [11] employ a supervised learning approach, where a classification decision, whether to segment or not, is made between each pair of tokens in the query. They use three groups of features: context features (preceding and following tokens in the query, if available), dependency features (POS tags), and statistical features (frequency counts on the Web). Another important contribution of this work is a manually annotated gold standard corpus (Bergsma-Wang-Corpus, BWC) comprising 500 queries sampled from the AOL query log dataset. BWC has been used as a standard test collection for query segmentation in subsequent work [15, 30, 47, 71]. Rather than using a supervised approach that requires training data, Tan and Peng [71] suggest an unsupervised method that uses n-gram frequencies from a large web corpus as well as from Wikipedia. Many other works adopt a similar rationale for segmentation, e.g., Huang et al. [36] use web-scale n-gram language models, while Mishra et al. [56] exploit n-gram frequencies from a large query log. Hagen et al. [30] present a simple frequency-based method, relying only on web n-gram frequencies and Wikipedia titles, that achieves comparable performance to state-of-the-art approaches while being less complex and more robust. Further, Hagen et al. [30] enrich the Bergsma-Wang-Corpus by means of crowdsourcing; they also introduce the Webis Query Segmentation Corpus, which is a larger sample from the AOL query log, consisting of 50k queries. Finally, there is a line of work on segmenting queries based on retrieval results, the idea being that it is hard to make a segmentation decision based on the query terms alone. Instead, one can bootstrap segmentation decisions based on the document corpus, using the top retrieved results [8] or snippets [15]. Note that these methods involve an extra retrieval round for obtaining the segmentation.

Evaluating against manual annotations "implicitly assumes that a segmentation technique that scores better against human annotations will also automatically lead to better IR performance" [65]. Instead, Saha Roy et al. [65] propose a framework for extrinsic evaluation of query segmentation. This involves assessing how well the given segmentation performs on the end-to-end retrieval task, while treating the retrieval algorithm as a black box. Their results confirm that segmentation indeed benefits retrieval performance. Further, they find human annotations to be a good proxy, yet "human notions of query segments may not be the best for maximizing retrieval performance, and treating them as the gold standard limits the scope for improvement for an algorithm" [65].

7.1.2.2   Query Tagging

It is possible to obtain semantically more meaningful segments than what is yielded by query segmentation techniques. *Query tagging* (or *semantic tagging*) refers to the process of annotating query terms with labels from a predefined label set. *Part-of-speech (POS) tagging* is one of the basic techniques in natural language processing to capture the meaning of text. The task is to label each word with a tag that describes its grammatical role, such as noun (NN), verb (VB), adjective (JJ), etc. POS tags may have different granularity; it is possible, e.g., to distinguish between singular (NN) and plural (NNS) nouns or between regular (RB), comparative (RBR), and superlative (RBS) adverbs. On regular text, POS tagging can be performed with high accuracy; e.g., the Stanford POS tagger [74], one of the most widely used systems, achieves 97% accuracy on the Penn Treebank-3 corpus [52]. Applied to short queries, that tend to lack proper grammar, punctuation, or capitalization, existing NLP techniques are much less successful [6]. Based on a manually annotated sample of queries from a commercial search engine, Barr et al. [6] show that the distribution of POS tag types in queries is rather different from that of standard (edited and published) text. Specifically, they find that among keyword queries the "most common tag is the proper noun, which constitutes 40% of all query terms, and proper nouns and nouns together constitute 71% of query terms" [6], while many standard POS tags (e.g., verbs and determiners) seldom appear in web search queries. In query annotation, therefore, a great emphasis is placed on detecting noun phrases and entities. Barr et al. [6] further show that tagger performance is severely affected by the lack of capitalization in queries.

Bendersky et al. [8] mark up queries using three types of annotations: capitalization, POS tags, and segmentation indicators. Rather than relying on the query itself, they draw on the (latent) information need behind the query and leverage the document corpus using pseudo relevance feedback techniques. In follow-up work, instead of solving the above annotation tasks in isolation, Bendersky et al. [9] perform them jointly and leverage the dependencies between the different types of markup.

Another line of work focuses on the identification of "key concepts" (i.e., the most important noun phrases) in verbose natural language queries. By assigning higher weights to these key concepts during document scoring, one can attain better retrieval effectiveness. Bendersky and Croft [7] use the noun phrases in the query as candidate concepts and use a supervised machine learning approach to classify each as being a key concept or not, and set a concept's importance based on the classifier's confidence score. Features include corpus-based frequency statistics, computed from the document collection, from an external collection (Google n-grams [14]), and also from a large query log.

Query tagging may also be performed with respect to an underlying domain-specific schema. It is often assumed that queries have already been classified onto a given domain (like movies, books, products, etc.). The task then is to assign each query term a label indicating which field it belongs to [46, 51]. A great deal of attention has been directed toward the product search domain, "since this is one representative domain where structured information can have a substantial influence on search experience" [46] (not to mention the obvious commercial value). Nevertheless, the proposed methods should be applicable to other domains as well. In the product search domain, the set of fields comprises type, brand, model, attribute, etc. It is possible to construct field-specific lexicons from a knowledge repository that enumerate all possible values for each field. A simple lexicon-based approach, however, is insufficient due to ambiguity (e.g., "*iphone*" may refer to model or to attribute) and the presence of out-of-vocabulary terms [46]. Li et al. [46] approach query tagging as a sequential labeling task and employ semi-supervised conditional random fields (CRF). A small amount of hand-labeled queries are combined with a large amount of queries with derived labels, which are obtained in an unsupervised fashion (by leveraging search click-through data in conjunction with a product database). Manshadi and Li [51] present a hybrid method, which consists of a generative grammar model and parser, and a discriminative re-ranking module. Li [44] further distinguishes between the semantic roles of the constituents in noun phrase queries and makes a distinction between *intent heads* (corresponding to attribute names) and *intent modifiers* (referring to attribute values). For example, given the query "*alice in wonderland 2010 cast*," "*cast*" is the intent head, while "*alice in wonderland*" and "*2010*" are intent modifiers, with labels *title* and *year*, respectively. Li [44] uses CRF-based models with transition, lexical, semantic (lexicon-based), and syntactic (POS tags) features. Pound et al. [61] annotate queries with semantic constructs, such as entity, type, attribute, value, or relationship. This mapping is learned from an annotated query log, using a CRF model with part-of-speech tags as features.

7.1.3   Query Interpretation

Instead of labeling individual query terms (or sequences of terms), *query interpretation* (a.k.a. *semantic query understanding*) aims to determine the meaning of the query as a whole, by figuring out how the segmented and annotated query tokens relate to each other and together form an "executable" expression.

NLP approaches to question answering assume that the user's input is a grammatically well-formed question, from which a logical form (λ-calculus representation) can be inferred via *semantic parsing* [10, 80]. Viewing the knowledge base as a graph, these logical forms are tree-like graph patterns; executing them is equivalent to finding a matching subgraph of the knowledge graph.

In contrast, the database and IR communities generally operate with "telegraphic" queries. Due to the inherent ambiguity of such short keyword queries, systems typically need to evaluate multiple possible interpretations. In databases, the objective is to map queries to structured data tables and attributes; this task may also be referred to as *structured query annotation* [66]. For example, the query *"50 inch LG lcd tv"* could be mapped to the table *"TVs,"* with attributes *diagonal="50 inch," brand="LG,"* and *TV_type="lcd."* We note that this is very similar to the problem of query tagging, which we have discussed in the previous subsection [44, 46, 51]. Nevertheless, there are two main differences. First, query tagging involves supervised or semi-supervised learning (typically using a CRF-based model), while the works on query interpretation presented here strive for an unsupervised solution. Second, query tagging does not describe how the various recognized semantic constructs interact; capitalizing on that structured data is organized in tables, the methods here do not allow for arbitrary combinations of attributes. Commonly, a two-tier architecture is employed, consisting of an offline and an online component [26, 59, 61, 66]. A collection of structured query templates are generated offline by mining patterns automatically from query logs. In the online component, all plausible interpretations are generated for an incoming query, where an interpretation is represented as a semantically annotated query and a query template. These interpretations are then scored to determine a single one that most likely captures the user's intent.

Very similar techniques are used in web search for querying specific verticals, by identifying the domain (vertical) of interest and recognizing specific attributes; we discuss some of these techniques in Sect. 7.4. Generally speaking, compared to NLP and databases, the goals in IR approaches to entity retrieval are somewhat more modest. The structural interpretation of queries consists of identifying mentions of entities and target types, with respect to an underlying knowledge base; see Sects. 7.2 and 7.3. The remaining "un-mapped" query words are used for scoring the textual descriptions of candidate entities. Notably, instead of aiming for a single "best" query interpretation, the system ranks responses for each possible interpretation, and then takes a weighted combination of scores over all interpretations [33, 67].

7.2   Identifying Target Entity Types

One way to understand the meaning behind a search query is to identify the entity types that are targeted by the query. In other words, we wish to map the query to a small set of target types (or categories), which boils down to the task of estimating the relevance of each type with respect to the query. The top-ranked types can then be leveraged in the entity ranking model (as we have seen earlier in this book, in Sect. 4.3) or offered to the user as facets. The latter form of usage is very typical, among others, on e-commerce sites, for filtering the search results; see Fig. 7.1 for an illustrative example. We note that the methods presented in this section are not limited to type taxonomies of knowledge bases, but are applicable to other type categorization systems as well (e.g., product categories of an e-commerce site).

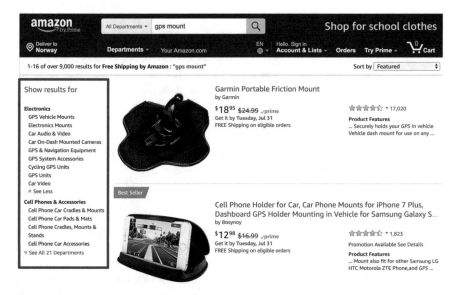

Fig. 7.1 Product categories displayed on Amazon in response to the query "gps mount"

7.2.1 Problem Definition

The problem of finding the target entity type(s) of the query can be formulated in different ways. The *entity type ranking* task, introduced by Vallet and Zaragoza [77], is as follows: Given an input query q, return a ranked list of entity types from a set $\mathcal{T}$ of possible types. Balog and Neumayer [5] investigate a variant of this task, called *hierarchical target type identification* (HTTI), which considers the hierarchical nature of entity type taxonomies. They aim to "find the single most specific type from an ontology [type taxonomy] that is general enough to cover all entities that are relevant to the query" [5]. Garigliotti et al. [27] refine this task definition (HTTIv2) to allow for a query to possibly have multiple "main" target types, provided they are sufficiently different, i.e., they lie on different branches in the type taxonomy. Furthermore, a query is also allowed to not have any target types, by assigning a special NIL-type. We adopt the revised task definition in [27], which is as follows:

Definition 7.1 *Target entity type identification* is the task of finding the target types of a given input query, from a type taxonomy, such that these types correspond to most specific types of entities that are relevant to the query. Target types cannot lie on the same branch in the taxonomy. If no matching entity type can be found in the taxonomy, then the query is assigned a special NIL-type.

Table 7.1 Notation used in Sect. 7.2

Symbol	Meaning
$c(t; x)$	Count of term t in x (x can be, e.g., an entity e or a type y)
e	Entity ($e \in \mathcal{E}$)
$\mathcal{E}$	Entity catalog (set of all entities)
$\mathcal{E}_q(k)$	Top-k ranked entities for query q
q	Query ($q = \langle q_1, \ldots, q_n \rangle$)
t	Term
$\mathcal{T}$	Type taxonomy
$\mathcal{T}_e$	Set of types assigned to entity e
y	Entity type ($y \in \mathcal{T}$)

Approached as a ranking problem, target entity type identification is the task of estimating the relevance of an entity type $y \in \mathcal{T}$ given a query q, expressed as $score(y; q)$. We present both unsupervised and supervised approaches in Sects. 7.2.2 and 7.2.3. Note that the NIL-type element of the problem is currently not considered. The different variants of the task definition essentially boil down to how the relevance assessments are obtained. We discuss this in Sect. 7.2.4. Table 7.1 summarizes the notation used in this section.

7.2.2   Unsupervised Approaches

According to one strategy, referred to as the *type-centric model* in [5], a term-based representation is built explicitly for each type; types can then be matched against the query using existing document retrieval models. Another strategy is to rank entities with respect to the query and then determine each type's relevance by considering the retrieval scores of entities of that type [41, 77]. Since types are not modeled directly, this approach is referred to as the *entity-centric model* [5].

An analogy can be drawn between these two approaches and the methods we have looked at earlier in the book for ranking entities. Specifically, the type-centric model corresponds to constructing term-based entity representations from documents mentioning those entities (Sect. 3.2), while the entity-centric model is akin to ranking entities without direct representations (Sect. 3.4). Essentially, one simply needs to replace entities with types and documents with entities in the equations. Zhang and Balog [82] further generalize these two retrieval strategies as design patterns for ranking arbitrary objects, called *early fusion* and *late fusion*, respectively.

7.2.2.1   Type-Centric Model

We construct a term-based representation ("type description") for each type, by aggregating descriptions of entities that are assigned that type. Then, those type representations can be ranked using conventional document retrieval methods. The term pseudo-counts for a type are computed using the following formula:

$$\tilde{c}(t; y) = \sum_{e \in \mathcal{E}} c(t; e)\, w(e, y)\,, \tag{7.1}$$

where $c(t; e)$ is the count of term t in the description of entity e and $w(e, y)$ denotes the entity-type association weight. This latter quantity may be interpreted as the importance of entity e for type y. A naïve but effective option is to set this weight uniformly across all entities that are typed with y:

$$w(e, y) = \begin{cases} \frac{1}{|\{e': y \in \mathcal{T}_{e'}\}|}, & y \in \mathcal{T}_e \\ 0, & y \notin \mathcal{T}_e\,, \end{cases} \tag{7.2}$$

where $\mathcal{T}_e$ denotes the set of types assigned to entity e. Substituting Eq. (7.2) back to Eq. (7.1) we get:

$$\tilde{c}(t; y) = \frac{1}{|\{e : y \in \mathcal{T}_e\}|} \sum_{e \in \mathcal{E}} \mathbb{1}(y \in \mathcal{T}_e)\, c(t; e)\,,$$

where $\mathbb{1}(y \in \mathcal{T}_e)$ returns 1 if entity e has y as one of its assigned types, otherwise returns 0. As the rewritten equation shows, this particular choice of entity-type association weighting merely serves as a normalization factor, so that types have comparable term pseudo-counts (representation lengths), irrespective of the number of entities that belong to them.

Given the term-based type representation, as defined by $\tilde{c}(t; y)$, types can be ranked using any standard retrieval method. Following [82], the final scoring may be formulated (for bag-of-words retrieval models) as:

$$score_{TC}(y; q) = \sum_{i=1}^{n} score(q_i; \tilde{c}, \varphi)\,, \tag{7.3}$$

where $score(q_i; \tilde{f}, \varphi)$ assigns a score to each query term q_i, based on the term pseudo-counts $\tilde{c}$, using some underlying term-based retrieval model (e.g., LM or BM25), which is parameterized by φ.

7.2.2.2 Entity-Centric Model

Instead of building a direct term-based representation of types, the *entity-centric model* works as follows. First, entities are ranked based on their relevance to the query. Then, the score for a given type y is computed by aggregating the relevance scores of the top-k entities with that type:

$$score_{EC}(y; q) = \sum_{e \in \mathcal{E}_q(k)} score(e; q)\, w(e, y)\,, \tag{7.4}$$

where $\mathcal{E}_q(k)$ denotes the set of top-k ranked entities for query q. The retrieval score of entity e is denoted by $score(q;e)$, which can be computed using any of the entity retrieval models discussed in Chaps. 3 and 4. The entity-type association weight, $w(e,y)$, is again as defined by Eq. (7.2). One advantage of this approach over the type-centric model is that it can directly benefit from improved entity retrieval. Also, out of the two unsupervised approaches, the entity-centric model is the more commonly used one [5, 41, 77].

7.2.3   Supervised Approach

Balog and Neumayer [5] observe that "the type-centric model tends to return more specific categories [types], whereas the entity-centric model rather assigns more general types." The complementary nature of these two approaches can be exploited by combining them. Additionally, one can also incorporate additional signals, including taxonomy-driven features [73] and various similarity measures between the type label and the query [78]. Table 7.2 presents a set of features for the target entity type identification task, compiled by Garigliotti et al. [27].

Here, we only detail two of the best performing features, both of which are based on distributional similarity between the type's label and the query. The first one, *simAggr*, considers the centroids of embedding vectors of terms in the type's label and terms in the query:

$$simAggr(y,q) = \cos(\mathbf{y}, \mathbf{q}) \,,$$

where $\mathbf{y}$ and $\mathbf{q}$ denote the respective centroid vectors.

The second one, *simMax*, takes the maximum pairwise similarity between terms in the type's label and in the query:

$$simMax(y,q) = \max_{t_y \in y, t_q \in q} \cos(\mathbf{t}_y, \mathbf{t}_q) \,,$$

where $\mathbf{t}_y$ and $\mathbf{t}_q$ are the embedding vectors corresponding to terms t_y and t_q, respectively. In both cases, pre-trained 300-dimensional Word2vec [55] vector embeddings were used. Further, the set of terms considered is limited to "content words" (i.e., nouns, adjectives, verbs, or adverbs) [27].

7.2.4   Evaluation

Next, we present evaluation measures and test collections for the target entity type identification task.

Table 7.2 Features for target entity type identification

Group	Feature	Description
Baseline features		
	$score_{TC}(y; q)$	Type-centric type score (cf. Eq. (7.3))[a]
	$score_{EC}(y; q)$	Entity-centric type score (cf. Eq. (7.4))[a,b]
Type taxonomy features		
	$depth(y)$	Hierarchical level of y, normalized by the taxonomy depth
	$children(y)$	Number of children of type y in the taxonomy
	$siblings(y)$	Number of siblings of type y in the taxonomy
	$numEntities(y)$	Number of entities in the KB with type y ($|\{e \in \mathcal{E} : y \in \mathcal{T}_e\}|$)
Type label features		
	l_y	Length of (the label of) type y in terms
	$sumIDF(y)$	Sum of IDF for terms in (the label of) type y ($\sum_{t \in y} IDF(t)$)
	$avgIDF(y)$	Avg. of IDF for terms in (the label of) type y ($\frac{1}{l_y} \sum_{t \in y} IDF(t)$)
	$sim_{JAC}(y, q)$	Jaccard similarity between terms of the type label and of the query
	$sim_{JACnouns}(y, q)$	Jaccard similarity between the type and the query, restricted to terms that are nouns
	$simAggr(y, q)$	Cosine sim. between the centroid embedding vectors of q and y
	$simMax(y, q)$	Max. cosine similarity of the embedding vectors between each pair of query and type label terms
	$simAvg(y, q)$	Avg. cosine similarity of the embedding vectors between each pair of query and type label terms

[a] Instantiated using both BM25 and LM as the underlying term-based retrieval model
[b] Instantiated with multiple cutoff thresholds $k \in \{5, 10, 20, 50, 100\}$

7.2.4.1 Evaluation Measures

Being a ranking task, type ranking is evaluated using standard rank-based measures. The ground truth annotations consist of a small set of types for each query, denoted as $\hat{\mathcal{T}}_q$. It might also be the case that a query has no target types ($\hat{\mathcal{T}}_q = \emptyset$); in that case the query might be annotated with a special NIL-type. Detecting NIL-types, however, is a separate task, which is still being researched, and which we do not deal with here. That is, we assume that $\hat{\mathcal{T}}_q \neq \emptyset$. The default setting is to take type relevance to be a binary decision; for each returned type y, it either matches one of the ground truth types (1) or not (0). The evaluation measures are mean average precision (MAP) and mean reciprocal rank (MRR). We shall refer to this as *strict* evaluation.

However, not all types of mistakes are equally bad. Imagine that the target entity type is *racing driver*. Then, returning a more specific type (*rally driver*) or a more general type (*athlete*) is less of a problem than returning a type from a completely different branch of the taxonomy (like *organization* or *location*). Balog and Neumayer [5] accommodate this by introducing *lenient* evaluation, where relevance is graded and near-misses are rewarded. Specifically, let $d(y, \hat{y})$ denote the distance between two types, the returned type y and the ground truth type $\hat{y}$, in the taxonomy. This distance is set to the number of steps between the two types, if they lie on the same branch (i.e., one of the types is a subtype of the other); otherwise, their distance is set to ∞. With the help of this distance function, the relevance level or gain of a type can be defined in a *linear* fashion:

$$r(y) = \max_{\hat{y} \in \hat{T}_q} \left(1 - \frac{d(y, \hat{y})}{h} \right),$$

where h is the depth of the type taxonomy. Notice that we consider the closest matching type from the set of ground types $\hat{T}_q$. Alternatively, distance may be turned into a relevance level using an *exponential* decay function:

$$r(y) = \max_{\hat{y} \in \hat{T}_q} \left(b^{-d(y, \hat{y})} \right),$$

where b is the base of the logarithm (set to 2 in [5]). If $d(y, \hat{y}) = \infty$, then the value of the exponential is taken to be 0. In the lenient evaluation mode, the final measure is normalized discounted cumulative gain (NDCG), using the above (linear or exponential) relevance gain values.

7.2.4.2 Test Collections

Balog and Neumayer [5] annotated 357 entity-oriented search queries with a single target type, using the DBpedia Ontology as the reference type taxonomy. These queries are essentially a subset of the ones in the DBpedia-Entity collection (cf. Sect. 3.5.2.7). According to their task definition, an "instance of" relation is required between the target type and relevant entities (as opposed to mere "relatedness," as in [77]). The guideline for the annotation process was to "pick a single type that is as specific as possible, yet general enough to cover all correct answers" [5]. For 33% of the queries, this was not possible because of one of the following three main reasons: (1) the query has multiple (legitimate) target types; e.g., "Ben Franklin" is a *Person* but may also refer to the ship (*MeanOfTransportation*) or the musical (*Work*); (2) there is no appropriate target type in the taxonomy for the given query; or (3) the intent of the query is not clear.

In follow-up work, Garigliotti et al. [27] annotated the complete set of queries from the DBpedia-Entity collection, in accordance with their revised task definition (cf. Sect. 7.2.1). Correspondingly, queries can have multiple target types or none (NIL-type option). The relevance assessments were obtained via crowdsourcing, using a newer (2015-10) version of the DBpedia Ontology as the type taxonomy. Around 58% of the queries in the collection have a single target type; the rest of the queries have multiple (mostly two or three) target types, including the NIL-type.

7.3 Entity Linking in Queries

Identifying entities in queries is another key technique that enables a better understanding of the underlying search intents. According to Guo et al. [28], over 70% of queries in web search contain named entities. The study by Lin et al. [49] reports a lower number, 43%, albeit using different annotation guidelines. The bottom line is that by being able to recognize entities in queries, the user experience can be improved for a significant portion of search requests (e.g., by enhanced result ranking or presentation). In Chap. 5, we have dealt in detail with the problem of *entity linking*, i.e., annotating documents with entities from a reference knowledge repository. Why can we not simply apply the same techniques to search queries? The reasons are at least threefold.

- One challenge is that queries are very short, typically consisting only of a few terms, and lack proper spelling and grammar. There is experimental evidence showing that methods perform substantially worse on very short and poorly composed texts (tweets) than on longer documents (news) [18, 50]. What is more, even methods that are designed in particular for short text perform significantly worse on queries than those that are specifically devised for queries [19].
- Another fundamental difference is that when documents are annotated with entities, "it is implicitly assumed that the text provides enough context for each entity mention to be resolved unambiguously" [32]. For queries, on the other hand, there is only limited context, or none at all.[4] It may be impossible to annotate entity mentions unambiguously in the case of queries. That is, a given query segment can possibly be linked to more than a single entity, leading to multiple legitimate interpretations of the query.
- Finally, obtaining entity annotations for queries is an online process that needs to happen during query-time, under serious time constraints. This is unlike annotating documents, which is typically performed offline. Therefore, we are not necessarily looking for the most effective solution, but for "one that represents the best trade-off between effectiveness and efficiency" [34].

The task of annotating queries with entities has been studied in a number of different flavors; we start with presenting an overview of these in Sect. 7.3.1. Then, in Sect. 7.3.2, we introduce a unified pipeline approach. The two main components of this pipeline are detailed in Sects. 7.3.3 and 7.3.4. Table 7.3 shows the notation used in this section.

---

[4]Search history information may provide contextual anchoring; this, however, is often unavailable. For example, if it is the first query in a search session, with no information about the previous searches of the user (which corresponds to the commonly studied ad hoc search scenario), then the query text is all we have.

Table 7.3 Notation used in
Sect. 7.4

Symbol	Meaning
$\mathcal{A}$	Annotation (set of mention-entity pairs)
C	Candidate entity annotations (ordered list)
e	Entity ($e \in \mathcal{E}$)
$\mathcal{E}$	Entity catalog (set of all entities)
$\mathcal{I}$	Query interpretation ($\mathcal{I} = \{\mathcal{A}_1, \ldots, \mathcal{A}_n\}$)
l_x	Length of x (number of terms)
m	Entity mention
$\mathcal{M}_q$	Set of entity mentions identified in query q
q	Query
t	Term

7.3.1   Entity Annotation Tasks

We distinguish between three entity annotation tasks, formulated for queries rather than for regular text. Table 7.4 highlights the differences between them, along with some illustrative examples.

- *Named entity recognition* is the task of identifying mentions of named entities and tagging the mentions with their respective types.
- *Semantic linking* seeks to find a ranked list of entities that are semantically related to the query string.
- *Interpretation finding* aims to discover all plausible meanings of the query; each interpretation consists of a set of non-overlapping and semantically compatible entity mentions, linked to a knowledge repository.

7.3.1.1   Named Entity Recognition

The task of *named entity recognition in queries* (NERQ) is analogous to the problem of named entity recognition in text (NER, cf. Sect. 5.1.1), namely: Identify named entities in the query text and classify them with respect to a set of predefined types from a taxonomy. NERQ was introduced and first studied by Guo et al. [28], who employed topic modeling techniques with weak supervision (WS-LDA). Pantel et al. [58] expanded upon this work by incorporating latent user intents and click signals.

Table 7.4 Comparison of various entity annotation tasks for queries

	Named entity recognition	Semantic linking	Interpretation finding
Result format	Set/ranked list	Ranked list	Sets of sets
Explicit entity mentions?	Yes	No	Yes
Mentions can overlap	No	Yes	No[b]
Evaluation criteria	Recognized entities[a]	Relevant entities	Interpretations
Evaluation measures	Set/rank-based	Rank-based	Set-based
Examples			
"obama mother"	*"obama"*/PER	BARACK OBAMA	{{BARACK OBAMA}}
		ANN DUNHAM	
"new york pizza	*"new york"*/LOC	NEW YORK CITY	{{NEW YORK CITY,
manhattan"	*"manhattan"*/LOC	NEW YORK-STYLE PIZZA	MANHATTAN},
		MANHATTAN	{NEW YORK-STYLE PIZZA,
		MANHATTAN PIZZA	MANHATTAN}}
		...	

[a] Along with their respective types
[b] Not within the same interpretation

7.3.1.2 Semantic Linking

Semantic linking refers to the task of identifying entities "that are intended or implied by the user issuing the query" [53]. This problem was introduced as *query mapping* by Meij et al. [53] and is also known as *semantic mapping* [32] and as *(ranked) concepts to Wikipedia* [18]. As the name we have adopted suggests, we seek to find entities that are semantically related to the query. Notably, the entities to be linked are meant for human and not for machine consumption (e.g., to help users acquire contextual information or to provide them with valuable navigational suggestions [53]). Therefore, we are not so much interested in detecting the specific entity mentions in the query, nor do we require the returned entities to form a coherent set. Further, an entity may be semantically related (i.e., relevant) even if it is not explicitly mentioned in the query. Take, e.g., the query *"charlie sheen lohan,"* for which ANGER MANAGEMENT (TV SERIES) would be a relevant entity. Mind that this is different from the task of entity retrieval; our goal is not to answer the user's underlying information need with a ranked list of entities, but to identify entities that are referenced (either explicitly or implicitly) in the query.

7.3.1.3 Interpretation Finding

Interpretation finding is the query counterpart to entity disambiguation, where the inherent ambiguity of queries is addressed head-on. A query "can legitimately have more than one interpretation" [17], where an interpretation is a set of "non-overlapping linked entity mentions that are semantically compatible with the query

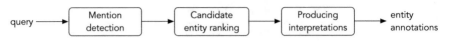

Fig. 7.2   Entity linking in queries pipeline

text" [17]. For example, the query *"new york pizza manhattan"* might be interpreted as the user wanting to eat a pizza in the MANHATTAN borough of NEW YORK CITY, or the user desiring a specific pizza flavor, NEW YORK-STYLE PIZZA, also in MANHATTAN. Interpretation finding aims at machine understanding of queries. The resulting annotations are utilized in the subsequent ranking process. A pioneering effort in this area was the Entity Recognition and Disambiguation (ERD) Challenge in 2014, organized by representatives of major web search engine companies [17]. Our ultimate interest in this section is on interpretation finding. In the next subsection, we shall present a pipeline architecture for addressing this task.

7.3.2   Pipeline Architecture for Interpretation Finding

We present a pipeline architecture for entity linking in queries, i.e., for the interpretation finding task, shown in Fig. 7.2. Notice that it is very similar to the one we employ for entity linking for documents (see Sect. 5.3). One important characteristic of this pipeline approach is it unifies the three tasks we discussed in the previous section under a common framework and shows how these tasks build on each other.

- The first step, *mention detection*, can be performed exactly the same way for queries as it is done for documents, i.e., using an extensive dictionary of entity surface forms; see Sect. 5.4.
- The *candidate entity ranking* step, as the name suggests, produces a ranking of candidate entities for the query. Specifically, given a set of mentions as input from the previous step, it emits a list of mention-entity pairs ordered by their degree of semantic relatedness to the query. Notice that this step directly translates to the task of *semantic linking*. One thing to point out here is that for the semantic linking task the actual mentions are ignored (i.e., for each entity only the highest scoring mention counts), while for interpretation finding the mentions also need to be passed along.
- Finally, *producing interpretations* is the query counterpart of the disambiguation component in conventional entity linking. The candidate entities identified in the previous step are used to form one or multiple query interpretations, where each interpretation consists of a set of semantically coherent entity linking decisions, with non-overlapping entity mentions.

In the next two subsections, we look at the candidate entity ranking and interpretation finding steps in detail. The details of mention detection are relatively
straightforward, as it is done analogously to entity linking in documents; see
Sect. 5.4.

7.3.3 Candidate Entity Ranking

Given a query q, the problem of *candidate entity ranking* is to return a ranked list
of entities $\langle e_1, \ldots, e_k \rangle$ from an entity catalog $\mathcal{E}$ that are semantically related to the
query. We shall assume that a set $\mathcal{M}_q$ of entity mentions has already been identified
in the query (see Sect. 5.4 for methods). For each mention $m \in \mathcal{M}_q$, let $\mathcal{E}_m$ denote
the set of candidate entities, i.e., entities that have a surface form matching m.
This candidate set may be further restricted to entities above a certain *commonness*
threshold (cf. Sect. 5.5). For example, in [32], a commonness threshold of 0.1 is
used. The goal, then, is to rank all candidate entities mentioned in q, $\mathcal{E}_q = \{e :
e \in \mathcal{E}_m, m \in \mathcal{M}_q\}$, based on $score(e; q, m)$, their semantic relatedness to the query.
Below, we present both unsupervised and supervised solutions for estimating this
score. Additionally, in Sect. 7.3.3.3, we introduce the "piggybacking" technique,
which is directed to alleviating the brevity of queries.

7.3.3.1 Unsupervised Approach

Hasibi et al. [32] propose to rank entities by combining term-based similarity score
with the commonness measure, using the following formula:

$$score(e; q, m) = P(e|q, m) \propto P(e|m) P(q|e) .$$

For the commonness computation, $P(e|m)$, we refer back to Eq. (5.3). While written
as a probability, the term-based similarity, $P(q|e)$, may in fact be computed using
any of the methods we presented in Chap. 3. What is important is that if the final
task is interpretation finding, these scores need to be comparable across queries.
One specific instantiation of this approach, referred to as *MLMcg* in [32], estimates
$P(q|e)$ using the query length normalized language model similarity [42]:

$$P(q|e) = \frac{\prod_{t \in q} P(t|\theta_e)^{P(t|q)}}{\prod_{t \in q} P(t|\mathcal{E})^{P(t|q)}} , \tag{7.5}$$

where $P(t|\theta_e)$ and $P(t|\mathcal{E})$ are the entity and collection language models, respec-
tively, computed using the mixture of language models (MLM) approach, cf.
Sect. 3.3.2.1. $P(t|q)$ is the term's relative frequency in the query $(c(t; q)/l_q)$.

Table 7.5 Features for candidate entity ranking

Group	Feature	Description
Mention		
	l_m	Length of mention m (number of terms)
	$|\mathcal{E}_m|$	Number of candidate entities for the mention
	$P(\text{link}|m)$	Link probability (cf. Eq. (5.2))
	$nameMatch(m)$	Number of entities with a surface form equal to mention m
	$partialMatch(m)$	Number of entities with a surface form partially matching m
Entity		
	$redirects(e)$	Number of Wikipedia redirect pages linking to the entity
	$links(e)$	Number of in/out-links of the entity in the knowledge graph
	$pageRank(e)$	PageRank of e in the knowledge graph
	$pageViews(e)$	Number of (Wikipedia) page views e received
Mention-entity		
	$P(e|m)$	Commonness (the probability of e being the link target of m)
	$contains(m, e)$	Whether the mention contains a surface form of the entity
	$contains(e, m)$	Whether a surface form of the entity contains the mention
	$equals(e, m)$	Whether a surface form of the entity equals the mention
	$editDist(m, e)$	Edit distance between the mention and the (best matching) surface form of the entity
	$firstPos(e, m)$	Position of the first occurrence of the mention in the entity's description in the knowledge repository
	$sim(m, f_e)$	Similarity between m and field f of the entity (cf. Sect. 3.3)
Query		
	$lenRatio(m, q)$	Mention to query length ratio (l_m / l_q)
	$contains(q, e)$	Whether the query contains a surface form of the entity
	$contains(e, q)$	Whether a surface form of the entity contains the query
	$equals(e, q)$	Whether a surface form of the entity equals the query
	$sim(q, e)$	Similarity between the query and entity (cf. Sect. 3.3)
	$sim(q, f_e)$	Similarity between q and field f of the entity (cf. Sect. 3.3)

7.3.3.2 Supervised Approach

Using a supervised learning approach, each (entity, query, mention) triple is described using a set of features. Table 7.5 displays a selection of features, assembled from the literature [19, 34, 54]. These are organized into four main groups:

- *Mention features* represent the characteristics of the specific mention.
- *Entity features* draw only from entity properties.
- *Mention-entity features* capture the binding between mentions and entities.
- *Query features* express query-mention and query-entity relationships.

We note that our assortment of features is by no means exhaustive. Also, we have limited ourselves to features that can be computed from publicly available resources; we refer to Blanco et al. [12] for additional query log-based features.

The supervised ranking model is trained on a set of labeled examples; for each mention and candidate entity pair, the target label is 1 if the entity is the correct link target of the given mention and is 0 otherwise.

7.3.3.3 Gathering Additional Context

One of the main challenges when annotating queries with entities is the lack of context. Meij et al. [53] develop features based on previous queries that the user issued in the same session. This method, however, is subject to the availability of session history. (Also, these features do not seem to make a significant contribution to the best results in [53].) Cornolti et al. [19] employ the so-called piggybacking technique (first introduced in [64]): Submitting the query to a web search engine API, and using the returned result snippets as additional context for entity disambiguation. The top-ranked result snippets are usually of very high quality, thanks to "sophisticated algorithms that leverage huge indexed document collections (the whole web), link graphs, and log analysis" [19]. The piggybacking technique has an additional benefit of being able to automatically correct spelling errors, by accessing the spelling correction feature of web search APIs. On the downside, it should be pointed out that the reliance on an external search service can seriously hinder the efficiency of the annotation process. Furthermore, there is no control over the underlying document ranking algorithm (which may change without notice).

Specifically, Cornolti et al. [19] retrieve the top 25 snippets using the original query. In addition to that, they also concatenate the original query with the string "*wikipedia*" and take the first 10 snippets, with the intention to boost results from Wikipedia. These search engine results are then used both for identifying candidate entities and for scoring them. Candidate entities are recognized in two ways:

1. Wikipedia articles occurring in the top-ranked results can be directly mapped to entities.
2. Annotating the result snippets with entities (using an entity linker designed for short text; Cornolti et al. [19] use WAT [60]) and keeping only those annotations that overlap with bold-highlighted substrings (reflecting query term matches) in the snippets.

In the candidate entity scoring phase, this information is utilized via various features, including the frequency and rank positions of mentions in snippets.

7.3.3.4 Evaluation and Test Collections

As we have explained earlier, the candidate entity ranking component of our pipeline corresponds to the semantic linking task. It is evaluated as a ranking problem, using standard rank-based measures, such as (mean) average precision and (mean) reciprocal rank. For each entity, only the highest scoring mention is considered:

$$score(e;q) = \arg\max_{m \in \mathcal{M}_q} score(e;q,m) .$$

Table 7.6 Test collections for evaluating semantic linking

Name	Reference KR	#Queries total	#Queries ≥1 annot.	Annotations can overlap	Sessions considered
YSQLE[a]	Wikipedia	2653	2583	Yes	Yes
GERDAQ [19][b]	Wikipedia	1000	889	No	No

[a] Yahoo! Webscope L24, https://webscope.sandbox.yahoo.com/
[b] http://acube.di.unipi.it/datasets/

At the time of writing, two publicly available test collections exist, which are summarized in Table 7.6.

Yahoo Search Query Log to Entities (YSQLE)   The dataset consists of 2635 web search queries, out of which 2583 are manually annotated with entities from Wikipedia. The annotations have been performed within the context of a search session (there are 980 sessions in total). Each linked entity is aligned with the specific mention (query span). Additionally, a single entity for each query may be labeled as "main," if it represents the main intent of the query. For example, the query *"France 1998 final"* is annotated with three entities: FRANCE NATIONAL FOOTBALL TEAM, FRANCE, and 1998 FIFA WORLD CUP FINAL, the last one being the main annotation.

GERDAQ   This collection, created by Cornolti et al. [19], consists of 1000 queries sampled from the KDD-Cup 2005 dataset [48]. The queries were annotated with Wikipedia entities via crowdsourcing, in two phases (first recall-oriented, then precision-oriented). The resulting dataset was then randomly split into training, development, and test sets, comprising 500, 250, and 250 queries, respectively. Each entity mention is linked to the highest scoring entity, according to the human annotators. Entity annotations do not overlap; entities below a given score threshold are discarded. On average, each query is annotated with two entities.

7.3.4   Producing Interpretations

In conventional entity linking, the generated annotations comprises a set of (semantically compatible) mention-entity pairs: $\mathcal{A} = \{(m_1, e_1), \ldots, (m_k, e_k)\}$, where e_i is the entity corresponding to mention m_i, and mention offsets must not overlap. In the context of queries, we shall refer to one such possible entity annotation as *interpretation*. Due to ambiguity, a query might have more than a single interpretation. Therefore, the objective of this component is to produce a set of query interpretations, $\mathcal{I} = \{\mathcal{A}_1, \ldots, \mathcal{A}_n\}$, where $\mathcal{A}_i$ is an interpretation.

We shall assume that all entity mentions in the query have been recognized and scored in a prior step (cf. Sect. 7.3.3). We shall refer to these as *candidate annotations*, and denote them as the list $C = \langle (m_1, e_1, s_1), \ldots, (m_k, e_k, s_k) \rangle$, where each annotation is a triple consisting of mention m_i, entity e_i, and score s_i. The list

---

Algorithm 7.1: Greedy interpretation finding [32]

Input: Candidate annotations C, score threshold τ
Output: Interpretations $\mathcal{I}$

```
1  C' ← prune(C, τ)
2  C' ← pruneContainmentMentions(C')
3  I ← createInterpretations(C')
4  return I
```

5 **Function** createInterpretations(C):
6 $\quad$ $\mathcal{I} \leftarrow \emptyset$
7 $\quad$ **for** $(m, e, s) \in C$ ordered by s **do**
8 $\quad\quad$ $h \leftarrow$ *False*
9 $\quad\quad$ **for** $\mathcal{A}_i \in \mathcal{I}$ **do**
10 $\quad\quad\quad$ **if** $\neg\, hasOverlap(m, \mathcal{A}_i)$ **then** /* Add to existing interpretation */
11 $\quad\quad\quad\quad$ $\mathcal{A}_i \leftarrow \mathcal{A}_i \cup \{(m, e)\}$
12 $\quad\quad\quad\quad$ $h \leftarrow$ *True*
13 $\quad\quad\quad$ **end**
14 $\quad\quad$ **end**
15 $\quad\quad$ **if** $\neg h$ **then** /* Create new interpretation */
16 $\quad\quad\quad$ $\mathcal{I} \leftarrow \mathcal{I} \cup \{(m, e)\}$
17 $\quad\quad$ **end**
18 $\quad$ **end**
19 $\quad$ **return** $\mathcal{I}$

C is ordered by decreasing score. Further, C may be truncated to a certain number of elements (top-k) or to annotations above a minimum score threshold. The task, then, is to form the set of interpretations $\mathcal{I}$, given the candidate annotations C as input.

We note that if one is to find only a single most likely interpretation, that can be done using existing entity linking methods (esp. using those that have been developed for annotating short text, such as TAGME [23] or WAT [60]). Even studies that address entity linking in queries often resort to the simpler problem of finding a single interpretation [12, 19]. Specifically, the top interpretation may be created greedily, by adding candidate annotations in decreasing order of score, as long as (1) they do not overlap and (2) the scores are above a given threshold (SMASH-S [19]). Alternatively, entity annotations may be selected using dynamic programming, such that they maximize the overall query likelihood [12]. In both cases, entity linking decisions are made individually, independently of each other (other than the constraint of non-overlapping). Our interest here, nevertheless, is focused on finding multiple interpretations, and we present both unsupervised (Sect. 7.3.4.1) and supervised (7.3.4.2) approaches for that.

7.3.4.1 Unsupervised Approach

Hasibi et al. [32] present the *greedy interpretation finding* (GIF) algorithm, shown in Algorithm 7.1, which consists of three steps: (1) pruning, (2) containment mention filtering, and (3) set generation. In the first step, the algorithm takes all candidate annotations and discards those with a score below the threshold τ. The threshold parameter is used to control the balance between precision and recall. In the second step, containment mentions are filtered out, by keeping only the highest scoring one. For instance, *"kansas city mo," "kansas city, "* and *"kansas"* are containment mentions; only a single one of these three is kept. Finally, interpretations are built iteratively, by processing the filtered candidate annotations C' in decreasing order of score. A given mention-entity pair (m, e) is added to an existing interpretation $\mathcal{A}_i$, if m does not overlap with the mentions already in $\mathcal{A}_i$; in the case multiple such interpretations exist, the pair (m, e) will be added to all of them. Otherwise, a new interpretation is created from the mention-entity pair. The GIF algorithm, despite its simplicity, is shown to be very effective and is on par with considerably more complex systems [34]. Its performance, however, crucially depends on that of the preceding candidate entity ranking step.

7.3.4.2 Supervised Approach

The main idea behind the *collective disambiguation* approach is to carry out a joint analysis of groups of mention-entity pairs, rather than greedily selecting the highest scoring entity for each mention. This idea can be realized by considering multiple possible candidate interpretations of the query and then applying supervised learning to select the most likely one(s). If the goal is to find multiple interpretations, then it is cast as a binary classification problem. If the objective is to find only the single most likely interpretation, then it may be approached as a ranking (regression) task.

As we have pointed out in Sect. 5.6.2.3, the joint optimization of entity annotations in text is an NP-hard problem. However, since queries are typically short, it is possible to enumerate all sensible combinations of entities. Specifically, Hasibi et al. [34] consider only the top-k candidate annotations for forming interpretations. The value of k can be chosen empirically, based on the effectiveness of the underlying candidate entity ranking component. In [34] $k = 5$ is used; less effective candidate entity ranking approaches may be compensated for by choosing a larger k value. Let $\mathcal{A}_1, \ldots, \mathcal{A}_n$ denote the candidate interpretations, which are generated by enumerating all possible valid (non-overlapping) combinations of mention-entity pairs from the candidate annotations C. For a given interpretation $\mathcal{A}_i$, we let $\mathcal{E}_i$ be the set of linked entities in that interpretation: $\mathcal{E}_i = \{e : (m, e) \in \mathcal{A}_i\}$.

The key challenge, then, is to design features that can capture the coherence of multiple entity annotations in the query. Two main groups of features are employed in [19, 34]:

- *Entity features* express the binding between the query and the entity and depend only on the individual entity (and not on other entities mentioned in the query). The top block of Table 7.7 lists entity features that are used in [34]. We note that other features that have been introduced for the candidate entity ranking step may also be used here (see Table 7.5). These features are computed for each entity that

Table 7.7 Features for producing interpretations

Group	Feature	Description
Entity		
	$links(e)$	Number of out-links of the entity in the knowledge graph
	$P(e\|m)$	Commonness (the probability of e being the link target of m)
	$score(e;q)$	Score from the candidate entity ranking step
	$iRank(e,q)$	Inverse rank from the candidate entity ranking step ($1/rank(e,q)$)
	$sim(q,e)$	Similarity between the query and entity (cf. Sect. 3.3)
	$contextSim(q,e)$	Contextual similarity between the query and entity, where context is the "rest" of the query, without the entity mention
Interpretation		
	$\min(\mathcal{R}_i)$	Minimum relatedness among entities in $\mathcal{E}_i$
	$\max(\mathcal{R}_i)$	Maximum relatedness among entities in $\mathcal{E}_i$
	$P_{co}(\mathcal{E}_i)$	Co-occurrence probability of entities in a Web corpus (Eq. (7.6))
	$H(\mathcal{E}_i)$	Entropy of $\mathcal{E}_i$ (Eq. (7.7))
	$sim(q,\mathcal{E}_i)$	Similarity between the query and $\mathcal{E}_i$ (Eq. (7.8))
	$coverage(\mathcal{A}_i,q)$	Mention coverage (Eq. (7.9))

is part of a given candidate interpretation ($e \in \mathcal{E}_i$), then aggregated by taking the minimum, maximum, or average of the individual values. Thus, each of these features is computed three times, using the three different aggregators.

- *Interpretation features* aim to capture the coherence of the set of linked entities in a given (candidate) interpretation ($\mathcal{A}_i$). These are computed collectively for all linked entities in the interpretation ($\mathcal{E}_i$). A selection of interpretation features are listed in the bottom block of Table 7.7. We briefly explain them below.

  – For relatedness features, we let $\mathcal{R}_i$ be the set of all pairwise relatedness scores between all entities in $\mathcal{E}_i$:

$$\mathcal{R}_i = \{rel(e_k,e_l) : e_k,e_l \in \mathcal{E}_i, e_k \neq e_l\},$$

  where $rel(e_k,e_l)$ is a measure of entity-relatedness; commonly WLM relatedness or Jaccard similarity is used (see Sect. 5.6.1.3 for other options). We take the minimum and the maximum of the values in $\mathcal{R}_i$ as two features.
  – The co-occurrence probability of all the entities in $\mathcal{E}_i$ may be estimated using a large web corpus:

$$P_{co}(\mathcal{E}_i) = \frac{|\bigcap_{e \in \mathcal{E}_i} \mathcal{D}_e|}{|\mathcal{D}|}, \qquad (7.6)$$

  where $\mathcal{D}_e$ is the set of documents in which e occurs, and $|\mathcal{D}|$ is the total number of documents in the corpus. With the help of this probability, the *entropy* of $\mathcal{E}_i$ may be calculated as:

$$H(\mathcal{E}_i) = -P_{co}(\mathcal{E}_i)\log P_{co}(\mathcal{E}_i) - (1 - P_{co}(\mathcal{E}_i))\log(1 - P_{co}(\mathcal{E}_i)). \qquad (7.7)$$

- The similarity between $\mathcal{E}_i$ and the query is calculated similarly to Eq. (7.5), but using a language model of all entities in the interpretation, instead of that of a single entity:

$$P(q|\mathcal{E}_i) = \frac{\prod_{t \in q} P(t|\theta_{\mathcal{E}_i})^{P(t|q)}}{\prod_{t \in q} P(t|\mathcal{E})^{P(t|q)}} , \qquad (7.8)$$

where the interpretation language model is estimated according to:

$$P(t|\theta_{\mathcal{E}_i}) = \frac{1}{|\mathcal{E}_i|} \sum_{e \in \mathcal{E}_i} P(t|\theta_e) .$$

- Mention coverage is the ratio of the query that is annotated, i.e., the length of all entity mentions over the length of the query:

$$coverage(\mathcal{A}_i, q) = \frac{\sum_{(m,e) \in \mathcal{A}_i} l_m}{l_q} , \qquad (7.9)$$

where l_m and l_q denote mention and query length, respectively.

7.3.4.3 Evaluation Measures

Let $\hat{\mathcal{I}} = \{\hat{\mathcal{A}}_1, \ldots, \hat{\mathcal{A}}_m\}$ denote the set of interpretations of query q according to the ground truth, and let $\mathcal{I} = \{\mathcal{A}_1, \ldots, \mathcal{A}_n\}$ denote the system-generated interpretations. For comparing these two sets, Carmel et al. [17] define *precision* and *recall* as:

$$P = \frac{|\mathcal{I} \cap \hat{\mathcal{I}}|}{|\mathcal{I}|} , \quad R = \frac{|\mathcal{I} \cap \hat{\mathcal{I}}|}{|\hat{\mathcal{I}}|} .$$

Hasibi et al. [32] point out that "according to this definition, if the query does not have any interpretations in the ground truth ($\hat{\mathcal{I}} = \emptyset$) then recall is undefined; similarly, if the system does not return any interpretations ($\mathcal{I} = \emptyset$), then precision is undefined." Therefore, they define precision and recall for *interpretation-based evaluation* as follows:

$$P_{int} = \begin{cases} |\mathcal{I} \cap \hat{\mathcal{I}}|/|\mathcal{I}|, & \mathcal{I} \neq \emptyset \\ 1, & \mathcal{I} = \emptyset, \hat{\mathcal{I}} = \emptyset \\ 0, & \mathcal{I} = \emptyset, \hat{\mathcal{I}} \neq \emptyset . \end{cases}$$

$$R_{int} = \begin{cases} |\mathcal{I} \cap \hat{\mathcal{I}}|/|\hat{\mathcal{I}}|, & \hat{\mathcal{I}} \neq \emptyset \\ 1, & \hat{\mathcal{I}} = \emptyset, \mathcal{I} = \emptyset \\ 0, & \hat{\mathcal{I}} = \emptyset, \mathcal{I} \neq \emptyset . \end{cases}$$

An interpretation is taken to be correct only if it matches all the entities of an interpretation in the ground truth exactly. For simplicity, the correctness of the mention offsets are not considered. Formally:

$$|\mathcal{I} \cap \hat{\mathcal{I}}| = \sum_{\mathcal{A}_i \in \mathcal{I}, \hat{\mathcal{A}}_j \in \hat{\mathcal{I}}} match(\mathcal{A}_i, \hat{\mathcal{A}}_j) ,$$

where

$$match(\mathcal{A}_i, \hat{\mathcal{A}}_j) = \begin{cases} 1, & \{e : e \in \mathcal{A}_i\} = \{e : e \in \hat{\mathcal{A}}_j\} \\ 0, & \text{otherwise} . \end{cases}$$

This evaluation is rather strict, as partial matches (for a given interpretation) are not given any credit. Alternatively, Hasibi et al. [32] propose a *lenient evaluation* that rewards partial matches. The idea is to combine interpretation-based evaluations (from above) with conventional entity linking evaluation, referred to as *entity-based evaluation*. Let $\mathcal{E}_\mathcal{I}$ denote the set of all entities from all interpretations returned by the entity linking system, $\mathcal{E}_\mathcal{I} = \cup_{i \in [1..n]} \{e : e \in \mathcal{A}_i\}$. Similarly, let the set $\hat{\mathcal{E}}_\mathcal{I}$ contain all entities from all interpretations in the ground truth, $\hat{\mathcal{E}}_\mathcal{I} = \cup_{j \in [1..m]} \{e : e \in \hat{\mathcal{A}}_j\}$. Then, precision and recall are defined as follows:

$$P_{ent} = \begin{cases} |\mathcal{E}_\mathcal{I} \cap \hat{\mathcal{E}}_\mathcal{I}| / |\mathcal{E}_\mathcal{I}|, & \mathcal{E}_\mathcal{I} \neq \emptyset \\ 1, & \mathcal{E}_\mathcal{I} = \emptyset, \hat{\mathcal{E}}_\mathcal{I} = \emptyset \\ 0, & \mathcal{E}_\mathcal{I} = \emptyset, \hat{\mathcal{E}}_\mathcal{I} \neq \emptyset . \end{cases}$$

$$R_{ent} = \begin{cases} |\mathcal{E}_\mathcal{I} \cap \hat{\mathcal{E}}_\mathcal{I}| / |\hat{\mathcal{E}}_\mathcal{I}|, & \hat{\mathcal{E}}_\mathcal{I} \neq \emptyset \\ 1, & \hat{\mathcal{E}}_\mathcal{I} = \emptyset, \mathcal{E}_\mathcal{I} = \emptyset \\ 0, & \hat{\mathcal{E}}_\mathcal{I} = \emptyset, \mathcal{E}_\mathcal{I} \neq \emptyset . \end{cases}$$

Finally, the overall precision and recall, in lenient evaluation, are defined as a linear combination of interpretation-based and entity-based precision and recall:

$$P = \frac{P_{int} + P_{ent}}{2} , \quad R = \frac{R_{int} + R_{ent}}{2} .$$

For simplicity, precision and recall are averaged with equal weights, but a weight parameter could also be introduced here. The F-measure (for any definition of precision and recall above) is computed according to Eq. (5.10).

So far, we have defined evaluation measures for a single query. For computing precision, recall, and F-measure over a set of queries, the (unweighed) average of the query-level scores is taken (i.e., macro-averaging is used).

Table 7.8 Test collections for interpretation finding

Name	Reference KR	#Queries total	#Queries ≥ 1 annot.	#Queries ≥ 1 interp.
ERD-dev [17]	Freebase	91	45	4
Y-ERD [32][a]	Freebase	2398	1256	9

[a] http://bit.ly/ictir2015-elq

7.3.4.4 Test Collections

There are two publicly available test collections for interpretation finding evaluation; see Table 7.8 for a summary.

ERD-dev The "short text" track of the Entity Recognition and Disambiguation (ERD) Challenge [17] provided a live evaluation platform for the interpretation finding task. A development set of 91 queries is made publicly available; of these, 45 queries have non-empty entity annotations. The test set comprises 500 queries; because of the live evaluation, the annotations for these queries are not available for traditional offline evaluation (we refer to Sect. 5.8.2.4 for a discussion on the live evaluation platform). The gold standard annotations are created manually, in accordance with the following three rules [17]: (1) for each entity, the longest mention is used, (2) only proper noun entities are annotated, and (3) overlapping mentions are not allowed within a single interpretation.

Y-ERD Hasibi et al. [32] created a larger test set based on the YSQLE collection (cf. Sect. 7.3.3.4). Following a set of guidelines, based on and expanding upon those of the ERD Challenge, they grouped independent entity annotations into semantically compatible sets of entity linking decisions, i.e., interpretations. Queries are annotated on their own, regardless of search sessions.

One observation that can be made from Table 7.8 is that very few queries actually have multiple interpretations. This explains why systems that returned only a single interpretation could end up being the best contenders at the ERD Challenge [17]. It remains an open question whether this is a limitation of currently available test collections, or if it is worth expending algorithmic effort toward finding multiple interpretations.

7.4 Query Templates

A large fraction of queries follow certain patterns. For instance, when people search for jobs, a frequently used query pattern is "*jobs in* ⟨*location*⟩," where ⟨location⟩ is a variable that can be instantiated, e.g., by a city ("*jobs in seattle*"), region ("*jobs in silicon valley*"), or country ("*jobs in the UK*"). From these patterns, templates may be inferred, which may be used for interpreting queries. As defined by Bortnikov et al. [13], "a template is a sequence of terms that are either text tokens or variables that can be substituted from some dictionary or taxonomy." To be consistent with

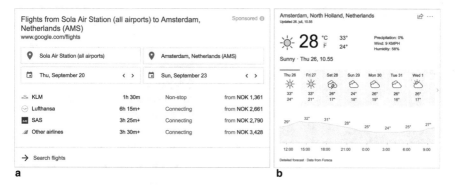

Fig. 7.3 (a) Flight search widget shown on Google in response to the query "*flights svg ams.*" (b) Weather widget on Bing for the query "*weather amsterdam*"

our earlier terminology we will use the expression *token*, which can be either a term (word) or an *attribute* (variable). In our examples, we mark attribute tokens as ⟨...⟩.

Templates provide structured interpretations of queries and have numerous advantages:

- Templates allow us not only to identify the target domain (vertical) of the query (such as flights or weather) but also to make parameterized requests to them, by mapping parts of the query to appropriate attributes of the given service. These services can then display various widgets or direct displays on the SERP, such as the ones shown in Fig. 7.3.
- Templates generalize well and can match queries that have not been observed in training data or search logs.
- Templates are very efficient in terms of online performance (only simple dictionary look-ups are required).

Such query templates may be crafted manually, e.g., by using regular expressions. Manual template building, however, has obvious limitations, due to the large variety of possible query formulations. A more scalable approach is to extract templates automatically from search logs. Based on Agarwal et al. [1], we formally introduce some concepts central to this problem in Sect. 7.4.1, followed by an explanation of methods that can be used for mining query templates from query logs in Sect. 7.4.2. Table 7.9 summarizes the notation used in this section.

7.4.1   Concepts and Definitions

Our objective is to direct queries to specific services or verticals, which will be referred to as *domains*. We begin by characterizing the schema of a given domain.

Definition 7.2 (Domain Schema) The schema of a given domain D is a pair $S_D = (\mathcal{A}, \mathcal{W})$, where $\mathcal{A} = \{a_1, \ldots, a_n\}$ is a set of *attributes* and $\mathcal{W} =$

Table 7.9 Notation used in Sect. 7.4

Symbol	Meaning
$\mathcal{A}$	Set of attributes
L	Search log ($L = (\mathcal{Q}, \mathcal{S}, \mathcal{C})$)
D	Domain
q	Query ($q \in \mathcal{Q}$)
$\mathcal{Q}$	Set of queries in the search log
$\mathcal{Q}_0$	Set of seed domain queries
$\mathcal{Q}_D$	Domain queries (queries relevant for D)
$\mathcal{Q}_s$	Set of click-through queries of site s
$\mathcal{Q}_u$	Set of queries instantiated by template u
s	Site ($s \in \mathcal{S}$)
$\mathcal{S}$	Set of sites
S_D	Domain schema ($S_D = (\mathcal{A}, \mathcal{W})$)
u	Query template ($u \in \mathcal{U}$)
$\mathcal{U}$	Template universe
$\mathcal{U}_q$	Set of templates generated by query q
$\mathcal{V}$	Vocabulary of terms
$\mathcal{W}$	Vocabulary of possible attribute values

$\{\mathcal{W}(a_1), \ldots, \mathcal{W}(a_n)\}$ is the *vocabulary* of the possible instances (i.e., values) of each of the attributes.

For example, attributes in the *jobs* domain may include $\mathcal{A} = \{company, location, category\}$. The vocabulary of the *company* attribute includes names of all entities that appear as possible values for that attribute: $\mathcal{W}(company) = \{$"Apple", "Microsoft", "Audi",... $\}$. Some attributes, like *category*, may require their own domain-specific dictionary.

Definition 7.3 (Query Template) A query template u is a sequence of tokens $u = \langle u_1, \ldots, u_n \rangle$, where each token u_i is either a term or an attribute: $u_i \in \mathcal{V} \cup \mathcal{A}$, where $\mathcal{V}$ is a vocabulary of terms and $\mathcal{A}$ is the set of possible attributes. We further require that at least one of the template tokens is an attribute: $\exists\, u_i \in \mathcal{A}$.

For example, "*jobs in* $\langle location \rangle$," consists of two terms and an attribute. This template can instantiate different queries. The inverse operation is template generation: Given a query, what templates can be generated from it?

Definition 7.4 (Template Instantiation and Generation) Given a query template $u = \langle u_1, \ldots, u_n \rangle$ and a query $q = \langle q_1, \ldots, q_m \rangle$, the template u *instantiates* q, or, equivalently, query q *generates* u, with respect to the domain schema S_D, if $n = m$ and for each token position $i \in [1, n]$,

- if token i in the template is an attribute, then query term q_i matches one of the possible instances of that attribute: $u_i \in \mathcal{A} \implies q_i \in \mathcal{W}(u_i)$,
- otherwise (if token i in the template is a term), the query and template tokens are equal: $u_i \notin \mathcal{A} \implies q_i = u_i \in \mathcal{V}$.

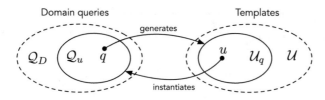

Fig. 7.4 Template generation and instantiation

The set of queries instantiated by template u is denoted by Q_u. The templates generated from query q are denoted by U_q. See Fig. 7.4 for a visual illustration.

Note that the vocabulary of a given attribute, according to the domain schema, may contain not only unigrams but n-grams as well (e.g., "new york"). When such an n-gram attribute instance is matched in the query, we treat it as a single query token.

To see an example of template generation, consider the query q = "*accounting jobs in new york*." According to our jobs domain schema, "*accounting*" is an instance of the *category* attribute and "*new york*" is an instance of the *location* attribute. This query can therefore generate the following three templates:

$$u_a = \text{``} \langle category \rangle \text{ } jobs \text{ } in \text{ } new \text{ } york, \text{''}$$

$$u_b = \text{``} accounting \text{ } jobs \text{ } in \text{ } \langle location \rangle, \text{''}$$

$$u_c = \text{``} \langle category \rangle \text{ } jobs \text{ } in \text{ } \langle location \rangle. \text{''}$$

Thus, $Q_q = \{u_a, u_b, u_c\}$.

7.4.2 Template Discovery Methods

Template discovery is the task of finding "good" templates (according to some quality measure, such as precision, recall, or F1-score) from a search log L, for a given domain D. Approaching this task as a ranking problem, the output is a ranked list of templates, sorted by a template score.

We shall assume that the search log $L = (Q, S, C)$ provides a set Q of queries with click-throughs C to a set S of sites. Specifically, let Q_s denote the click-through queries of site $s \in S$, i.e., $Q_s = \{q : q \in Q, clicks(q,s) > 0\}$, where $clicks(q,s)$ is the number of times a result originating from site s was clicked in response to q (over some time period). We shall further assume that we are given a seed set Q_0 of domain queries, i.e., queries that resulted in clicks on target pages.

Since our interest is in discovering templates for a given domain D, we shall generate templates that match *domain queries*, i.e., queries that are relevant for that domain. We write Q_D to denote the set of domain queries ($Q_D \subseteq Q$). Let U denote the template universe, i.e., templates that can be generated by at least one $q \in Q_D$. Formally: $U = \{u : \exists q \in Q_D, u \in Q_q\}$.

Algorithm 7.2: Classify & match [1]

Input: $L = (\mathcal{Q}, \mathcal{S}, \mathcal{C}), S_D, \mathcal{Q}_0, \tau$
Output: Templates u, ranked by $score(u)$

1 classify $\mathcal{Q}$ to $\mathcal{Q}_D$, trained on $\mathcal{Q}_0$, thresholded by τ
2 $\mathcal{U} \leftarrow \{u : u \in \mathcal{Q}_q, \forall q \in \mathcal{Q}_D\}$
3 **foreach** $u \in \mathcal{U}$ **do**
4 $\quad\mid$ compute $score(u)$
5 **end**
6 **return** $\mathcal{U}$ sorted by $score(u)$ /* P_u, R_u, or $F1_u$ */

7.4.2.1 Classify&Match

Agarwal et al. [1] introduce a natural baseline algorithm, called *Classify&Match*, which operates in two stages. First, domain queries $\mathcal{Q}_D$ are separated from all queries $\mathcal{Q}$ in the query log using automatic classification. Specifically, a query classifier is trained on the seed domain queries $\mathcal{Q}_0$ using the method from [45], with a threshold τ applied on the results. In the second stage, for each template $u \in \mathcal{U}$ is scored against the (estimated) set of domain queries $\mathcal{Q}_D$, using precision, recall, or F-measure as $score(u)$. See Algorithm 7.2.

Quality Measures To be able to measure the quality of a given template u, we establish *precision* and *recall*, analogously to the standard retrieval measures. The "target" set is the collection of domain queries, $\mathcal{Q}_D$. We shall assume that this set is clearly identified (e.g., by taking all queries that resulted in clicks on a set of target pages). The "matched" set is $\mathcal{Q}_u$, i.e., queries that are instantiated by the template. The *precision* of template u is the fraction of $\mathcal{Q}_u$ that falls within $\mathcal{Q}_D$:

$$P_u = \frac{\mathcal{Q}_D \cap \mathcal{Q}_u}{\mathcal{Q}_u} . \tag{7.10}$$

The *recall* of template u is the fraction of $\mathcal{Q}_D$ that is covered by $\mathcal{Q}_u$:

$$R_u = \frac{\mathcal{Q}_D \cap \mathcal{Q}_u}{\mathcal{Q}_D} . \tag{7.11}$$

Ultimately, the overall quality of a template needs to be measured by a combination of precision and recall, e.g., by using the F-measure ($F1_u$, cf. Eq. (5.10)).

7.4.2.2 QueST

While simple and intuitive, the above naïve baseline approach suffers from two shortcomings. First, queries are ambiguous and click-throughs are noisy by nature; deterministically separating domain and non-domain queries is problematic. Instead, probabilistic modeling is needed that can encapsulate the fuzziness

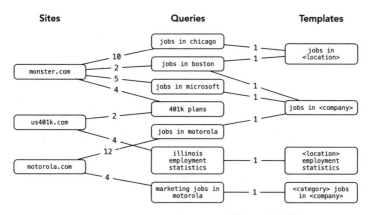

Fig. 7.5 QST graph based on a toy-sized search log. Example is taken from [1]

of domain relevance. Second, search logs are not only noisy but also sparse. By separating template mining into two separate stages, an important (indirect) connection between sites and templates is lost. To overcome these issues, Agarwal et al. [1] present an iterative inferencing framework, called *QueST*, over a tripartite graph of queries, sites, and templates. Precision and recall is defined for each type of node. The process of template discovery can then be seen as propagating precision and recall across the query-site-template graph using random walks. We shall elaborate on the details below.

QST Graph Queries, sites, and templates are represented as a tripartite graph (*QST graph*) $G_{QST} = (V, E)$, where the vertices are $V = \mathcal{Q} \cup \mathcal{S} \cup \mathcal{U}$, and there are two types of edges E, with $w(x, y)$ denoting the edge weight between nodes x and y:

- *Query-site edges*: For each query q that clicks to site s, there is an edge with the click-through frequency as weight: $\forall s, \forall q \in \mathcal{Q}_s : w(q, s) = clicks(q, s)$.
- *Query-template edges*: For each query q that instantiates template u, there is an edge in between with weight 1: $\forall u \in \mathcal{U}, \forall q \in \mathcal{Q}_u : w(q, u) = 1$.

Figure 7.5 displays the QST graph for a toy-sized example.

Probabilistic Modeling In practice, due to query ambiguity and noisy click-throughs, the crisp separation of queries into target domains is rather problematic (e.g., the query "*microsoft*" might be in *job* or in *product*). Therefore, we shall generalize the deterministic notions of precision and recall from the previous subsection to probabilistic measures. Let $match(q, x)$ denote the event that query q and x are semantically matching, where x may be a template u, a site s, or a domain D. Further, we let $P(match(q, x))$ be the probability of *semantic relevance* between q and x. Precision in Eq. (7.10) can then be rewritten in the "match" notation as the

---

Algorithm 7.3: QueST [1]

Input: $L = (\mathcal{Q}, \mathcal{S}, \mathcal{C})$, S_D, $\mathcal{Q}_0$, P_0, R_0
Output: Templates u, ranked by $score(u)$

1 $\mathcal{U} \leftarrow \{u : u \in \mathcal{Q}_q, \forall q \in \mathcal{Q}_D\}$
2 construct G_{QST} given $\mathcal{Q}, S, \mathcal{U}, \mathcal{C}$
3 $R_u, R_q, R_s \leftarrow$ QuestR on G_{QST} with R_0 /* inference recall */
4 $P_u, P_q, P_s \leftarrow$ QuestP on G_{QST} with P_0 /* inference precision */
5 **return** $\mathcal{U}$ sorted by $score(u)$ /* P_u, R_u, or Fl_u */

---

following conditional probability:

$$P_u = \frac{P\,(match(q,D), match(q,u))}{P\,(match(q,u))} = P\,(match(q,D)|match(q,u))\ .$$

Similarly, recall in Eq. (7.11) is rewritten as:

$$R_u = \frac{P\,(match(q,D), match(q,u))}{P\,(match(q,D))} = P\,(match(q,u)|match(q,D))\ .$$

Next, we extend the notions of precision and recall to sites and queries. This is needed for being able to perform integrated inferencing on queries, sites, and templates. The precision and recall of a site is modeled analogously to that of templates. That is, precision measures how likely queries match domain D given that they match site s; for recall, it is the other way around. Formally:

$$P_s = P\,(match(q,D)|match(q,s))\ ,$$

$$R_s = P\,(match(q,s)|match(q,D))\ .$$

The precision of a query q is simply its probability of matching the domain. The recall of q is the fraction of domain queries that are actually q.

$$P_q = P\,(match(q,D))\ ,$$

$$R_q = P\,\big(q' = q|match(q',D)\big)\ .$$

Inference Framework Let us remember our ultimate goal, which is to estimate precision P_u and recall R_u for each template $u \in \mathcal{U}$. As part of the integrated inferencing process, we will also estimate precision and recall for other types of vertices in the QST graph, i.e., for queries (P_q, R_q, $\forall q \in \mathcal{Q}$) and for sites ($P_s$, R_s, $\forall s \in \mathcal{S}$). The QueST algorithm, shown in Algorithm 7.3, infers precision and recall for each vertex, then ranks templates by precision, recall, or the combined F-measure. This process of propagating precision and recall may be interpreted as random walks in opposite directions on the QST-graph. See Fig. 7.6 for an illustration.

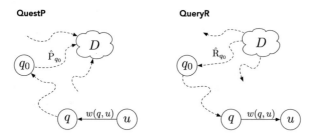

Fig. 7.6 Inferencing precision and recall. QuestP is for precision and propagates backward. QuestR is for recall and propagates forward. Dashed arrows signify random walks

Key to this semi-supervised learning process is to have a small set of seed queries $\mathcal{Q}_0$. Each of these queries $q_0 \in \mathcal{Q}_0$ is labeled with precision $\hat{P}_{q_0}$, indicating how likely it is to fall within the target domain D. As it is infeasible for users to provide seed recall, it is estimated as precision normalized across all seed queries:

$$\hat{R}_{q_0} = \frac{\hat{P}_{q_0}}{\sum_{q' \in \mathcal{Q}_0} \hat{P}_{q'}} .$$

These initial precision and recall values get propagated in the QST-graph through a set of inferencing equations, which we shall present below. For the derivation of these equations, we refer to [1]. The algorithm is reported to converge in four to five iterations [1].

QuestP Estimating P_u may be thought of as the probability of reaching the target domain D in a (backward) random walk starting from u; see Fig. 7.6 (left). We take as input the initial precision estimates for the seed queries, $\hat{P}_{q_0}, q_0 \in \mathcal{Q}_0$. Based on this seed knowledge, the precision P_v of every vertex $v \in V_G$ is determined by propagating precision from its neighboring vertices. The following three equations specify inference for template vertices ($\mathcal{Q} \to \mathcal{U}$), site vertices ($\mathcal{Q} \to \mathcal{S}$), and query vertices ($\mathcal{U} \to \mathcal{Q} \wedge \mathcal{S} \to \mathcal{Q}$), respectively.

$$P_u = \sum_{q \in \mathcal{Q}_u} \frac{w(q,u)}{\sum_{q'} w(q',u)} P_q ,$$

$$P_s = \sum_{q \in \mathcal{Q}_s} \frac{w(q,s)}{\sum_{q'} w(q',s)} P_q ,$$

$$P_q = \begin{cases} \hat{P}_q, & \text{if } q \in \mathcal{Q}_0 \\ \alpha \sum_{u \in \mathcal{U}} \frac{w(q,u)}{\sum_{u'} w(q,u')} P_u + (1-\alpha) \sum_{s \in \mathcal{S}} \frac{w(q,s)}{\sum_{s'} w(q,s')} P_s, & \text{otherwise} . \end{cases}$$

The parameter α controls the relative importance of templates and sites in inferring the precision of queries ($\alpha = 0.5$ in [1]). Note that for seed queries $q \in \mathcal{Q}_0$, the initial precision $\hat{P}_q$ is taken as "ground truth" and will not change.

QuestR The inference of probabilistic recall follows a very similar process. We may think of R_u as the probability of arriving at node u in a (forward) random walk starting from given seeds $q_0 \in \mathcal{Q}_0$ originating from a hidden domain D; see Fig. 7.6 (right). Recall is distributed to neighboring nodes using the following inference equations:

$$R_u = \sum_{q \in \mathcal{Q}_u} \frac{w(q,u)}{\sum_{u'} w(q,u')} R_q \,,$$

$$R_s = \sum_{q \in \mathcal{Q}_s} \frac{w(q,s)}{\sum_{s'} w(q,s')} R_q \,,$$

$$R_q = \beta_1 \hat{R}_q + \beta_2 \sum_{u \in \mathcal{U}} \frac{w(q,u)}{\sum_{q'} w(q',u)} R_u + (1 - \beta_1 - \beta_2) \sum_{s \in \mathcal{S}} \frac{w(q,s)}{\sum_{q'} w(q',s)} R_s \,.$$

Notice that, unlike for precision, the recall of seed queries will also get re-estimated. Parameters β_1 and β_2 specify the relative importance of the different sources, $\hat{R}_q$, R_u, and R_s, when estimating R_q (with $\beta_1 = 0.1$ and $\beta_2 = 0.45$ in [1]).

7.5   Summary

The question driving this chapter has been how to obtain a semantically enriched representation of the user's information need from a keyword query. We have looked at three specific forms of enrichment, all of which are semantic annotations performed with the help of a knowledge repository. First, we have discussed how to annotate queries with target types from a type taxonomy. Second, we have performed entity linking on queries in a number of flavors, from merely recognizing entity mentions to forming coherent interpretation sets. Third, we have generated query templates, which provide structured interpretations of queries by mapping parts of the query to the specific entity attributes. These structured interpretations can then be used to make parameterized requests to particular search services (e.g., verticals). Having thus enriched the user's query with inferred information about their underlying information need, the response to that query can be made more effectively.

7.6 Further Reading

There is a rich and diverse body of research on understanding search queries that was not possible to compress into this chapter. One important practical issue that we have not discussed is *spell checking*. According to Cucerzan and Brill [20], roughly 10–15% of Web search queries contain spelling errors. It is therefore strongly recommended to perform spelling correction before commencing any of the query analysis steps. For example, Blanco et al. [12] report on a 3% improvement in entity linking performance, ascribed to spelling correction. *Query refinement* (also known as *query modification*) refers to the automated process of changing ill-formed queries submitted by users before scoring results. It includes tasks such as spelling error correction, word splitting, word merging, phrase segmentation, word stemming, and acronym expansion [29].

Han et al. [31] address the limitations of machine-based methods for query interpretation by utilizing crowdsourcing in a hybrid crowd-machine framework.

The methods we have presented in this chapter do not make use of the searcher's context, such as age, gender, topic, or location. To a large extent, it is because this type of information is unavailable in public test collections. Contemporary web search engines leverage contextual information by serving personalized search results according to the user's profile [69, 72]. This, however, is typically done by designing specific ranking features [2]. For example, Murnane et al. [57] utilize *personal context* for named entity disambiguation by modeling user interests with respect to a personal knowledge context (using Wikipedia).

References

1. Agarwal, G., Kabra, G., Chang, K.C.C.: Towards rich query interpretation: Walking back and forth for mining query templates. In: Proceedings of the 19th international conference on World Wide Web, WWW '10, pp. 1–10. ACM (2010). doi: 10.1145/1772690.1772692

2. Agichtein, E., Brill, E., Dumais, S., Ragno, R.: Learning user interaction models for predicting web search result preferences. In: Proceedings of the 29th Annual International ACM SIGIR Conference on Research and Development in Information Retrieval, SIGIR '06, pp. 3–10. ACM (2006). doi: 10.1145/1148170.1148175

3. Arguello, J., Diaz, F., Callan, J., Crespo, J.F.: Sources of evidence for vertical selection. In: Proceedings of the 32nd international ACM SIGIR conference on Research and development in information retrieval, SIGIR '09, pp. 315–322. ACM (2009). doi: 10.1145/1571941.1571997

4. Ashkan, A., Clarke, C.L.A.: Characterizing commercial intent. In: Proceedings of the 18th ACM Conference on Information and Knowledge Management, CIKM '09, pp. 67–76. ACM (2009). doi: 10.1145/1645953.1645965

5. Balog, K., Neumayer, R.: Hierarchical target type identification for entity-oriented queries. In: Proceedings of the 21st ACM international conference on Information and knowledge management, CIKM '12, pp. 2391–2394. ACM (2012). doi: 10.1145/2396761.2398648

6. Barr, C., Jones, R., Regelson, M.: The linguistic structure of English web-search queries. In: Proceedings of the Conference on Empirical Methods in Natural Language Processing, EMNLP '08, pp. 1021–1030 (2008)

7. Bendersky, M., Croft, W.B.: Discovering key concepts in verbose queries. In: Proceedings of the 31st Annual International ACM SIGIR Conference on Research and Development in Information Retrieval, SIGIR '08, pp. 491–498. ACM (2008). doi: 10.1145/1390334.1390419

8. Bendersky, M., Croft, W.B., Smith, D.A.: Structural annotation of search queries using pseudo-relevance feedback. In: Proceedings of the 19th ACM international conference on Information and knowledge management, CIKM '10, pp. 1537–1540. ACM (2010). doi: 10.1145/1871437.1871666

9. Bendersky, M., Croft, W.B., Smith, D.A.: Joint annotation of search queries. In: Proceedings of the 49th Annual Meeting of the Association for Computational Linguistics: Human Language Technologies - Volume 1, pp. 102–111. Association for Computational Linguistics (2011)

10. Berant, J., Chou, A., Frostig, R., Liang, P.: Semantic parsing on freebase from question-answer pairs. In: Empirical Methods in Natural Language Processing, EMNLP '13, pp. 1533–1544. Association for Computational Linguistics (2013)

11. Bergsma, S., Wang, Q.I.: Learning noun phrase query segmentation. In: Proceedings of the 2007 Joint Conference on Empirical Methods in Natural Language Processing and Computational Natural Language Learning, EMNLP-CoNLL '07, pp. 819–826. Association for Computational Linguistics (2007)

12. Blanco, R., Ottaviano, G., Meij, E.: Fast and space-efficient entity linking for queries. In: Proceedings of the Eighth ACM International Conference on Web Search and Data Mining - WSDM '15, pp. 179–188. ACM (2015). doi: 10.1145/2684822.2685317

13. Bortnikov, E., Donmez, P., Kagian, A., Lempel, R.: Modeling transactional queries via templates. In: Proceedings of the 34th European Conference on Advances in Information Retrieval, ECIR '12, pp. 13–24. Springer (2012). doi: 10.1007/978-3-642-28997-2_2

14. Brants, T., Franz, A.: Web 1T 5-gram Version 1 LDC2006T13 (2006)

15. Brenes, D.J., Gayo-Avello, D., Garcia, R.: On the fly query entity decomposition using snippets. CoRR **abs/1005.5** (2010)

16. Broder, A.: A taxonomy of web search. SIGIR Forum **36**(2), 3–10 (2002)

17. Carmel, D., Chang, M.W., Gabrilovich, E., Hsu, B.J.P., Wang, K.: ERD'14: Entity recognition and disambiguation challenge. SIGIR Forum **48**(2), 63–77 (2014). doi: 10.1145/2701583.2701591

18. Cornolti, M., Ferragina, P., Ciaramita, M.: A framework for benchmarking entity-annotation systems. In: Proceedings of the 22nd International Conference on World Wide Web, WWW '13, pp. 249–260 (2013). doi: 10.1145/2488388.2488411

19. Cornolti, M., Ferragina, P., Ciaramita, M., Rüd, S., Schütze, H.: A piggyback system for joint entity mention detection and linking in web queries. In: Proceedings of the 25th International Conference on World Wide Web, WWW '16, pp. 567–578. International World Wide Web Conferences Steering Committee (2016). doi: 10.1145/2872427.2883061

20. Cucerzan, S., Brill, E.: Spelling correction as an iterative process that exploits the collective knowledge of web users. In: Proceedings of the 2004 Conference on Empirical Methods in Natural Language Processing, EMNLP '04 (2004)

21. Dai, H.K., Zhao, L., Nie, Z., Wen, J.R., Wang, L., Li, Y.: Detecting online commercial intention (OCI). In: Proceedings of the 15th International Conference on World Wide Web, WWW '06, pp. 829–837. ACM (2006). doi: 10.1145/1135777.1135902

22. Diaz, F.: Integration of news content into web results. In: Proceedings of the Second ACM International Conference on Web Search and Data Mining, WSDM '09, pp. 182–191. ACM (2009). doi: 10.1145/1498759.1498825

23. Ferragina, P., Scaiella, U.: TAGME: on-the-fly annotation of short text fragments (by Wikipedia entities). In: Proceedings of the 19th ACM International Conference on Information and Knowledge Management, CIKM '10, pp. 1625–1628. ACM (2010). doi: 10.1145/1871437.1871689

24. Gabrilovich, E., Broder, A., Fontoura, M., Joshi, A., Josifovski, V., Riedel, L., Zhang, T.: Classifying search queries using the web as a source of knowledge. ACM Trans. Web 3(2), 5:1–5:28 (2009). doi: 10.1145/1513876.1513877

25. Gabrilovich, E., Markovitch, S.: Computing semantic relatedness using Wikipedia-based explicit semantic analysis. In: Proceedings of the 20th International Joint Conference on Artificial Intelligence, IJCAI'07, pp. 1606–1611. Morgan Kaufmann Publishers Inc. (2007)

26. Ganti, V., He, Y., Xin, D.: Keyword++: A framework to improve keyword search over entity databases. Proc. VLDB Endow. 3(1–2), 711–722 (2010). doi: 10.14778/1920841.1920932

27. Garigliotti, D., Hasibi, F., Balog, K.: Target type identification for entity-bearing queries. In: Proceedings of the 40th International ACM SIGIR Conference on Research and Development in Information Retrieval, SIGIR '17. ACM (2017). doi: 10.1145/3077136.3080659

28. Guo, J., Xu, G., Cheng, X., Li, H.: Named entity recognition in query. In: Proceedings of the 32nd international ACM SIGIR conference on Research and development in information retrieval, SIGIR '09, pp. 267–274. ACM (2009). doi: 10.1145/1571941.1571989

29. Guo, J., Xu, G., Li, H., Cheng, X.: A unified and discriminative model for query refinement. In: Proceedings of the 31st Annual International ACM SIGIR Conference on Research and Development in Information Retrieval, SIGIR '08, pp. 379–386 (2008). doi: 10.1145/1390334.1390400

30. Hagen, M., Potthast, M., Stein, B., Bräutigam, C.: Query segmentation revisited. In: Proceedings of the 20th International Conference on World Wide Web, WWW '11, pp. 97–106 (2011). doi: 10.1145/1963405.1963423

31. Han, J., Fan, J., Zhou, L.: Crowdsourcing-assisted query structure interpretation. In: Proceedings of the Twenty-Third International Joint Conference on Artificial Intelligence, IJCAI '13, pp. 2092–2098. AAAI Press (2013)

32. Hasibi, F., Balog, K., Bratsberg, S.E.: Entity linking in queries: Tasks and evaluation. In: Proceedings of the 2015 International Conference on The Theory of Information Retrieval, ICTIR '15, pp. 171–180. ACM (2015). doi: 10.1145/2808194.2809473

33. Hasibi, F., Balog, K., Bratsberg, S.E.: Exploiting entity linking in queries for entity retrieval. In: Proceedings of the 2016 ACM on International Conference on the Theory of Information Retrieval, ICTIR '16, pp. 209–218. ACM (2016). doi: 10.1145/2970398.2970406

34. Hasibi, F., Balog, K., Bratsberg, S.E.: Entity linking in queries: Efficiency vs. effectiveness. In: Proceedings of the 39th European conference on Advances in Information Retrieval, ECIR '17, pp. 40–53. Springer (2017). doi: 10.1007/978-3-319-56608-5_4

35. Hu, J., Wang, G., Lochovsky, F., Sun, J.t., Chen, Z.: Understanding user's query intent with Wikipedia. In: Proceedings of the 18th International Conference on World Wide Web, WWW '09, pp. 471–480. ACM (2009). doi: 10.1145/1526709.1526773

36. Huang, J., Gao, J., Miao, J., Li, X., Wang, K., Behr, F., Giles, C.L.: Exploring web scale language models for search query processing. In: Proceedings of the 19th International Conference on World Wide Web, WWW '10, pp. 451–460. ACM (2010). doi: 10.1145/1772690.1772737

37. Huston, S., Croft, W.B.: Evaluating verbose query processing techniques. In: Proceedings of the 33rd International ACM SIGIR Conference on Research and Development in Information Retrieval, SIGIR '10, pp. 291–298. ACM (2010). doi: 10.1145/1835449.1835499

38. Jansen, B.J., Booth, D.: Classifying web queries by topic and user intent. In: Proceedings of the 28th of the international conference extended abstracts on Human factors in computing systems, CHI EA '10, pp. 4285–4290. ACM (2010)

39. Jansen, B.J., Booth, D.L., Spink, A.: Determining the informational, navigational, and transactional intent of web queries. Inf. Process. Manage. **44**(3), 1251–1266 (2008). doi: 10.1016/j.ipm.2007.07.015

40. Jones, R., Rey, B., Madani, O., Greiner, W.: Generating query substitutions. In: Proceedings of the 15th International Conference on World Wide Web, WWW '06, pp. 387–396. ACM (2006). doi: 10.1145/1135777.1135835

41. Kaptein, R., Serdyukov, P., De Vries, A., Kamps, J.: Entity ranking using Wikipedia as a pivot. In: Proceedings of the 19th ACM international conference on Information and knowledge management, CIKM '10, pp. 69–78. ACM (2010). doi: 10.1145/1871437.1871451

42. Kraaij, W., Spitters, M.: Language models for topic tracking. In: Croft, W., Lafferty, J. (eds.) Language Modeling for Information Retrieval, *The Springer International Series on Information Retrieval*, vol. 13, pp. 95–123. Springer (2003)

43. Lee, U., Liu, Z., Cho, J.: Automatic identification of user goals in web search. In: Proceedings of the 14th International Conference on World Wide Web, WWW '05, pp. 391–400. ACM (2005). doi: 10.1145/1060745.1060804

44. Li, X.: Understanding the semantic structure of noun phrase queries. In: Proceedings of the 48th Annual Meeting of the Association for Computational Linguistics, ACL '10, pp. 1337–1345. Association for Computational Linguistics (2010)

45. Li, X., Wang, Y.Y., Acero, A.: Learning query intent from regularized click graphs. In: Proceedings of the 31st annual international ACM SIGIR conference on Research and development in information retrieval, SIGIR '08, pp. 339–346. ACM (2008). doi: 10.1145/1390334.1390393

46. Li, X., Wang, Y.Y., Acero, A.: Extracting structured information from user queries with semi-supervised conditional random fields. In: Proceedings of the 32Nd International ACM SIGIR Conference on Research and Development in Information Retrieval, SIGIR '09, pp. 572–579. ACM (2009). doi: 10.1145/1571941.1572039

47. Li, Y., Hsu, B.J.P., Zhai, C., Wang, K.: Unsupervised query segmentation using clickthrough for information retrieval. In: Proceedings of the 34th International ACM SIGIR Conference on Research and Development in Information Retrieval, SIGIR '11, pp. 285–294. ACM (2011). doi: 10.1145/2009916.2009957

48. Li, Y., Zheng, Z., Dai, H.K.: KDD CUP-2005 report: facing a great challenge. SIGKDD Explor. Newsl. **7**(2), 91–99 (2005)

49. Lin, T., Pantel, P., Gamon, M., Kannan, A., Fuxman, A.: Active objects. In: Proceedings of the 21st international conference on World Wide Web, WWW '12, pp. 589–598. ACM (2012). doi: 10.1145/2187836.2187916

50. Liu, X., Zhang, S., Wei, F., Zhou, M.: Recognizing named entities in tweets. In: Proceedings of the 49th Annual Meeting of the Association for Computational Linguistics: Human Language Technologies - Volume 1, HLT '11, pp. 359–367. Association for Computational Linguistics (2011)

51. Manshadi, M., Li, X.: Semantic tagging of web search queries. In: Proceedings of the Joint Conference of the 47th Annual Meeting of the ACL and the 4th International Joint Conference on Natural Language Processing of the AFNLP: Volume 2 - Volume 2, ACL '09, pp. 861–869. Association for Computational Linguistics (2009)

52. Marcus, M.P., Marcinkiewicz, M.A., Santorini, B.: Building a large annotated corpus of English: The Penn Treebank. Comput. Linguist. **19**(2), 313–330 (1993)

53. Meij, E., Bron, M., Hollink, L., Huurnink, B., de Rijke, M.: Mapping queries to the linking open data cloud: A case study using DBpedia. Web Semant. **9**(4), 418–433 (2011)

54. Meij, E., Weerkamp, W., De Rijke, M.: Adding semantics to microblog posts. In: Proceedings of the Fifth ACM International Conference on Web Search and Data Mining, WSDM '12, pp. 563–572. ACM (2012). doi: 10.1145/2124295.2124364

55. Mikolov, T., Sutskever, I., Chen, K., Corrado, G., Dean, J.: Distributed representations of words and phrases and their compositionality. In: Proceedings of the 26th International Conference on Neural Information Processing Systems, NIPS'13, pp. 3111–3119. Curran Associates Inc. (2013)

56. Mishra, N., Saha Roy, R., Ganguly, N., Laxman, S., Choudhury, M.: Unsupervised query segmentation using only query logs. In: Proceedings of the 20th International Conference Companion on World Wide Web, WWW '11, pp. 91–92. ACM (2011). doi: 10.1145/1963192.1963239

57. Murnane, E.L., Haslhofer, B., Lagoze, C.: RESLVE: leveraging user interest to improve entity disambiguation on short text. In: Proceedings of the 22nd International Conference on World Wide Web, WWW '13 Companion, pp. 81–82. ACM (2013). doi: 10.1145/2487788.2487823

58. Pantel, P., Lin, T., Gamon, M.: Mining entity types from query logs via user intent modeling. In: Proceedings of the 50th Annual Meeting of the Association for Computational Linguistics: Long Papers - Volume 1, ACL '12, pp. 563–571. Association for Computational Linguistics (2012)

59. Paparizos, S., Ntoulas, A., Shafer, J., Agrawal, R.: Answering web queries using structured data sources. In: Proceedings of the 2009 ACM SIGMOD International Conference on Management of Data, SIGMOD '09, pp. 1127–1130. ACM (2009). doi: 10.1145/1559845.1560000

60. Piccinno, F., Ferragina, P.: From TagME to WAT: A new entity annotator. In: Proceedings of the First International Workshop on Entity Recognition & Disambiguation, ERD '14, pp. 55–62. ACM (2014). doi: 10.1145/2633211.2634350

61. Pound, J., Hudek, A.K., Ilyas, I.F., Weddell, G.: Interpreting keyword queries over web knowledge bases. In: Proceedings of the 21st ACM International Conference on Information and Knowledge Management, CIKM '12, pp. 305–314. ACM (2012). doi: 10.1145/2396761.2396803

62. Risvik, K.M., Mikolajewski, T., Boros, P.: Query segmentation for web search. In: Proceedings of the 12th International Conference on World Wide Web, WWW '03 (2003)

63. Rose, D.E., Levinson, D.: Understanding user goals in web search. In: Proceedings of the 13th International Conference on World Wide Web, WWW '04, pp. 13–19 (2004). doi: 10.1145/988672.988675

64. Rüd, S., Ciaramita, M., Müller, J., Schütze, H.: Piggyback: Using search engines for robust cross-domain named entity recognition. In: Proceedings of the 49th Annual Meeting of the Association for Computational Linguistics: Human Language Technologies, pp. 965–975 (2011)

65. Saha Roy, R., Ganguly, N., Choudhury, M., Laxman, S.: An IR-based evaluation framework for web search query segmentation. In: Proceedings of the 35th International ACM SIGIR Conference on Research and Development in Information Retrieval, SIGIR '12, pp. 881–890. ACM (2012). doi: 10.1145/2348283.2348401

66. Sarkas, N., Paparizos, S., Tsaparas, P.: Structured annotations of web queries. In: Proceedings of the 2010 ACM SIGMOD International Conference on Management of Data, SIGMOD '10, pp. 771–782 (2010). doi: 10.1145/1807167.1807251

67. Sawant, U., Chakrabarti, S.: Learning joint query interpretation and response ranking. In: Proceedings of the 22nd International Conference on World Wide Web, WWW '13, pp. 1099–1109 (2013). doi: 10.1145/2488388.2488484

68. Shen, D., Sun, J.T., Yang, Q., Chen, Z.: Building bridges for web query classification. In: Proceedings of the 29th Annual International ACM SIGIR Conference on Research and Development in Information Retrieval, SIGIR '06, pp. 131–138. ACM (2006). doi: 10.1145/1148170.1148196

69. Speretta, M., Gauch, S.: Personalized search based on user search histories. In: Proceedings of the 2005 IEEE/WIC/ACM International Conference on Web Intelligence, WI '05, pp. 622–628. IEEE Computer Society (2005). doi: 10.1109/WI.2005.114

70. Srba, I., Bielikova, M.: A comprehensive survey and classification of approaches for community question answering. ACM Trans. Web **10**(3), 18:1–18:63 (2016). doi: 10.1145/2934687

71. Tan, B., Peng, F.: Unsupervised query segmentation using generative language models and Wikipedia. In: Proceedings of the 17th international conference on World Wide Web, WWW '08, pp. 347–356. ACM (2008). doi: 10.1145/1367497.1367545
72. Teevan, J., Dumais, S.T., Horvitz, E.: Personalizing search via automated analysis of interests and activities. In: Proceedings of the 28th Annual International ACM SIGIR Conference on Research and Development in Information Retrieval, SIGIR '05, pp. 449–456. ACM (2005). doi: 10.1145/1076034.1076111
73. Tonon, A., Catasta, M., Prokofyev, R., Demartini, G., Aberer, K., Cudré-Mauroux, P.: Contextualized ranking of entity types based on knowledge graphs. Web Semant. **37–38**, 170–183 (2016). doi: 10.1016/j.websem.2015.12.005
74. Toutanova, K., Klein, D., Manning, C.D., Singer, Y.: Feature-rich part-of-speech tagging with a cyclic dependency network. In: Proceedings of the 2003 Conference of the North American Chapter of the Association for Computational Linguistics on Human Language Technology - Volume 1, NAACL '03, pp. 173–180. Association for Computational Linguistics (2003). doi: 10.3115/1073445.1073478
75. Tsur, G., Pinter, Y., Szpektor, I., Carmel, D.: Identifying web queries with question intent. In: Proceedings of the 25th International Conference on World Wide Web, WWW '16, pp. 783–793. International World Wide Web Conferences Steering Committee (2016). doi: 10.1145/2872427.2883058
76. Ullegaddi, P.V., Varma, V.: Learning to rank categories for web queries. In: Proceedings of the 20th ACM International Conference on Information and Knowledge Management, CIKM '11, pp. 2065–2068. ACM (2011). doi: 10.1145/2063576.2063891
77. Vallet, D., Zaragoza, H.: Inferring the most important types of a query: A semantic approach. In: Proceedings of the 31st annual international ACM SIGIR conference on Research and development in information retrieval, SIGIR '08, pp. 857–858. ACM (2008). doi: 10.1145/1390334.1390541
78. Voskarides, N., Meij, E., Tsagkias, M., de Rijke, M., Weerkamp, W.: Learning to explain entity relationships in knowledge graphs. In: Proceedings of the 53rd Annual Meeting of the Association for Computational Linguistics and the 7th International Joint Conference on Natural Language Processing (Volume 1: Long Papers), pp. 564–574. Association for Computational Linguistics (2015)
79. Wei, X., Peng, F., Dumoulin, B.: Analyzing web text association to disambiguate abbreviation in queries. In: Proceedings of the 31st Annual International ACM SIGIR Conference on Research and Development in Information Retrieval, SIGIR '08, pp. 751–752 (2008). doi: 10.1145/1390334.1390485
80. Yih, S.W.t., Chang, M.W., He, X., Gao, J.: Semantic parsing via staged query graph generation: Question answering with knowledge base. In: Proceedings of the Joint Conference of the 53rd Annual Meeting of the ACL and the 7th International Joint Conference on Natural Language Processing of the AFNLP. ACL - Association for Computational Linguistics (2015)
81. Yin, X., Shah, S.: Building taxonomy of web search intents for name entity queries. In: Proceedings of the 19th International Conference on World Wide Web, WWW '10, pp. 1001–1010. ACM (2010). doi: 10.1145/1772690.1772792

82. Zhang, S., Balog, K.: Design patterns for fusion-based object retrieval. In: Proceedings of the 39th European conference on Advances in Information Retrieval, ECIR '17. Springer (2017). doi: 10.1007/978-3-319-56608-5_66
83. Zhou, K., Cummins, R., Halvey, M., Lalmas, M., Jose, J.M.: Assessing and predicting vertical intent for web queries. In: Proceedings of the 34th European conference on Advances in Information Retrieval, ECIR'12, pp. 499–502. Springer (2012). doi: 10.1007/978-3-642-28997-2_50

Chapter 8
Leveraging Entities in Document Retrieval

This chapter focuses on the classic IR task of *ad hoc document retrieval* and discusses how entities may be leveraged to improve retrieval performance. At their core, all document retrieval methods compare query and document representations. Traditionally, these representations are based on terms (words). Entities facilitate a semantic understanding of both the user's information need, as expressed by the keyword query, and of the document's content. Entities thus may be used to improve query and/or document representations. As a first step of that process, entities that are related to the query need to be identified, thereby establishing a mapping between queries and entities. As we shall explain in Sect. 8.1, one may go beyond considering only those entities that are explicitly mentioned in the query. We next present three different families of approaches, which are illustrated in Fig. 8.1.

- *Expansion-based* methods utilize entities as a source of expansion terms to enrich the representation of the query (Sect. 8.2).
- *Projection-based* methods treat entities as a latent layer, while leaving the original document/query representations intact (Sect. 8.3).
- *Entity-based* methods consider explicitly the entities that are recognized in documents, as first-class citizens, and embrace entity-based representations in "duet" with traditional term-based representations in the retrieval model (Sect. 8.4).

This particular order corresponds to the temporal evolution of research in this area, where the tendency toward more and more explicit entity semantics is clearly reflected. Throughout this chapter, we shall assume that both queries and documents have been annotated with entities, using entity linking techniques we have discussed before (see Chap. 5 for documents and Sect. 7.3 for queries). A particular challenge involved here is how to deal with the uncertainty of these automatic annotations.

In practice, necessitated by efficiency considerations, all methods described in this chapter are implemented as re-ranking mechanisms. The details are found in Sect. 8.5. Finally, we present standard datasets and useful resources in Sect. 8.6. We refer to Table 8.1 for the notation used throughout this chapter.

© The Author(s) 2018
K. Balog, *Entity-Oriented Search*, The Information Retrieval Series 39,
https://doi.org/10.1007/978-3-319-93935-3_8

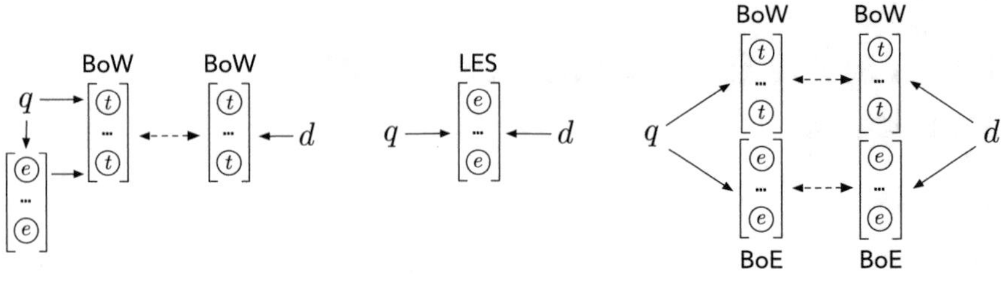

(a) Expansion-based   (b) Projection-based   (c) Entity-based

Fig. 8.1 Three main ways to leverage entities for improved document retrieval. The representations are bag-of-words (BoW), bag-of-entities (BoE), and latent entity space (LES). The main difference between projection-based (**b**) and entity-based (**c**) methods is that the former treats entities as a latent layer between queries and documents, while the latter explicitly models the entities mentioned in the document and complements the traditional bag-of-words (BoW) representations with bag-of-entities (BoE) representations

Table 8.1 Notation used in this chapter

Symbol	Description
$c(t;x)$	Count (raw frequency) of term t in the x
d	Document ($d \in \mathcal{D}$)
$\mathcal{D}$	Collection of documents
$\mathcal{D}_q(k)$	Top-k ranked documents in response to query q
e	Entity ($e \in \mathcal{E}$)
$\mathcal{E}$	Entity catalog (set of all entities)
$\mathcal{E}_q(k)$	Top-k ranked entities in response to query q
$\mathcal{E}_q$	Set of query entities
l_x	Length of the description of x ($l_x = \sum_{t \in x} c(t;x)$)
q	Query ($q = \langle q_1, \ldots, q_n \rangle$)
t	Term ($t \in \mathcal{V}$)
$\mathcal{T}$	Type taxonomy
$\mathcal{T}_e$	Set of types assigned to e
$\mathcal{V}$	Vocabulary of terms

8.1 Mapping Queries to Entities

A common component that is shared by all approaches that follow later in this chapter is the mapping of queries to entities. The goal is to identify a set of entities that may be semantically related to the query.[1] We shall refer to this set $\mathcal{E}_q$ of related entities as *query entities*. Naturally, not all query entities are equally strongly related to the query. Therefore, we use the probability $P(e|q)$ to express the likelihood of

---

[1] Notice that this is highly related to the task of *ad hoc entity retrieval* (cf. Chap. 3), as well as to the *candidate entity ranking* and *semantic linking* subtasks in query entity annotation (cf. Sect. 7.3).

entity e being related to query q. The estimation of this probability may be based on (1) entities mentioned in the query, (2) entities retrieved directly from a knowledge base, and/or (3) entities retrieved indirectly, through pseudo-relevant documents. Let us look at these in order.

- *Entities mentioned in the query.* The presence of an entity mention in the query provides a unique opportunity for improving the understanding of the user's information need [5]. Entities can be identified and disambiguated using entity linking techniques, cf. Sect. 7.3. Let $\mathcal{E}_q$ be the set of entities that have been identified in query q. For each of the query entities $e \in \mathcal{E}_q$, we let $score_{ELQ}(e;q)$ be the associated confidence score. We note that the annotated entity mentions may overlap (i.e., we are not concerned with forming interpretation sets). For queries that are associated with a specific entity (i.e., $|\mathcal{E}_q| = 1$), it makes sense to use that entity's description as pseudo-feedback information [56]. More generally, we consider a single entity that has the highest annotation score:

$$P(e|q) = \begin{cases} 1, & e = \arg\max_{e \in \mathcal{E}_q} score_{ELQ}(e;q) \\ 0, & \text{otherwise} . \end{cases} \tag{8.1}$$

Further generalization to an arbitrary number of entities can easily be done, by introducing a minimum confidence threshold on the annotations:

$$P(e|q) = \begin{cases} \frac{1}{Z} score_{ELQ}(e;q), & score_{ELQ}(e;q) > \gamma \\ 0, & \text{otherwise} , \end{cases}$$

where γ is a score threshold parameter, and Z is a normalization factor, such that $0 \leq P(e|q) \leq 1$. Additionally, entities relevant to those mentioned in the query may also be considered [34].

- *Entities retrieved from a knowledge base.* An alternative route may be taken by querying a knowledge base directly for relevant entities [18]. We let $score_{ER}(e;q)$ be the relevance score of entity e given q. This score may be computed using any of the entity retrieval methods introduced in Chap. 3. For pragmatic considerations, only the top-k entities are considered, denoted as $\mathcal{E}_q(k)$. $P(e|q)$ then becomes:

$$P(e|q) = \begin{cases} \frac{1}{Z} score_{ER}(e;q), & e \in \mathcal{E}_q(k) \\ 0, & \text{otherwise} , \end{cases}$$

where Z is a normalization coefficient.

- *Entities from pseudo-relevant documents.* The third method uses the top-ranked documents retrieved in response to the query, in the spirit of pseudo relevance feedback [40]. This corresponds to the setting of ranking entities without direct representations (cf. Sect. 3.4). Formally:

$$P(e|q) \propto \sum_{d \in \mathcal{D}_q(k)} P(e|d)P(d|q), \tag{8.2}$$

where $\mathcal{D}_q(k)$ denotes the set of top-k highest scoring documents retrieved in response to query q, $P(d|q)$ corresponds to document d's relevance to the query, and $P(e|d)$ is the probability of observing the entity in d. $P(e|d)$ may be taken as a maximum-likelihood estimate:

$$P(e|d) = \frac{c(e;d)}{\sum_{e' \in d} c(e';d)} \, ,$$

where $c(e;d)$ is the number of times entity e occurs in document d. Additionally, the frequency of e across the document collection may also be taken into account (to demote entities that occur in too many documents) by adding an IDF-like component, see Eq. (3.6). One alternative to relying on maximum-likelihood estimation is presented by Meij et al. [40], who re-estimate the probability mass of the entities using parsimonious language models.

We note that entity relevance may be estimated selectively or jointly using the above methods, depending on the type of the query. For example, Xu et al. [56] employ Eq. (8.1) for queries that can be associated with a single entity; for other queries, the top-k ranked documents are considered, i.e., Eq. (8.2) is used.

8.2 Leveraging Entities for Query Expansion

Keyword queries are typically too short to describe the underlying information need accurately. Query expansion is one of the classical techniques used in document retrieval, dating all the way back to the 1970s [43]. The idea is to supplement the keyword query with additional terms, thereby having a more elaborate expression of the underlying information need. These additional terms may be extracted from documents that are deemed relevant. In most cases, however, there is no explicit feedback from the user as to which documents are relevant and which are not. Instead, one may "blindly" assume that the top-ranked documents are relevant, and extract expansion terms from these. This technique is known as *pseudo* (or *blind*) *relevance feedback* and has been thoroughly investigated in the past. In general, pseudo relevance feedback helps more queries than it hurts [39]. Clearly, it can only be effective when the initial set of retrieved documents is good, otherwise it merely introduces noise. Prior work has demonstrated the benefits of exploiting external collections for query expansion [2, 20, 49]. In this section, we will leverage a knowledge base as an external resource, and utilize entities for query expansion. This can bring in external semantic signals that may not be available within feedback documents.

First, in Sect. 8.2.1, we describe how traditional document-based feedback works. The aim of that section is to show how an expanded query model $\hat{\theta}_q$ can be constructed and subsequently used for retrieval. Next, in Sect. 8.2.2, we present a general framework for performing entity-centric query expansion, i.e., estimating $\hat{\theta}_q$ with the help of entities. A core component of this framework is *term selection*, which may be approached using either unsupervised or supervised methods. These are discussed in Sects. 8.2.3 and 8.2.4, respectively.

8.2.1 Document-Based Query Expansion

To give an idea of how traditional (term-based) pseudo relevance feedback works, we present one of the most popular approaches, the *relevance model* by Lavrenko and Croft [33]. This method assumes that there exists some underlying relevance model R, which generates both the query and the relevant documents. Then, based on the observed query q (which is a sample from R), we attempt to learn the parameters of R. The probability of drawing a term t from R is approximated with the probability of observing that term, given the query:

$$P(t|R) \approx P(t|q_1, \ldots, q_n) \, .$$

Lavrenko and Croft [33] present two methods for estimating this conditional probability. The better performing of the two, referred to as *RM1*, assumes that the terms in the query and in relevant documents are sampled identically and independently from the relevance model (i.i.d. sampling):

$$P(t|R) \propto \sum_{d \in \mathcal{D}_q(m)} P(d) P(t|\theta_d) \prod_{i=1}^{n} P(q_i|\theta_d) \, , \tag{8.3}$$

where $\mathcal{D}_q(m)$ is the set of top-m highest ranked documents for the original query, according to some retrieval model (commonly query likelihood, i.e., $P(q|\theta_d)$).[2] These are used as evidence for estimating the relevance model, with m typically set between 10 and 50 [1, 17, 37, 56]. The prior document probability, $P(d)$, is usually assumed to be uniform. The term probabilities $P(t|\theta_d)$ and $P(q_i|\theta_d)$ are smoothed term probabilities from the document's language model (cf. Sect. 3.3.1.1).

The number of expansion terms has a direct impact on retrieval efficiency. Therefore, in practice, only the top-k expansion terms with the highest probability are used for expansion, with k typically ranging between 10 and 50 (see, e.g., [1, 5, 17]). Thus, the top-k highest scoring terms according to Eq. (8.3) are taken to form the expanded query model $\hat{\theta}_q$ (with the probabilities renormalized such that $\sum_t P(t|\hat{\theta}_q) = 1$).

To avoid the query shifting too far away from the user's original intent (an issue known as *topic drift*), it is common to define the final query model θ_q as a linear combination of the maximum likelihood and expanded query models [59]:

$$P(t|\theta_q) = (1 - \lambda)\frac{c(t;q)}{l_q} + \lambda P(t|\hat{\theta}_q) \, , \tag{8.4}$$

---

[2] We use m to denote the number of feedback documents, as the variable k will be used for the number of feedback terms.

Fig. 8.2 Entity-based query
expansion. Query entities
$(\mathcal{E}_q)$ are utilized to add and
re-weigh terms in the original
query q, resulting in an
expanded query model θ_q

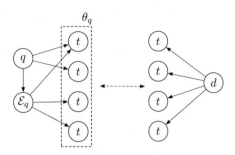

where $c(t; q)$ is the number of times t occurs in q, l_q is the total number of terms
in the query, and λ is a mixture parameter. This parameter controls the influence of
the expanded query model and is typically in the range $[0.4, 0.6]$ (see, e.g., [1, 56]).
The combination of RM1 with the original query in Eq. (8.4) is commonly referred
to in the literature as RM3 [37]. RM3 is widely regarded as a state-of-the-art model
for pseudo relevance feedback.

Finally, the combined query model θ_q is used in a second retrieval round to obtain
the final document ranking. For example, using the query likelihood retrieval model
(a.k.a. the standard language modeling approach) the scoring of documents is done
according to:[3]

$$\log P(q|\theta_d) = \sum_{t \in q} P(t|\theta_q) \log P(t|\theta_d) ,$$

where $P(t|\theta_d)$ is the probability of term t in the (smoothed) language model of
document d. Note that any other retrieval model can be used for obtaining the final
document ranking by using θ_q as a (weighted) query.

8.2.2 Entity-Centric Query Expansion

Given that our focus is on entities, the question we ask is this: Can we use
entities, instead of documents, for estimating an expanded query model? The idea
of entity-centric query expansion is illustrated in Fig. 8.2, where the expanded query
model θ_q is estimated by utilizing the set of query entities $\mathcal{E}_q$.

As a general framework, we follow the method proposed by Meij et al.
[40], where query expansion is formulated as a double translation process: first,
translating the query to a set of relevant entities, then considering the vocabulary
of terms associated with those entities as possible expansion terms to estimate the

---

[3]This formula can be derived by replacing the query term count $c(t; q)$ in Eq. (3.8) with the
probability of the term given the query language model, $P(t|\theta_q)$. Note that this scoring is rank-
equivalent to measuring the Kullback–Leibler divergence between the document and the query
term distributions $(KL(\theta_q||\theta_d))$ [1].

expanded query model. Formally:

$$P(t|\hat{\theta}_q) \propto \sum_{e \in \mathcal{E}_q} P(t|e,q)P(e|q) . \tag{8.5}$$

Most existing approaches can be instantiated into Eq. (8.5). The first component, $P(t|e,q)$, expresses how strongly term t is associated with entity e given q. Meij et al. [40] further impose a conditional independence assumption between the term and the query. That is, once an entity is selected for a query, the probability of the expansion term depends only on that entity: $P(t|e,q) \cong P(t|e)$. This distribution governs how terms are sampled from the description of e. We shall refer to it as *term selection*. The second component, $P(e|q)$, identifies entities to be used for query expansion and corresponds to the relevance of e given the query. This latter component we have already discussed in Sect. 8.1. Therefore, we shall now focus on the estimation of $P(t|e,q)$, considering both unsupervised and supervised methods.

8.2.3 Unsupervised Term Selection

The probability distribution $P(t|e)$ is estimated by selecting expansion terms from the description or surface forms of a given query entity ($e \in \mathcal{E}_q$). (Notice that term selection here depends only on the entity, and not on the original query.)

Entity Description One of the simplest ways to perform term selection is to pick the most important terms from the entity's term-based representation, which we refer to as the entity description. Following our notation from before and assuming a single-field entity representation, we write $c(t; e)$ to denote the count (raw frequency) of term t in the description of e. Further, we introduce the shorthand notation $w(t, e)$ for the importance of term t given e. Commonly, this is estimated using the TF-IDF (here: TF-IEF) weighting scheme:

$$w(t,e) = TF(t,e) \times IEF(t) ,$$

where term frequency *TF* and inverse entity frequency *IEF* have been defined in Eqs. (3.1) and (3.2), respectively. Another popular choice is to use entity language models, i.e., $w(t,e) = P(t|\theta_e)$. (We refer back to Sect. 3.3.1.1 for the construction of entity language models.) Once term scores are computed, $P(t|e)$ is formed by taking the top-k terms and re-normalizing their scores:

$$P(t|e) = \begin{cases} \frac{1}{Z} w(t,e) , & t \in \mathcal{V}(k) \\ 0 , & \text{otherwise} , \end{cases}$$

where $\mathcal{V}(k)$ is the set of top-k terms with the highest score, and Z is a normalization coefficient. Xu et al. [56] further exploit the structured nature of entity descriptions, by taking a linear combination of term scores across multiple entity fields. When the entity does not have a ready-made description available in the knowledge repository, terms may be sampled from documents that mention the entity [40]. Instead of considering entity-term co-occurrences on the document level, they may be restricted to smaller contexts, such as the sentence mentioning the entity or a fixed sized window of words around the entity mention [18]. The key difference between the above methods is how the entity description is obtained; we have covered all these variants earlier in the book, in Chap. 3.

Surface Forms   Another approach is to use the various surface forms (aliases) of the query entities as expansion terms [18, 34]. The main intuition behind doing so is that by including the different ways an entity can be referred to (i.e., its "synonyms"), documents relevant to that entity can be retrieved more effectively. Name variants of an entity can be conveniently enumerated as a separate field in its description. Then, the estimation of $P(t|e)$ is performed as before, except that the term importance score is based on $c(t; f_n)$, instead of $c(t; e)$, where f_n is the field holding the name variants of e.

8.2.4   Supervised Term Selection

So far, it has been implicitly assumed that all expansion terms are useful and benefit retrieval performance, when added to the original query. This assumption was challenged by Cao et al. [8], who showed that not all expansion terms are actually valuable. Some terms are neutral (do not affect performance), while others are in fact harmful. They further demonstrated that it is difficult to tell apart good expansion terms from bad ones based solely on term distributions. One needs to incorporate additional signals to be able to select useful expansion terms. Therefore, Cao et al. [8] propose to combine multiple features using supervised learning to predict the usefulness of expansion terms. This task may be formulated as a binary classification problem (separating good expansion terms from bad ones) [8] or cast as a ranking problem (ranking expansion terms based on their predicted utility) [5]. We follow the latter approach and specifically focus on ranking expansion terms given a particular query entity e and the query q. The resulting term importance score $score(t; e, q)$ may be plugged into Eq. (8.5) as an estimate of $P(t|e, q)$. Note that the dependence of the expansion term on the original query is kept.

8.2.4.1   Features

Brandão et al. [5] present five specific feature functions for entity-based query expansion. The order in which we discuss them corresponds to their usefulness (from most to least useful).

The first two features are simple statistical measures of term frequency, which rely on fielded entity descriptions. *Term frequency* is the total number of times t occurs across the set $\mathcal{F}_e$ fields of the entity:

$$TF(t,e) = \sum_{f_e \in \mathcal{F}_e} c(t; f_e) \, .$$

The *term spread* feature measures the spread of a term across multiple fields, by counting in how many different fields the term occurs:

$$TS(t,e) = \sum_{f_e \in \mathcal{F}_e} \mathbb{1}(c(t; f_e) > 0) \, ,$$

where $\mathbb{1}()$ is a binary indicator function.

Term proximity accounts for the proximity between an expansion term and the original query terms in the description of the entity:

$$TP(t,q,e) = \sum_{i=1}^{n} \sum_{w=1}^{m} \frac{c_w(t,q_i;e)}{2^{w-1}} \, ,$$

where $c_w(t,q_i;e)$ is the total number of co-occurrences of terms t and q_i, within an unordered window size of w, in the description of e. In [5], entities are represented by their Wikipedia pages, and windows are measured in terms of sentences, up to $m = 5$.

The last two features are taxonomic, utilizing the types of entities. Let $\mathcal{E}_t$ denote the set of entities that contain the term t: $\mathcal{E}_t = \{e' \in \mathcal{E} : c(t;e') > 0\}$. Further, let $\mathcal{E}_e$ be the set of entities that share at least one type with the query entity e: $\mathcal{E}_e = \{e' \in \mathcal{E} : \mathcal{T}_e \cap \mathcal{T}_{e'} \neq \emptyset\}$. The similarity between the sets of related entities $\mathcal{E}_t$ and $\mathcal{E}_e$ may be measured using *Dice's coefficient*:

$$DC(t,e) = 2 \, \frac{|\mathcal{E}_t \cap \mathcal{E}_e|}{|\mathcal{E}_t| + |\mathcal{E}_e|} \, .$$

Another option is to use *mutual information*:

$$MI(t,e) = \begin{cases} |\mathcal{E}_t \cap \mathcal{E}_e| \, \log \frac{|\mathcal{E}_t \cap \mathcal{E}_e|}{|\mathcal{E}_t| \times |\mathcal{E}_e|}, & |\mathcal{E}_t \cap \mathcal{E}_e| > 0 \\ 0, & \text{otherwise} \, . \end{cases}$$

All the above features score candidate expansion terms with respect to a given query entity. It is also possible to leverage information associated with entities in a knowledge base, without utilizing query entities directly. A specific example of such an approach is given by Xiong and Callan [52], who select expansion terms that have similar type distributions with that of the query.

The type distribution of a term is estimated according to:

$$P(y|\theta_t) = \frac{P(t|y)}{\sum_{y' \in \mathcal{T}} P(t|y')} \, ,$$

where $\mathcal{T}$ is the type taxonomy and the term probability of a type $P(t|y)$ is approximated based on the term's frequency in the descriptions of all entities with that type:

$$P(t|y) = \frac{\sum_{e \in y} c(t;e)}{\sum_{e \in y} l_e} .$$

Similarly, the type distribution of the query is estimated according to:

$$P(y|\theta_q) = \frac{P(q|y)}{\sum_{y' \in \mathcal{T}} P(q|y')} ,$$

where $P(q|y)$ is the product of the query terms' likelihood given type y:

$$P(q|y) = \prod_{q_i} P(q_i|y) .$$

Then, expansion terms are selected based on the similarity of their type distributions θ_t to the type distribution of the query θ_q, measured by negative *Jensen–Shannon divergence*:

$$score_{JSD}(t;q) = -\frac{1}{2} KL(\theta_q||\theta_{q,t}) - \frac{1}{2} KL(\theta_t||\theta_{q,t}) , \qquad (8.6)$$

where

$$P(y|\theta_{q,t}) = \frac{1}{2}\big(P(y|\theta_q) + P(y|\theta_t)\big) .$$

Notice that the estimate in Eq. (8.6) depends only on the query and not on the query entities. Further note that all unsupervised term importance estimates from Sect. 8.2.3 can also be used as features in supervised term selection.

8.2.4.2 Training

To be able to apply supervised learning, target labels are required. The question is: How to measure if a term is a good expansion term? Cao et al. [8] propose to identify the ground truth labels of terms according to their direct impact on retrieval effectiveness. Formally, the *gain* attained by appending the candidate expansion term t to the original query q (denoted by the $\oplus$ operator) is measured as:

$$\delta(t) = \frac{\zeta(q \oplus t) - \zeta(q)}{\zeta(q)} ,$$

where ζ can be any standard IR evaluation measure, such as MAP or NDCG. Then, terms above a certain threshold (0.005 in [8]) may be considered as good expansion terms (target label $+1$), while the rest being bad terms (target label -1).

Instead of measuring the direct impact of terms with respect to some retrieval measure, Xiong and Callan [52] use their influence on ranking scores. We write

$\mathcal{R}^+$ and $\mathcal{R}^-$ to denote the set of relevant and irrelevant documents for query q, respectively. Further, $score(d;q)$ denotes the retrieval score of document d for q. The gain from a term over the retrieved documents is then calculated as:

$$\delta(t) = \frac{1}{|\mathcal{R}^+|} \sum_{d \in \mathcal{R}^+} \big(score(d;q \oplus t) - score(d;q)\big)$$

$$- \frac{1}{|\mathcal{R}^-|} \sum_{d \in \mathcal{R}^-} \big(score(d;q \oplus t) - score(d;q)\big) \, .$$

Xiong and Callan claim that this latter formulation "reflects an expansion term's effectiveness more directly" [52][4] and performs better experimentally.

8.3 Projection-Based Methods

Traditional keyword-based IR models have an inherent limitation of not being able to retrieve (relevant) documents that have no explicit term matches with the query. While query expansion can remedy this to some extent, the limitation still remains. Concept-based retrieval methods attempt to tackle this challenge by relying on auxiliary structures to obtain semantic representations of queries and documents in a higher-level concept space. Such structures include controlled vocabularies (dictionaries and thesauri) [28, 47], ontologies [9], and entities from a knowledge repository [23]. Our interest here is in the latter group.

The overall idea is "to construct a high-dimensional latent entity space, in which each dimension corresponds to one entity, and map both queries and documents to the latent space accordingly" [35]. The relevance between a query and a document is then estimated based on their projections to this latent entity space. This approach allows to uncover hidden (latent) semantic relationships between queries and documents. See Fig. 8.1b for an illustration.

This idea is related to that of topic modeling, as developed in *latent semantic indexing* [19] and *latent Dirichlet allocation* [3]. While topic models can now be computed on web-scale [32], their utility to improve retrieval effectiveness is limited. For example, Yi and Allan [57] have demonstrated that relevance models (cf. Sect. 8.2.1) consistently outperform more elaborate topic modeling methods. Latent entity representations, on the other hand, may be obtained at a relatively low cost, are easy to interpret, and have been clearly shown to improve retrieval effectiveness.

In this section, we present three specific approaches for ranking documents using latent entity representations.

---

[4]It is more direct in the sense that changes in a given evaluation metric only happen if a given expansion term manages to affect the ranking scores to an extent that documents exchange positions.

8.3.1 Explicit Semantic Analysis

Explicit semantic analysis (ESA) is an influential concept-based retrieval method by Gabrilovich and Markovitch [26], where "the semantics of a given word are described by a vector storing the word's association strengths to Wikipedia-derived concepts" [23]. Unlike in latent semantic analysis (LSA) [21], the use of a knowledge repository gives meaningful interpretation to each element (concept) in the vector representation, hence the name "explicit." Work on ESA has primarily focused on using Wikipedia as the underlying knowledge repository [22, 23, 25, 26]. Nevertheless, it may be used with any other knowledge repository, provided that it has a sufficient coverage of concepts and concepts have textual descriptions associated with them. For terminological consistency, we will continue to use the term "entity" instead of "concept," when referring to entries of a knowledge repository,[5] but follow the terminology "concept vector" and "concept space" from the original work.

8.3.1.1 ESA Concept-Based Indexing

The semantic representation of a given term t is a *concept vector* of length $|\mathcal{E}|$:

$$\mathbf{t} = \langle w(e_1, t), \ldots, w(e_{|\mathcal{E}|}, t) \rangle,$$

where each element of the vector corresponds to an entity in the knowledge repository and its value quantifies the strength of the association between term t and the given entity. For a given term-entity pair, $w(e, t)$ is computed by taking the TF-IDF weight of t in the description of e (in ESA, the Wikipedia article of e). Further, cosine normalization is applied to disregard differences in entity representation length:

$$w(e, t) = \frac{TFIDF(t, e)}{\sqrt{\sum_{t' \in \mathcal{V}} TFIDF(t', e)^2}}.$$

The semantic representation of a given piece of text (bag of terms) is computed by taking the centroid of the individual terms' concept vectors. Formally, the concept vector corresponding to input text z is given by $\mathbf{z} = \langle w(e_1, z), \ldots, w(e_{|\mathcal{E}|}, z) \rangle$. Each element of this vector represents the relatedness of the corresponding entity to the input text. The value of the jth vector element is calculated as:

$$w(e_j, z) = \frac{1}{l_z} \sum_{t \in z} c(t; z) \, w(e_j, t),$$

where l_z is the length of z and $c(t; z)$ is the number of times term t appears in z. See Fig. 8.3 for an illustration.

---

[5]We refer back to Sect. 1.1.1 for the difference between concepts and entities.

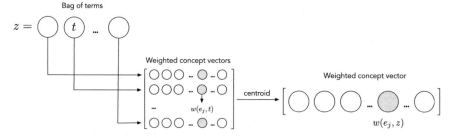

Fig. 8.3 Semantic representation of a piece of text (z) using explicit semantic analysis (ESA) [26]

Given that these concept-based vectors are sparse, with most weights being zero, they can be efficiently represented using an inverted index. These inverted index representations of concept vectors are further pruned by retaining only the top-k entities with the highest weights. In [23], this cutoff is applied to both term and text concept vectors ($\mathbf{t}$ and $\mathbf{z}$, respectively), with k set to 50. This helps to eliminate spurious and insignificant entity associations and also reduces index size.

8.3.1.2 ESA Concept-Based Retrieval

The semantic similarity between query q and document d may be computed by mapping both to the ESA concept space, and taking the cosine similarity of their concept vectors. That is, $score(q;d) = \cos(\mathbf{q},\mathbf{d})$, where $\mathbf{q}$ and $\mathbf{d}$ denote the concept vectors corresponding to q and d, respectively.

One issue that arises here is that long documents are difficult to map to the concept space. As Egozi et al. [23] explain, "a small part of a long document might be relevant to the current query, but the semantics of this part may be underrepresented in the concepts vector for the full document." The proposed solution, motivated by prior work on passage-based retrieval [7], is to break up the document into shorter passages. In [23], passages are of fixed length and overlapping. Each passage, $s \in d$, is represented by its own concept vector, $\mathbf{s}$, and matched against the query. The final retrieval score combines the full document's similarity score with that of the best performing passage:

$$score_{ESA}(q;d) = \cos(\mathbf{q},\mathbf{d}) + \max_{s \in d} \cos(\mathbf{q},\mathbf{s}) \ .$$

Due to the fact that queries are short and noisy, the initially generated query concept vector needs further refinement. Egozi et al. [23] propose to utilize the idea of pseudo relevance feedback for this purpose. First, keyword-based retrieval (using q) is performed on the passage level. Then, the top-k passages are treated as pseudo-relevant (i.e., positive) examples, while the bottom-k passages are taken to be pseudo-non-relevant (i.e., negative) examples. Then, a subset of the initial query entities is selected, based on the positive and negative example passages, resulting in a modified query $\mathbf{q}'$. Finally, documents are ranked using the refined concept-based query $\mathbf{q}'$. We refer to [23] for further details on how to select which entities to retain in the modified query concept vector.

8.3.2   Latent Entity Space Model

Liu and Fang [35] present the *latent entity space* (LES) model, which is based on a generative probabilistic framework. The document's retrieval score is taken to be a linear combination of the latent entity space score and the original query likelihood score:[6]

$$score_{LES}(q;d) = \alpha \underbrace{\sum_{e \in \mathcal{E}} P(q|e)P(e|d)}_{\text{LES score}} + (1 - \alpha)\,P(q|d)\,. \tag{8.7}$$

In the absence of labeled training data, Liu and Fang [35] suggest to set interpolation parameter α to a value between 0.5 and 0.7. The latent entity space score is calculated as a linear combination over latent entities ($e \in \mathcal{E}$) and involves the estimation of two components: *query projection*, $P(q|e)$, and *document projection*, $P(e|d)$. Next, we detail the estimation of these two probabilities.

Query Projection   The probability $P(q|e)$ may be interpreted as the likelihood of the query q being generated by entity e. A straightforward option is to base this estimate on the language model of the entity, θ_e. This language model may be constructed from the entity's description (cf. Sect. 3.3.1.1). The query projection probability is then computed by taking a product over the query terms (which is essentially the query likelihood score of the entity):

$$P(q|e) = \prod_{t \in q} P(t|\theta_e)^{c(t;q)}\,.$$

Another approach to estimating this probability is to leverage the set of query entities $\mathcal{E}_q$ (which we have obtained in Sect. 8.1) in a pairwise manner:

$$P(q|e) \propto \sum_{e' \in \mathcal{E}_q} sim(e,e')\,P(e'|q)\,, \tag{8.8}$$

where $sim(e,e')$ may be any symmetric pairwise similarity measure (in [35] the cosine similarity between θ_e and $\theta_{e'}$ is used; see Sect. 4.5.1 for other possibilities), and $P(e'|q)$ is the query association probability for e'.

Liu and Fang [35] find that the latter, entity-similarity-based method works better than the former, unigram-based approach. Using Eq. (8.8) implies that the summation in Eq. (8.7) can be restricted to the set of query entities $\mathcal{E}_q$ as opposed to

---

[6]In [35], the scores of the two components are re-normalized to make them compatible.

Table 8.2 Features used in EsdRank [51]

Group	Description
Query-entity features	
$P(e\|q)$	Query-entity association probability (cf. Sect. 8.1)
$score_{ER}(e; q)$	Entity retrieval score (relevance of e given q)
$sim(\mathcal{T}_e, \mathcal{T}_q)$	Type-based similarity between the entity ($\mathcal{T}_e$) and the query ($\mathcal{T}_q$)
$maxSim(e, \mathcal{E}_q)$	Max. pairwise similarity between e and other query entities $e' \in \mathcal{E}_q$
$avgSim(e, \mathcal{E}_q)$	Avg. pairwise similarity between e and other query entities $e' \in \mathcal{E}_q$
Entity-document features	
$sim(e, d)$	Textual similarity between e and d
$sim(\mathcal{T}_e, \mathcal{T}_d)$	Type-based similarity between the entity ($\mathcal{T}_e$) and the document ($\mathcal{T}_d$)
$numMentioned(\mathcal{E}_e^i, d)$	Number of related entities (at most i hops from e) mentioned in d
Other features	
$IDF(e)$	IDF score of the entity (based on the number of documents that are annotated with e)
$quality(d)$	Quality score of d (document length, SPAM score, PageRank, etc.)

Note that many of these features are instantiated multiple times, using different similarity methods or parameter configurations

the entire entity catalog $\mathcal{E}$, which will have a significant positive effect on efficiency. It is further shown in [35] that the quality of entity language models θ_e can have a significant impact on end-to-end retrieval performance.

Document Projection The probability $P(e|d)$ may be interpreted as the projection of document d to the latent entity space. It may be estimated using existing document retrieval models, e.g., by computing entity likelihood (i.e., the probability of e generated from the language model of d). Liu and Fang [35] estimate $P(e|d)$ based on the negative cross-entropy between the document and entity language models:

$$P(e|d) = \exp\left(-CE(\theta_e \parallel \theta_d)\right) = \exp\left(\sum_{t \in \mathcal{V}} P(t|\theta_e) \log P(t|\theta_d)\right).$$

8.3.3 EsdRank

The idea of using entities as a bridge between documents and queries may also be expressed in a discriminative learning framework. Xiong and Callan [51] introduce *EsdRank* for ranking documents, using a combination of query-entity and entity-document features. These correspond to the notions of query projection and document projection components of LES, respectively, from before. Using a discriminative learning framework, additional signals can also be incorporated easily, such as entity popularity or document quality. Next, we present these main groups of features, which are summarized in Table 8.2. We then continue by briefly discussing the learning-to-rank algorithm used in EsdRank.

8.3.3.1 Features

Query-entity features include the following:

- *Query entity probability*, which can be computed by different methods in the query-to-entity mapping step (cf. Sect. 8.1).
- *Entity retrieval score*, which may be computed by using any standard retrieval model (such as LM, BM25, SDM) to score the query against the entity's description (in the case of unstructured entity representations) or a given entity field (in the case of structured entity representations).
- *Type-based similarity* between the target types of the query, $\mathcal{T}_q$, and the types of the entity, $\mathcal{T}_e$. The former may be computed using the target type detection methods we presented in Sect. 7.2, while the latter is provided in the knowledge base. Specifically, Xiong and Callan [51] consider the top three types identified for the query.
- *Entity similarity* considers the similarity between the candidate entity and other query entities in a pairwise manner. Then, these pairwise similarities are aggregated by taking their maximum or average.

Entity-document features comprise the following:

- *Text-based similarity* is measured between the document and the entity description (or fields thereof). These may be computed, e.g., by using cosine similarity or by applying standard retrieval models (LM, BM25, SDM, etc.) to score documents by treating the entity description as the search query.
- *Type-based similarity* may be computed between documents and entities, similarly to how it is done for queries and entities. Assigning types to the document may be approached as a multiclass classification problem or as a ranking problem (as it was done for queries in Sect. 7.2). Ultimately, the top-k types, i.e., with the highest confidence score, are considered for the document ($k = 3$ in [51]).
- *Graph-based similarity* considers the relationships of the entity. Let $\mathcal{E}_e^i$ denote the set of entities that are reachable from entity e in i hops, where $i \in [0..2]$ and $\mathcal{E}_e^0 = \{e\}$. Then, graph-based similarity in [51] is measured by the number of entities in $\mathcal{E}_e^i$ that are mentioned in d.

Other features may include, among others:

- *Entity frequency*, which reflects the popularity of the entity within the corpus and can be measured, e.g., using IDF.
- *Document quality indicators*, such as document length, URL length, SPAM score, PageRank score, number of inlinks, etc.

8.3.3.2 Learning-to-Rank Model

Xiong and Callan [51] introduce Latent-ListMLE, which extends the ListMLE [50] method. ListMLE is a listwise learning-to-rank algorithm that uses a parametric model to estimate the probability of a document ranking being generated by a query. Then, it employs maximum likelihood estimation (MLE) to find the parameters that maximize the likelihood of the best ranking. One important assumption that ListMLE makes, to keep the optimization problem tractable, is that the probability of a document being ranked at a given position i is independent from all those documents that are ranked at earlier positions, $1 \ldots i - 1$.

Latent-ListMLE extends ListMLE by adding a latent layer of candidate entities in the generation process. Similarly to ListMLE, it is assumed that the probabilities of selecting entities and documents at each position are independent of those that have been selected at earlier positions. However, instead of conditioning the document ranking probability directly on the query, first entities are sampled based on the query, then the probability of a document ranking is conditioned on the sampled entities. The probability of a document ranking $\mathbf{d} = \langle d_1, \ldots, d_k \rangle$ given q is:

$$
P(\mathbf{d}|q; w, \theta) = \prod_{i=1}^{k} \sum_{e \in \mathcal{E}_q} P\left(d_i | e, \mathcal{D}_q(i,k)\right) P(e|q),
$$

where $\mathcal{D}_q(i,k) = \{d_i, \ldots, d_k\}$ is the set of documents that were not ranked in positions $1, \ldots, i - 1$. The parameters of the model, w and θ, are learned using MLE and the EM algorithm. We refer to Xiong and Callan [51] for the details.

8.4 Entity-Based Representations

The main difference between the approaches in the previous section and those that will follow below is that instead of projecting documents to a *latent* entity layer, we will make use of *explicit* entity annotations of documents. We shall assume that the document has been annotated by some entity linking tool. The resulting set $\mathcal{E}_d$ of entities will be referred to as *document entities*. Entities may be blended with terms in a single representation layer, such as it is done in entity-based language models (Sect. 8.4.1). Alternatively, a separate bag-of-entities representation may be introduced and combined with the traditional bag-of-terms representation (Sect. 8.4.2).

8.4.1 Entity-Based Document Language Models

Raviv et al. [42] introduce entity-based language models (ELM), which consider individual terms as well as term sequences that have been annotated as entities (both in documents and in queries). They implement this idea by extending the vocabulary of terms ($\mathcal{V}$) with entities ($\mathcal{E}$). We shall write x to denote a vocabulary token, which here may be a term or an entity, $x \in \mathcal{V} \cup \mathcal{E}$. Further, we write $c(x; d)$ to denote the (pseudo) count of x in document d. The representation length of the document is then given by $l_d = \sum_{x \in d} c(x; d)$. The maximum likelihood estimate of token x given d is defined as:

$$P(x|d) = \frac{c(x;d)}{l_d} . \tag{8.9}$$

This maximum likelihood estimate is then smoothed with a background (collection-level) language model analogously to how it is done for unigram language models, e.g., using Dirichlet smoothing:

$$P(x|\theta_d) = \frac{c(x;d) + \mu P(x|\mathcal{D})}{l_d + \mu} , \tag{8.10}$$

where μ is a smoothing parameter, and the collection language model is also a maximum likelihood estimate, computed over the set $\mathcal{D}$ of documents:

$$P(x|\mathcal{D}) = \frac{\sum_{d \in \mathcal{D}} c(x;d)}{\sum_{d \in \mathcal{D}} l_d} .$$

What remains to be defined is how the token pseudo-counts are computed. Raviv et al. [42] propose two alternatives:

- **Hard confidence-level thresholding** Only those entity annotations are considered in the document that are above a given (pre-defined) threshold $\tau \in [0, 1]$. That is, the pseudo-count of token x is (1) the raw frequency of the term in the document, if the token is a term, and (2) the total number of mentions of the entity in the document with a minimum annotation confidence of τ, if x is an entity:

$$\tilde{c}(x;d) = \begin{cases} \lambda c(x;d) , & x \in \mathcal{V} \\ (1 - \lambda) \sum_{i=1}^{l_d} \mathbb{1}(x_i = x, score_{EL}(x_i;d) \geq \tau) , & x \in \mathcal{E} , \end{cases}$$

where x_i refers to the token at position i in the document and $score_{EL}(x_i; d)$ is the entity linking confidence associated with that token. The binary indicator function $\mathbb{1}()$ returns 1 if its argument evaluates to true, otherwise returns 0. The λ parameter controls the relative importance given to term vs. entity tokens.

- **Soft confidence-level thresholding** Instead of considering only entity annotations above a given threshold and treating them uniformly, the second method recognizes all entities that are linked in the document and weighs them by their corresponding confidence levels:

$$\tilde{c}(x;d) = \begin{cases} \lambda c(x;d) , & x \in \mathcal{V} \\ (1 - \lambda) \sum_{i=1}^{l_d} \mathbb{1}(x_i = x) \, score_{EL}(x_i;d) , & x \in \mathcal{E} . \end{cases}$$

The ranking of documents is based on the negative cross-entropy (CE) between the query and document language models:[7]

$$score_{ELM}(d;q) = -CE(\theta_q||\theta_d) = \sum_{x \in \mathcal{V} \cup \mathcal{E}} P(x|\theta_q) \log P(x|\theta_d) \,,$$

where the query language model θ_q is a maximum-likelihood estimate (as in Eq. (8.9), but by replacing q with d). The document language model θ_d is instantiated by Eq. (8.10).

8.4.2 Bag-of-Entities Representation

Entity-based language models use a single representation layer, in which terms and entities are mixed together. We shall now discuss a line of work by Xiong et al. [53–55], where the term-based and entity-based representations are kept apart and are used in "duet." That is, queries and documents are represented in the term space as well as in the entity space. The latter is referred to as the *bag-of-entities* representation. Recall that we have already discussed this idea in the context of the ad hoc entity retrieval task in Sect. 4.2.2.[8]

8.4.2.1 Basic Ranking Models

Xiong et al. [53] present two basic ranking models based on bag-of-entities representations.

- **Coordinate Match** ranks documents based on the number of query entities they mention:

$$score_{CM}(d;q) = \sum_{e \in \mathcal{E}_q} \mathbb{1}\big(c(e;d) > 0\big) \,. \tag{8.11}$$

- **Entity Frequency** also considers the frequency of query entities in documents:

$$score_{EF}(d;q) = \sum_{e \in \mathcal{E}_q} c(e;q) \log c(e;d) \,. \tag{8.12}$$

---

[7]Note that scoring based on cross-entropy $CE(\theta_q||\theta_d)$ is rank-equivalent to scoring based on Kullback–Leibler divergence $KL(\theta_q||\theta_d)$ [58].

[8]Interestingly, the idea of a bag-of-entities representation was proposed independently and published at the same conference by Hasibi et al. [30] and Xiong et al. [53] for entity retrieval and document retrieval, respectively.

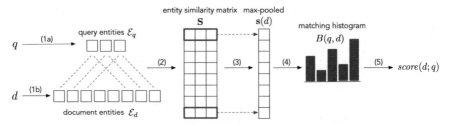

Fig. 8.4 Overview of the explicit semantic ranking (ESR) model [55]. The steps are: (1) entity linking in queries (1a) and in documents (1b); (2) computing pairwise entity similarities; (3) max pooling along the query dimension; (4) bin-pooling; (5) ranking documents using the histogram counts as features

These ranking functions are used to re-rank the top-k documents retrieved by a standard term-based retrieval model ($k = 100$ in [53]). Despite their simplicity, both models were shown to significantly outperform conventional term-based retrieval models [53].

8.4.2.2 Explicit Semantic Ranking

The *explicit semantic ranking* (ESR) [55] model incorporates relationship information from a knowledge graph to enable "soft matching" in the entity space. Figure 8.4 depicts an overview of the approach.

ESR first creates a query-document *entity similarity matrix* **S**. Each element $\mathbf{S}(e, e')$ in this matrix represents the similarity between a query entity $e \in \mathcal{E}_q$ and a document entity $e' \in \mathcal{E}_d$:

$$\mathbf{S}(e, e') = \cos(\mathbf{e}, \mathbf{e}') \, ,$$

where $\mathbf{e}$ is the embedding vector of entity e. In [55], entity embeddings are trained based on neighboring entities (i.e., entity relationships) in a knowledge graph.

ESR performs two pooling steps. The first one is max-pooling along the query dimension:

$$\mathbf{s}(d) = \max_{e \in \mathcal{E}_q} \mathbf{S}(e, \mathcal{E}_d) \, .$$

The second step is bin-pooling (introduced as *matching histogram mapping* in [29]), to group and count the number of document entities according to the strength of their matches to the query:

$$B_i(q, d) = \log \sum_j \mathbb{1}(st_i \leq \mathbf{s}_j(d) < ed_i) \, , \tag{8.13}$$

Fig. 8.5 Query-document
matching in the word-entity
duet framework [54]

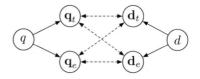

where $[st_i, ed_i)$ is the score range for the ith bin, and $B_i(q,d)$ is the number of
entities that fall into that bin. The bin ranges in [55] are $[0, 0.25)$, $[0.25, 0.5)$,
$[0.5, 0.75)$, $[0.75, 1)$, $[1, 1]$. Negative bins are discarded. The rightmost bin with
range $[1, 1]$ counts the *exact matches* in the entity space, while the other bins
correspond to various degrees of *soft matches*. The resulting bin scores B_i are fed
as features to a standard learning-to-rank model.

8.4.2.3 Word-Entity Duet Framework

Most recently, Xiong et al. [54] present the *word-entity duet framework*, which
also incorporates cross-space interactions between term-based and entity-based
representations, leading to four types of matches. The idea is illustrated on Fig. 8.5,
where $\mathbf{q}_t$ and $\mathbf{d}_t$ denote the bag-of-words, while $\mathbf{q}_e$ and $\mathbf{d}_e$ denote the bag-of-entities
representations of the query and the document, respectively. Each element in these
vectors corresponds to the frequency of a given term/entity in the query/document.
Based on these representations, query-document matching may be computed in four
different ways:

- **Query terms to document terms** ($match(\mathbf{q}_t, \mathbf{d}_t)$): This corresponds to traditional
  term-based matching between a query and a document, and can be computed
  using standard retrieval models (e.g., LM or BM25) on various document fields
  (title and body in [54]).
- **Query entities to document terms** ($match(\mathbf{q}_e, \mathbf{d}_t)$): Relevance matching is
  performed by using the names or descriptions of query entities as (pseudo-
)queries, and employing standard retrieval models to score them against the
  document (title or body fields).
- **Query terms to document entities** ($match(\mathbf{q}_t, \mathbf{d}_e)$): Similar in spirit to the
  previous kind of matching, the relevance between the query text and document
  entities is estimated by considering the names and descriptions of those entities.
  However, since the document may mention numerous entities, only the top-k
  ones with the highest relevance to the query are considered. Specifically, Xiong
  et al. [54] consider the top three entities from the document's title field and the
  top five entities from the document's body.
- **Query entities to document entities** ($match(\mathbf{q}_e, \mathbf{d}_e)$): Matches in the entity space
  can be measured using the coordinate match and entity frequency methods, cf.
  Eqs. (8.11) and (8.12). Additionally, matches can also be considered by using
  entity embeddings from a knowledge graph. In particular, Xiong et al. [54] learn

entity embeddings using the TransE model [4] and then use the ESR matching histogram scores (cf. Eq. (8.13)) as query-document ranking features.

The four-way matching scores from above are combined in a feature-based ranking framework.

8.4.2.4   Attention-Based Ranking Model

A main challenge with entity-based representations is the inherent uncertainty of automatic query entity annotations. It is inevitable that some entity mentions will be mistakenly linked, especially in short queries. Consequently, documents that match these (erroneous) entities would end up being promoted in the ranking. Xiong et al. [54] address this problem by developing an attention mechanism that can effectively demote noisy query entities.

A total of four attention features are designed, which are extracted for each query entity. *Entity ambiguity* features are meant to characterize the risk associated with an entity annotation. These are: (1) the entropy of the probability of the surface form being linked to different entities (e.g., in Wikipedia), (2) whether the annotated entity is the most popular sense of the surface form (i.e., has the highest commonness score, cf. Eq. (5.3))), and (3) the difference in commonness scores between the most likely and second most likely candidates for the given surface form. The fourth feature is *closeness*, which is defined as the cosine similarity between the query entity and the query in an embedding space. Specifically, a joint entity-term embedding is trained using the skip-gram model [41] on a corpus, where entity mentions are replaced with the corresponding entity identifiers. The query's embedding is taken to be the centroid of the query terms' embeddings.

We write $\Phi_{\mathbf{q}_t, \mathbf{d}_t}$, $\Phi_{\mathbf{q}_e, \mathbf{d}_t}$, $\Phi_{\mathbf{q}_t, \mathbf{d}_e}$, and $\Phi_{\mathbf{q}_e, \mathbf{d}_e}$ to refer to the four-way query-document features in the word-entity duet framework (cf. Sect. 8.4.2.3). Attention features are denoted as Φ_{Attn}. Using these five groups of features, the *AttR-Duet* model aims to learn a ranking function $score(d;q)$ that will be used for re-ranking an initial set of candidate documents.

The architecture of *AttR-Duet* is shown in Fig. 8.6. The model takes four matrices as input: $\mathbf{R}_t$, $\mathbf{R}_e$, $\mathbf{A}_t$, and $\mathbf{A}_e$. In the following, we will suppose that the query contains n words $q = \langle q_1, \ldots, q_n \rangle$ and there are m query entities $\mathcal{E}_q = \{e_1, \ldots, e_m\}$. $\mathbf{R}_t$ and $\mathbf{R}_e$ are ranking features for terms and entities, respectively. The rows of these matrices are made up of the word-duet feature vectors corresponding to each query term/entity:

$$\mathbf{R}_t(q_i, :) = \Phi_{\mathbf{q}_t, \mathbf{d}_t}(q_i) \sqcup \Phi_{\mathbf{q}_t, \mathbf{d}_e}(q_i)$$
$$\mathbf{R}_e(e_j, :) = \Phi_{\mathbf{q}_e, \mathbf{d}_t}(e_j) \sqcup \Phi_{\mathbf{q}_e, \mathbf{d}_e}(e_j),$$

where $\sqcup$ is a vector concatenation operator. $\mathbf{A}_t$ and $\mathbf{A}_e$ are attention features for terms and entities, respectively. Recall that the main objective is to handle the

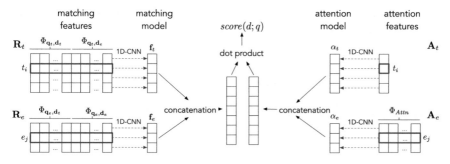

Fig. 8.6 Architecture of the AttR-Duet model [54]. The left side models query-document matching using the four-way term-entity features. The right side models the importance of query entities via attention features. The combination of these two yields the final document score

uncertainty of query entity annotations. Therefore, for terms, uniform attention is used; for entities, we employ the attention features introduced above:

$$\mathbf{A}_t(q_i, :) = 1$$

$$\mathbf{A}_e(e_j, :) = \Phi_{Attn}(e_j) \, .$$

The matching part (left side of Fig. 8.6) consists of two *convolutional neural networks* (CNNs), one for matching query terms ($\mathbf{R}_t$) and another for matching query entities ($\mathbf{R}_e$) against the document d. The convolution is applied on the term/entity dimension, "assuming that the ranking evidence from different query words [terms] or entities should be treated the same" [54]. Using a single CNN layer, one filter, and a linear activation function, the matching scores of terms and entities can be written as the following linear models:

$$\mathbf{f}_t(q_i) = \mathbf{w}_t^m \cdot \mathbf{R}_t(q_i, :) + b_t^m$$

$$\mathbf{f}_e(e_j) = \mathbf{w}_e^m \cdot \mathbf{R}_e(e_j, :) + b_e^m \, ,$$

where $\cdot$ is the dot product; $\mathbf{f}_t$ and $\mathbf{f}_e$ are n- and m-dimensional vectors, respectively; $\{\mathbf{w}_t^m, \mathbf{w}_e^m, b_t^m, b_e^m\}$ are the matching parameters to learn.

The attention part (right side of Fig. 8.6) also contains two CNNs, one for query terms ($\mathbf{A}_t$) and one for query entities ($\mathbf{A}_e$), using the same convolution idea as before. Using a single CNN layer and ReLU activation (to ensure non-negative attention weights), the attention weights on terms and entities can be written as:

$$\alpha_t(t) = ReLU \left(\mathbf{w}_t^a \cdot \mathbf{A}_t(t, :) + b_t^a \right)$$

$$\alpha_e(e) = ReLU \left(\mathbf{w}_e^a \cdot \mathbf{A}_e(e, :) + b_e^a \right) \, ,$$

where $\{\mathbf{w}_t^a, \mathbf{w}_e^a, b_t^a, b_e^a\}$ are the attention parameters to learn.

The final model, AttR-Duet, combines the matching and attention scores as:

$$score_{AD}(d; q) = \mathbf{f}_t \cdot \alpha_t + \mathbf{f}_e \cdot \alpha_e \, .$$

The matching part and attention parts of the model are learned simultaneously, by optimizing pairwise hinge loss:

$$L(q, \mathcal{R}^+, \mathcal{R}^-) = \sum_{d \in \mathcal{R}^+} \sum_{d' \in \mathcal{R}^-} [1 - score(d; q) + score(d'; q)]_+ \, ,$$

where $\mathcal{R}^+$ and $\mathcal{R}^-$ denote the set of relevant and non-relevant documents, respectively, and $[\,]_+$ is the hinge loss.

8.5   Practical Considerations

Efficiency is a key concern when serving search results. Compared to traditional term-based approaches, the computational overhead involved with the presented approaches stems from two components: (1) identifying the query entities, and (2) leveraging query entities in document scoring. Modern entity linking tools can already handle (1) with low latency, cf. Sect. 7.3. As for (2), document scoring is typically implemented as a re-ranking mechanism. That is, an initial retrieval is performed using the original query and a standard retrieval method, such as BM25 or LM. Then, the top-k scoring documents are re-ranked using the more advanced retrieval method. This is the same standard practice as in learning-to-rank [38]. Using a smaller k can result in markedly improved efficiency compared to a larger k. At the same time, using lower k values limits the scope, and hence potential, of the advanced method. A typical k value used in published work is around 100 [35, 51].

The efficiency of query expansion methods (Sect. 8.2) can be strongly affected by the number of expansion terms used (e.g., Meij et al. [40] consider maximum ten expansion terms). For approaches that operate with entity-based representations of documents (Sects. 8.4 and 8.3) entity annotations of documents can be performed offline and the linked entities can be stored in an inverted index structure. Similarly, entity descriptions can be constructed and indexed offline.

8.6   Resources and Test Collections

Experimental evaluation is commonly conducted using the test suites of the TREC 2009–2014 Web track [11–16], which employ the ClueWeb09 and ClueWeb12 collections. Additionally, the Robust04 newswire collection has also been used, with a set of topics from the ad hoc task in TREC 6–8 (#301–450) and topics developed for the TREC 2003–2004 Robust track (#601–700) [48]. See Table 8.3 for an overview. The reference knowledge base is typically Freebase, due to the availability of Freebase-annotated versions of the ClueWeb corpora, released by Google, referred to as the FACC1 dataset [27]; see Sect. 5.9.2. In addition to document annotations, the FACC1 dataset also

Table 8.3 Test collections for evaluating document retrieval methods that leverage entities

Document collection	#Documents	TREC topics	#Queries
Robust04	528k	Ad hoc #301–450, Robust #601–700	250
ClueWeb09-B	50M	Web #1–#200	200
ClueWeb12-B13	52M	Web #201–#300	100

contains manual entity annotations for the TREC 2009–2012 Web track queries. These annotations are limited to explicitly mentioned entities; 94 out of the 200 queries contain an entity. Dalton et al. [18] provide manually revised query annotations to improve recall, resulting in 191 of the 200 queries containing an entity.[9] For obtaining automatic entity annotations, TAGME [24] is a popular choice.

8.7 Summary

In this chapter, we have focused on leveraging entities for ad hoc document retrieval. The guiding principle behind these approaches is to obtain a semantically richer representation of the user's information need by identifying entities that are related to the query. This knowledge can then be utilized in the document retrieval process in various ways. In particular, we have discussed three families approaches: (1) *expansion-based*, which uses entities as a source for expanding the query with additional terms; (2) *projection-based*, where the relevance matching between a query and a document is performed by projecting them to a latent space of entities; and (3) *entity-based*, where explicit semantic representations of queries and documents are obtained in the entity space to augment the term-based representations. Moving from (1) to (2) and then from (2) to (3) corresponds to making increasingly more explicit use of entities, which, as it turns out, also translates to increasingly higher retrieval effectiveness. Entity-based representations, according to the current state of the art, can outperform a language modeling baseline by over 80% and a strong learning-to-rank baseline by over 20% in terms of NDCG@20, measured on the ClueWeb09-B collection [54].

8.8 Further Reading

It is also possible to combine the different perspectives of the discussed methods in a hybrid approach. For example, the EQFE method by Dalton et al. [18] uses explicit entity annotations of documents and performs query expansion based on entities and their properties (types and categories). Thereby, it bears some characteristics of

---

[9]http://ciir.cs.umass.edu/downloads/eqfe/.

both entity-based representations and expansion-based methods. However, they do not use query expansion in the conventional sense (i.e., creating an expanded query model), but rather expand ranking features which are combined in a learning-to-rank approach.

Entity-based text representation may be utilized in many other tasks, e.g., computing document similarity [44], text classification [10], or question answering [6, 45]. Medical search is another prominent example for the use of controlled vocabulary representations, with a lot of work conducted in the context of the TREC Genomics track [31, 36, 46].

References

1. Balog, K., Weerkamp, W., de Rijke, M.: A few examples go a long way: Constructing query models from elaborate query formulations. In: Proceedings of the 31st annual international ACM SIGIR conference on Research and development in information retrieval, SIGIR '08, pp. 371–378. ACM (2008). doi: 10.1145/1390334.1390399
2. Bendersky, M., Metzler, D., Croft, W.B.: Effective query formulation with multiple information sources. In: Proceedings of the Fifth ACM International Conference on Web Search and Data Mining, WSDM '12, pp. 443–452 (2012). doi: 10.1145/2124295.2124349
3. Blei, D.M., Ng, A.Y., Jordan, M.I.: Latent Dirichlet allocation. J. Mach. Learn. Res. 3, 993–1022 (2003)
4. Bordes, A., Usunier, N., Garcia-Durán, A., Weston, J., Yakhnenko, O.: Translating embeddings for modeling multi-relational data. In: Proceedings of the 26th International Conference on Neural Information Processing Systems, NIPS'13, pp. 2787–2795. Curran Associates Inc. (2013)
5. Brandão, W.C., Santos, R.L.T., Ziviani, N., de Moura, E.S., da Silva, A.S.: Learning to expand queries using entities. J. Am. Soc. Inf. Sci. Technol. pp. 1870–1883 (2014)
6. Cai, L., Zhou, G., Liu, K., Zhao, J.: Large-scale question classification in cQA by leveraging Wikipedia semantic knowledge. In: Proceedings of the 20th ACM International Conference on Information and Knowledge Management, CIKM '11, pp. 1321–1330. ACM (2011). doi: 10.1145/2063576.2063768
7. Callan, J.P.: Passage-level evidence in document retrieval. In: Proceedings of the 17th Annual International ACM SIGIR Conference on Research and Development in Information Retrieval, SIGIR '94, pp. 302–310. Springer (1994)
8. Cao, G., Nie, J.Y., Gao, J., Robertson, S.: Selecting good expansion terms for pseudo-relevance feedback. In: Proceedings of the 31st Annual International ACM SIGIR Conference on Research and Information Retrieval, SIGIR '08, pp. 243–250. ACM (2008). doi: 10.1145/1390334.1390377
9. Castells, P., Fernandez, M., Vallet, D.: An adaptation of the vector-space model for ontology-based information retrieval. IEEE Trans. on Knowl. and Data Eng. 19(2), 261–272 (2007). doi: 10.1109/TKDE.2007.22
10. Chang, M.W., Ratinov, L., Roth, D., Srikumar, V.: Importance of semantic representation: Dataless classification. In: Proceedings of the 23rd National Conference on Artificial Intelligence - Volume 2, AAAI'08, pp. 830–835. AAAI Press (2008)
11. Clarke, C.L.A., Craswell, N., Soboroff, I.: Overview of the TREC 2009 Web track. In: The Eighteenth Text REtrieval Conference Proceedings, TREC '09. NIST Special Publication 500-278 (2010)
12. Clarke, C.L.A., Craswell, N., Soboroff, I., V. Cormack, G.: Overview of the TREC 2010 Web track. In: The Nineteenth Text REtrieval Conference Proceedings, TREC '10. NIST Special Publication 500-294 (2011)

13. Clarke, C.L.A., Craswell, N., Soboroff, I., Voorhees, E.M.: Overview of the TREC 2011 Web track. In: The Twentieth Text REtrieval Conference Proceedings, TREC '11. NIST Special Publication 500-296 (2012)
14. Clarke, C.L.A., Craswell, N., Voorhees, E.M.: Overview of the TREC 2012 Web track. In: The Twenty-First Text REtrieval Conference Proceedings, TREC '12. NIST Special Publication 500-298 (2013)
15. Collins-Thompson, K., Bennett, P., Diaz, F., Clarke, C.L.A., Voorhees, E.M.: TREC 2013 Web track overview. In: The Twenty-Second Text REtrieval Conference Proceedings, TREC '13. NIST Special Publication 500-302 (2014)
16. Collins-Thompson, K., Macdonald, C., Bennett, P., Diaz, F., Voorhees, E.M.: TREC 2014 Web track overview. In: The Twenty-Third Text REtrieval Conference Proceedings, TREC '14. NIST Special Publication 500-308 (2015)
17. Croft, B., Metzler, D., Strohman, T.: Search Engines: Information Retrieval in Practice. 1st edn. Addison-Wesley Publishing Co. (2009)
18. Dalton, J., Dietz, L., Allan, J.: Entity query feature expansion using knowledge base links. In: Proceedings of the 37th International ACM SIGIR Conference on Research and Development in Information Retrieval, SIGIR '14, pp. 365–374. ACM (2014). doi: 10.1145/2600428.2609628
19. Deerwester, S., Dumais, S.T., Furnas, G.W., Landauer, T.K., Harshman, R.: Indexing by latent semantic analysis. J. Am. Soc. Inf. Sci. Technol. 41(6), 391–407 (1990)
20. Diaz, F., Metzler, D.: Improving the estimation of relevance models using large external corpora. In: Proceedings of the 29th Annual International ACM SIGIR Conference on Research and Development in Information Retrieval, SIGIR '06, pp. 154–161. ACM (2006). doi: 10.1145/1148170.1148200
21. Dumais, S.T.: Latent semantic analysis. Ann. Rev. Info. Sci. Tech. 38(1), 188–230 (2004). doi: 10.1002/aris.1440380105
22. Egozi, O., Gabrilovich, E., Markovitch, S.: Concept-based feature generation and selection for information retrieval. In: Proceedings of the 23rd National Conference on Artificial Intelligence - Volume 2, AAAI'08, pp. 1132–1137. AAAI Press (2008)
23. Egozi, O., Markovitch, S., Gabrilovich, E.: Concept-based information retrieval using explicit semantic analysis. ACM Trans. Inf. Syst. 29(2), 8:1–8:34 (2011)
24. Ferragina, P., Scaiella, U.: TAGME: On-the-fly annotation of short text fragments (by Wikipedia entities). In: Proceedings of the 19th ACM International Conference on Information and Knowledge Management, CIKM '10, pp. 1625–1628. ACM (2010). doi: 10.1145/1871437.1871689
25. Gabrilovich, E., Markovitch, S.: Overcoming the brittleness bottleneck using Wikipedia: Enhancing text categorization with encyclopedic knowledge. In: Proceedings of the 21st National Conference on Artificial Intelligence - Volume 2, AAAI'06, pp. 1301–1306. AAAI Press (2006)
26. Gabrilovich, E., Markovitch, S.: Wikipedia-based semantic interpretation for natural language processing. J. Artif. Int. Res. 34(1), 443–498 (2009)
27. Gabrilovich, E., Ringgaard, M., Subramanya, A.: FACC1: Freebase annotation of Clueweb corpora, version 1. Tech. rep., Google, Inc. (2013)
28. Gonzalo, J., Verdejo, F., Chugur, I., Cigarrin, J.: Indexing with WordNet synsets can improve text retrieval. In: Proceedings of the COLING/ACL'98 Workshop on Usage of WordNet for NLP, pp. 38–44 (1998)
29. Guo, J., Fan, Y., Ai, Q., Croft, W.B.: A deep relevance matching model for ad-hoc retrieval. In: Proceedings of the 25th ACM International on Conference on Information and Knowledge Management, CIKM '16, pp. 55–64. ACM (2016). doi: 10.1145/2983323.2983769
30. Hasibi, F., Balog, K., Bratsberg, S.E.: Exploiting entity linking in queries for entity retrieval. In: Proceedings of the 2016 ACM on International Conference on the Theory of Information Retrieval, ICTIR '16, pp. 209–218. ACM (2016). doi: 10.1145/2970398.2970406

31. Hersh, W., Voorhees, E.: TREC genomics special issue overview. Inf. Retr. **12**(1), 1–15 (2009). doi: 10.1007/s10791-008-9076-6
32. Jagerman, R., Eickhoff, C., de Rijke, M.: Computing web-scale topic models using an asynchronous parameter server. In: Proceedings of the 40th International ACM SIGIR Conference on Research and Development in Information Retrieval, SIGIR '17, pp. 1337–1340. ACM (2017). doi: 10.1145/3077136.3084135
33. Lavrenko, V., Croft, W.B.: Relevance based language models. In: Proceedings of the 24th annual international ACM SIGIR conference on Research and development in information retrieval, SIGIR '01, pp. 120–127. ACM (2001). doi: 10.1145/383952.383972
34. Liu, X., Chen, F., Fang, H., Wang, M.: Exploiting entity relationship for query expansion in enterprise search. Inf. Retr. **17**(3), 265–294 (2014). doi: 10.1007/s10791-013-9237-0
35. Liu, X., Fang, H.: Latent entity space: A novel retrieval approach for entity-bearing queries. Inf. Retr. **18**(6), 473–503 (2015). doi: 10.1007/s10791-015-9267-x
36. Lu, Z., Kim, W., Wilbur, W.J.: Evaluation of query expansion using mesh in pubmed. Inf. Retr. **12**(1), 69–80 (2009). doi: 10.1007/s10791-008-9074-8
37. Lv, Y., Zhai, C.: A comparative study of methods for estimating query language models with pseudo feedback. In: Proceedings of the 18th ACM Conference on Information and Knowledge Management, CIKM '09, pp. 1895–1898. ACM (2009). doi: 10.1145/1645953.1646259
38. Macdonald, C., Santos, R.L., Ounis, I.: The whens and hows of learning to rank for web search. Inf. Retr. **16**(5), 584–628 (2013). doi: 10.1007/s10791-012-9209-9
39. Manning, C.D., Raghavan, P., Schütze, H.: Introduction to Information Retrieval. Cambridge University Press (2008)
40. Meij, E., Trieschnigg, D., de Rijke, M., Kraaij, W.: Conceptual language models for domain-specific retrieval. Inf. Process. Manage. **46**(4), 448–469 (2010). doi: http://dx.doi.org/10.1016/j.ipm.2009.09.005
41. Mikolov, T., Sutskever, I., Chen, K., Corrado, G., Dean, J.: Distributed representations of words and phrases and their compositionality. In: Proceedings of the 26th International Conference on Neural Information Processing Systems, NIPS'13, pp. 3111–3119. Curran Associates Inc. (2013)
42. Raviv, H., Kurland, O., Carmel, D.: Document retrieval using entity-based language models. In: Proceedings of the 39th International ACM SIGIR Conference on Research and Development in Information Retrieval, SIGIR '16, pp. 65–74. ACM (2016). doi: 10.1145/2911451.2911508
43. Rocchio, J.: Relevance feedback in information retrieval. In: Salton, G. (ed.) The SMART Retrieval System—Experiments in Automatic Document Processing. Prentice-Hall, Inc. (1971)
44. Schuhmacher, M., Ponzetto, S.P.: Knowledge-based graph document modeling. In: Proceedings of the 7th ACM International Conference on Web Search and Data Mining, WSDM '14, pp. 543–552. ACM (2014). doi: 10.1145/2556195.2556250
45. Srba, I., Bielikova, M.: A comprehensive survey and classification of approaches for community question answering. ACM Trans. Web **10**(3), 18:1–18:63 (2016). doi: 10.1145/2934687
46. Stokes, N., Li, Y., Cavedon, L., Zobel, J.: Exploring criteria for successful query expansion in the genomic domain. Inf. Retr. **12**(1), 17–50 (2009). doi: 10.1007/s10791-008-9073-9
47. Voorhees, E.M.: Using wordnet to disambiguate word senses for text retrieval. In: Proceedings of the 16th Annual International ACM SIGIR Conference on Research and Development in Information Retrieval, SIGIR '93, pp. 171–180. ACM (1993). doi: 10.1145/160688.160715
48. Voorhees, E.M.: The TREC Robust retrieval track. SIGIR Forum **39**(1), 11–20 (2005). doi: 10.1145/1067268.1067272
49. Weerkamp, W., Balog, K., de Rijke, M.: Exploiting external collections for query expansion. ACM Trans. Web **6**(4), 18:1–18:29 (2012). doi: 10.1145/2382616.2382621
50. Xia, F., Liu, T.Y., Wang, J., Zhang, W., Li, H.: Listwise approach to learning to rank: Theory and algorithm. In: Proceedings of the 25th International Conference on Machine Learning, ICML '08, pp. 1192–1199. ACM (2008). doi: 10.1145/1390156.1390306
51. Xiong, C., Callan, J.: Esdrank: Connecting query and documents through external semi-structured data. In: Proceedings of the 24th ACM International on Conference on

Information and Knowledge Management, CIKM '15, pp. 951–960. ACM (2015a). doi: 10.1145/2806416.2806456

52. Xiong, C., Callan, J.: Query expansion with freebase. In: Proceedings of the 2015 International Conference on The Theory of Information Retrieval, ICTIR '15, pp. 111–120. ACM (2015b). doi: 10.1145/2808194.2809446

53. Xiong, C., Callan, J., Liu, T.Y.: Bag-of-entities representation for ranking. In: Proceedings of the 2016 ACM on International Conference on the Theory of Information Retrieval, ICTIR '16, pp. 181–184. ACM (2016). doi: 10.1145/2970398.2970423

54. Xiong, C., Callan, J., Liu, T.Y.: Word-entity duet representations for document ranking. In: Proceedings of the 40th International ACM SIGIR Conference on Research and Development in Information Retrieval, SIGIR '17, pp. 763–772. ACM (2017a). doi: 10.1145/3077136.3080768

55. Xiong, C., Power, R., Callan, J.: Explicit semantic ranking for academic search via knowledge graph embedding. In: Proceedings of the 26th International Conference on World Wide Web, WWW '17, pp. 1271–1279. International World Wide Web Conferences Steering Committee (2017b). doi: 10.1145/3038912.3052558

56. Xu, Y., Jones, G.J.F., Wang, B.: Query dependent pseudo-relevance feedback based on Wikipedia. In: Proceedings of the 32nd International ACM SIGIR Conference on Research and Development in Information Retrieval, SIGIR '09, pp. 59–66 (2009). doi: 10.1145/1571941.1571954

57. Yi, X., Allan, J.: A comparative study of utilizing topic models for information retrieval. In: Proceedings of the 31th European Conference on IR Research on Advances in Information Retrieval, ECIR '09, pp. 29–41. Springer-Verlag (2009). doi: 10.1007/978-3-642-00958-7_6

58. Zhai, C.: Statistical language models for information retrieval A critical review. Found. Trends Inf. Retr. 2(3), 137–213 (2008)

59. Zhai, C., Lafferty, J.: Model-based feedback in the language modeling approach to information retrieval. In: Proceedings of the 10th international conference on Information and knowledge management, CIKM '01, pp. 403–410. ACM (2001). doi: 10.1145/502585.502654

Chapter 9
Utilizing Entities for an Enhanced Search Experience

Over the past decade, search engines have not only improved the quality of search results, but also evolved in how they interact with users. Modern search engines provide assistance to users throughout the entire search process, from formulating their information needs to presenting results and recommending additional content. This chapter presents a selection of topics, where entities are utilized with the overall aim of improving users' search experiences.

First, in Sect. 9.1, we discuss techniques for assisting users with articulating their information needs. These include query assistance services, such as query auto-completion and query suggestions, and specialized query building interfaces. Next, in Sect. 9.2, we turn to the question of result presentation. In conventional document retrieval, the standard way of serving results is to display a *snippet* for each document, consisting of its title and a short summary. This summary is automatically extracted from the document with the aim of explaining why that particular document is relevant to the query at hand. Moving from documents to entities as the unit of retrieval, the question we need to ask is: How can one generate dynamic summaries of entities when displaying them as search results? Finally, in Sect. 9.3, we describe entity recommendation methods that present users with contextual suggestions, encourage exploration, and allow for serendipitous discoveries. We study the related entity retrieval problem in different flavors, depending on what kind of input is available to the recommendation engine. Furthermore, we address the question of explaining the relationship in natural language between entities presented to the user. We refer to Table 9.1 for the notation used in this chapter.

9.1 Query Assistance

Chapter 7 dealt with query understanding from the machine's point of view. In this section, we bring the user's perspective to the forefront. How can a semantic search

© The Author(s) 2018   299
K. Balog, *Entity-Oriented Search*, The Information Retrieval Series 39,
https://doi.org/10.1007/978-3-319-93935-3_9

Table 9.1 Notation used in this chapter

Symbol	Meaning
e	Entity ($e \in \mathcal{E}$)
E	Graph edges
$\mathcal{E}_q$	Set of entities linked in query q ($\mathcal{E}_q \subset \mathcal{E}$)
$\mathcal{E}_s$	Set of entities clicked in session s ($\mathcal{E}_s \subset \mathcal{E}$)
$\mathcal{F}_e$	Knowledge base facts about entity e
q	Query
$\mathcal{Q}$	Query log (set of unique queries)
$\mathcal{Q}_s$	Set of queries issued within search session s
s	Search session ($s \in \mathcal{S}$)
$\mathcal{S}$	Set of search sessions
t	Term ($t \in \mathcal{V}$)
$\mathcal{T}$	Type taxonomy
$\mathcal{T}_e$	Types of entity e
u	Query template ($u \in \mathcal{U}$)
$\mathcal{U}$	Set of templates
$\mathcal{U}_q$	Set of templates for query q
V	Graph vertices
y	Entity type ($t \in \mathcal{T}$)

system assist the user in the process of articulating and expressing her information need? First, we discuss automatic methods that provide users with query suggestions while they are typing their query (Sect. 9.1.1) or after the query has been submitted (Sect. 9.1.2). Then, we present examples of specialized query building interfaces that enable users to formulate semantically enriched (keyword++) queries, by explicitly marking entities, types, or relationships (Sect. 9.1.3).

9.1.1   Query Auto-completion

Query auto-completion (QAC) provides users with query suggestions as they enter terms in the search box. Query auto completion is a common feature in modern search engines; see Fig. 9.1 for an illustration. It helps users to express their search intent as well as to avoid possible spelling mistakes [15].

Most systems rely on the wisdom of the crowds, i.e., suggest completions (matching the entered prefix) that have been most popular among users in the past, based on query logs [5]. Typically, QAC is viewed as a ranking problem, where the aim is "to return the user's intended query at the top position of a list of [candidate] query completions" [15].

Formally, we let q_0 be the incomplete query that the user has typed in so far and q_s be a suggested candidate query suffix. Let $c(q_0 \oplus q_s)$ be the number of times

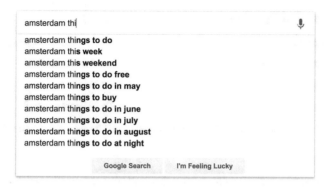

Fig. 9.1 Query auto-completion in the Google search engine

we observe the query $q_0 \oplus q_s$ issued in the query log $\mathcal{Q}$, where $\oplus$ is the string concatenation operator. The baseline approach to QAC, referred to as *most popular completion* in [5], is to rank suggestions according to:

$$score(q_s; q_0) = P(q_s|q_0) = \frac{c(q_0 \oplus q_s)}{\sum_{q_0 \oplus q_{s'} \in \mathcal{Q}} c(q_0 \oplus q_{s'})} \; .$$

Given that many information needs revolve around entities, entities can be leveraged to improve query auto-completion.

9.1.1.1 Leveraging Entity Types

Meij et al. [32] focus on a specific subset of queries that can be decomposed into entity and refiner components, such as "ASPIRIN *side effects*" or "BRITNEY SPEARS *video*" (refiners typeset in italics). They show that exploiting the type of entities being sought can improve QAC for rare queries. Specifically, we let $q_0 = e$, where the *query entity e* is recognized using entity linking techniques (cf. Sect. 7.3). Further, let $\mathcal{T}_e$ denote the entity types assigned to e in the knowledge base. Their best performing model (called M1 in [32]) looks at the most likely completion for a given entity type $y \in \mathcal{T}_e$:

$$score(q_s; q_0) = P(q_s|y) = \frac{c(q_s, y)}{\sum_{q_0 \oplus q_{s'} \in \mathcal{Q}} c(q_{s'}, y)} \; ,$$

where $c(q_s, y)$ is the number of times we can observe completion q_s with an entity of type y in the query log. Note that entities commonly have multiple types assigned to them. Selecting a single "best" type y out of the types of the query entity is an

Related searches for amsterdam things to see

best time to **go** to amsterdam	amsterdam **netherlands** things to **do**
visiting amsterdam **for the first time**	amsterdam **tourist attractions**
must do in amsterdam **first timers**	**best way** to see amsterdam
amsterdam **attractions for adults**	amsterdam **tourist information**

Fig. 9.2   Query suggestions offered by the Bing search engine for the query "*amsterdam things to see*"

open issue that is not dealt with in [32]. Instead, they evaluate performance for all possible types and then choose the type that led to the best performance on the training set.

9.1.2   Query Recommendations

Unlike "as-you-type" query auto-completion, which assists users during the articulation of their information need, *query recommendations* (a.k.a. *query suggestions*) are presented on the SERP once an initial query has been issued. The idea is to help users formulate more effective queries by providing suggestions for the next query. These suggestions are semantically related to the submitted query and either dive deeper into the current search direction or move to a different aspect of the search task [37]. Query suggestions are an integral feature of major web search engines and an active area of research [10, 11, 18, 25, 37, 43, 46]. Figure 9.2 shows query recommendations in action in a major web search engine.

Generating query recommendations is commonly viewed as a ranking problem, where given an input query q, the task is to assign a score to each candidate suggestion q', $score(q'; q)$. Like QAC, query recommendation also relies on the wisdom of the crowds by exploiting query co-occurrences and/or click-through information mined from search logs. While log-based methods work well for popular queries, it is a challenge for them to ensure *coverage* (i.e., provide meaningful suggestions) for rare queries. Most query assistance services perform poorly, or are not even triggered, on long-tail queries, simply because there is little to no historical data available for them. In this section, we will discuss methods that alleviate this problem by utilizing entities in a knowledge base.

We start by presenting the query-flow graph (QFG) method [10] in Sect. 9.1.2.1, which is a seminal approach for generating query recommendations. Then, in Sects. 9.1.2.2–9.1.2.4, we introduce various extensions to the QFG approach that tap into specific characteristics of entities, such as types and relationships. All these methods rely on the ability to recognize entities mentioned in queries. We refer back to Sect. 7.3 for the discussion of techniques for entity linking in queries. The set $\mathcal{E}_q$ denotes the entities identified in query q.

9.1.2.1 Query-Flow Graph

The query-flow graph (QFG), proposed by Boldi et al. [10], is a compact representation of information contained in a query log. It is a directed graph $G = (V, E, W)$, where the set V of vertices is the distinct set Q of queries contained in a query log, plus two special vertices, v_s and v_t, representing the start and terminal states of sessions: $V = Q \cup \{v_s, v_t\}$. A *session* is defined as a "sequence of queries of one particular user within a specific time limit" [10]. Commonly, the session time limit is taken to be 30 min. Further, $E \subseteq V \times V$ is a set of directed edges and $W : E \rightarrow (0, 1]$ is a weighting function assigning a weight $w(q_i, q_j)$ to each edge $(q_i, q_j) \in E$. Two queries, q_i and q_j, are connected with a directed edge $(q_i \rightarrow q_j)$, if there is at least one session in the log in which q_i and q_j are consecutive.

The key aspect of the construction of the query-flow graph is how the weighing function connecting two queries, $w(q_i, q_j)$, is defined. A simple and effective solution is to base it on relative frequencies of the query pair appearing in the query log. Specifically, the weight of the edge connecting two queries is computed as:

$$w(q_i, q_j) = \begin{cases} \frac{1}{Z} c(q_i, q_j), & (c(q_i, q_j) > \tau) \vee (q_i = v_s) \vee (q_j = v_t) \\ 0, & \text{otherwise}, \end{cases}$$

where $c(q_i, q_j)$ is the number of times query q_j follows immediately q_i in a session. The normalization coefficient Z is set to such that the sum of outgoing edge weights equals to 1 for each vertex, i.e., $\sum_j w(q_i, q_j) = 1$. Thus, weight $w(q_i, q_j)$ may be seen as the probability that q_i is followed by q_j in the same search session. This normalization may be viewed as the transition matrix of a Markov chain. Figure 9.3 shows an excerpt from a query-flow graph.

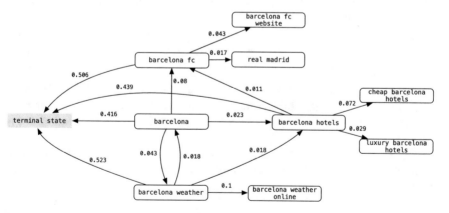

Fig. 9.3 Example of a query-flow graph, based on [10]. Note that not all outgoing edges are reported (thus, the sum of outgoing edges from each vertex does not reach 1). The terminal state vertex (v_t) is distinguished using a gray background

Based on this graph-based representation, query recommendations can be obtained by performing proximity-based top-k vertex retrieval, either neighborhood-based or path-based. A simple recommendation scheme is to pick, for an input query q, the top-k vertices connected with the largest edge weights. However, as observed by Boldi et al. [10], this method tends to drift off toward popular but unrelated queries. A better recommendation would be to pick the most important query q' relative to the initial query q. The recommendation algorithm proposed in [10] is a random walk with restart to a single vertex: a random surfer starts at the initial query q, and at each step either (1) follows one of the outlinks from the current vertex with probability α or (2) jumps back to q with probability $1 - \alpha$. This process may also be viewed as applying "a form of personalized PageRank, where the preference vector is concentrated in a single node [vertex]" [10]. More formally, the process is described as computing the transition matrix $\mathbf{A}$ of a Markov chain:

$$\mathbf{A} = \alpha \mathbf{P} + (1 - \alpha) \mathbf{1} \mathbf{i}_q^\mathsf{T} \, ,$$

where $\mathbf{P}$ is the row-normalized weight matrix of the query-flow graph, $\mathbf{1}$ is the identity matrix, and $\mathbf{i}_j$ is a "one-hot" vector whose entries are all zeroes, except for the jth vector whose value is 1. The parameter α is chosen to be 0.85 in [10]. $\mathbf{A}$ has a unique stationary distribution vector $\mathbf{v}$, such that $\mathbf{v}^\mathsf{T} \mathbf{A} = \mathbf{v}$. This distribution, called the *random-walk score relative to q*, can be computed using the power iteration method. Then, the highest scoring queries can be returned as the most relevant query suggestions for q. Notably, if the top-scoring query is the termination vertex v_t, then it means that the query chain is most likely to end at that point. In that case, it may be wise not to offer any query suggestions to the user.

Instead of using the raw random walk scores, Boldi et al. [10] propose to use a weighting scheme, so as to avoid returning very common queries as suggestions. The variant that yields the best recommendations in their experiments is given by the following formula:

$$score(q'; q) = \frac{rw_q(q')}{\sqrt{rw(q')}} \, , \tag{9.1}$$

where $rw_q(q')$ is the random walk score of the query q' with respect to q (personalized PageRank [26]) and $rw(q')$ is the random walk score of q' computed using a uniform preference vector (no personalization, i.e., starting at random at any vertex).

Having introduced the QFG approach, we shall next look at a number of extensions that utilize entities in a knowledge base, in order to improve recommendations for long-tail queries.

9.1.2.2   Exploiting Entity Aspects

Reinanda et al. [39] define *entity aspects* as a set of refiners that represent the same search intent in the context of an entity. For example, for the entity BARCELONA F.C., the refiners "live," "live streaming," and "live stream" represent the same search intent, i.e., they amount to one particular aspect of that entity. For each entity e, Reinanda et al. [39] mine a set of aspects $\mathcal{A}_e = \{a_1, \ldots, a_m\}$ from a search log. First, the set of queries mentioning that entity, $\mathcal{Q}_e \subset \mathcal{Q}$, is identified using entity linking. Then, refiners r (referred to as *query contexts* in [39]) are extracted by removing the mention of the entity from the queries. Next, refiners that express the same intent are clustered together. Clustering relies on the pairwise similarity between two refiners, $sim(r_i, r_j)$, which is taken to be the maximum of the following three types of similarity:

- *Lexical similarity*, estimated by the Jaro-Winkler distance between r_i and r_j.
- *Semantic similarity*, using the cosine similarity of word embeddings of the refiners, $\cos(\mathbf{r}_i, \mathbf{r}_j)$. Specifically, the vector representation of a refiner, $\mathbf{r}_i$, is computed as the sum of Word2vec [33] vector of each term within.
- *Click similarity*, obtained using the cosine similarity between the vectors of clicked sites.

For clustering refiners, *hierarchical agglomerative clustering* with complete linkage is employed. Refiners are placed in the same cluster if their similarity is above a given threshold. By the end of this step, each cluster of refiners corresponds to an entity aspect.

The mined aspects are used for query recommendations as follows. During the construction of the query-flow graph, a distinction is made between queries that contain a mention of an entity (i.e., are *entity-bearing*) and those that are not. For an entity-bearing query, the mentioned entity e and the refiner r are extracted.[1] Then, r is matched against the appropriate entity aspect, by finding the aspect $a_i \in \mathcal{A}_e$ that contains r as its cluster member, $r \in a_i$. This way, semantically equivalent queries are collapsed into a single entity aspect vertex in the modified query-flow graph. Non-entity-bearing queries are handled as in the regular query-flow graph, i.e., each unique query corresponds to a vertex.

For an incoming new query, the process of generating recommendations works as follows. First, entity linking is performed on the query to decide if it is entity-bearing. Then, the query is matched against the appropriate graph vertex (using the same procedure that was used during the construction of the query-flow graph). Finally, recommendations are retrieved from the query-flow graph.[2] Note that entity aspect vertices may contain multiple semantically equivalent queries. In that case, a single one of these is to be selected (for each vertex), e.g., based on query popularity. This approach (referred to as QFG+A) is shown to achieve small but consistent and significant improvements over generic QFG query recommendations.

---

[1]In this work, a single "main" entity is selected from each query (due to the fact that the average query length is short, most queries mention just one entity or none).

[2]Reinanda et al. [39] use the simple recommendation scheme, based on raw edge weights. However, it is also possible to apply random walks using the weighting scheme proposed in [10], as is given in Eq. (9.1).

9.1.2.3   Entity Types

An obvious limitation of query-flow graphs is that they cannot make recommendations for long-tail or previously unseen queries. Szpektor et al. [46] alleviates this limitation by enhancing query-flow graphs with *query templates*.[3] Templates are defined by replacing the entity mention in the query by a placeholder, which is an entity type. For example, the queries "*New York hotels*," "*Los Angeles hotels*," and "*Paris hotels*" may be abstracted into a "$\langle city\rangle$ *hotels*" query template. Then, general recommendation rules, like "$\langle city\rangle$ *hotels*" $\rightarrow$ "$\langle city\rangle$ *restaurants*," may be extracted from query logs. Using such rules, it is possible to generate the recommendation "*Yancheng restaurants*" for the input query "*Yancheng hotels*," even if none of those queries have been observed before. Next, we detail the elements of this approach.

Generating Query Templates   Each query q in the query log $\mathcal{Q}$ is considered for template construction. Given a query, every word n-gram (up to length 3 in [46]) is checked whether it refers to an entity. If yes, then that query segment of the query is replaced with the type(s) of the corresponding entity to produce the corresponding template(s). Here, entity types are not taken to be a flat set but are considered to exist in a hierarchical type taxonomy. We let $\tilde{\mathcal{T}}_e$ denote the most specific type(s) that entity e is assigned to in the knowledge base. For example, using the DBpedia Ontology as the type taxonomy, the entity ALBERT EINSTEIN has a single most-specific type $\tilde{\mathcal{T}}_e = \{Scientist\}$. By definition, the distance between an entity and (one of) its most specific type(s) is set to 1. Since the type taxonomy is a subsumption hierarchy (cf. Sect. 2.3.1), if entity e is an instance of type y, then it will also be an instance of all supertypes of y. We let $\hat{\mathcal{T}}_e$ denote the set of all supertypes of the types in $\tilde{\mathcal{T}}_e$. For the example entity ALBERT EINSTEIN, $\hat{\mathcal{T}}_e = \{Person, Agent\}$ (since *Scientist* is a subtype of *Person*, which is a subtype of *Agent*). Thus, the type assignments of an entity are partitioned into most specific types and their supertypes (i.e., $\mathcal{T}_e = \tilde{\mathcal{T}}_e \cup \hat{\mathcal{T}}_e$ and $\tilde{\mathcal{T}}_e \cap \hat{\mathcal{T}}_e = \emptyset$). Further, $d(y, y')$ is defined as the shortest path distance between types y and y' in the type taxonomy. The distance function $d(e, y)$ between entity e and type y is then defined as follows:

$$d(e, y) = \begin{cases} 1, & y \in \tilde{\mathcal{T}}_e \\ 1 + \min_{y' \in \tilde{\mathcal{T}}_e} d(y, y'), & y \in \hat{\mathcal{T}}_e \\ \infty, & \text{otherwise} . \end{cases}$$

For example, $d(\text{ALBERT EINSTEIN}, Scientist) = 1$, $d(\text{ALBERT EINSTEIN}, Person) = 2$, and $d(\text{ALBERT EINSTEIN}, Agent) = 3$.

---

[3]The notion of query templates is similar to that in [1] (cf. Sect. 7.4), with two main differences: (1) Templates here are defined over a taxonomy of entity types as opposed to attributes and (2) they are defined globally, i.e., are not restricted to any particular domain/vertical (such as travel or weather).

The set of templates constructed from a given query q is denoted as $\mathcal{U}_q$. Each template $u \in \mathcal{U}_q$ is associated with a confidence score $w(u, q)$, which expresses how well u generalizes q. Intuitively, the higher a type is located in the type hierarchy, the higher the risk of the corresponding template over-generalizing the query. Thus, the confidence score can be set to be exponentially decaying with the distance between entity e and type y:

$$w(q, u) = \alpha^{d(e, y)}, \tag{9.2}$$

where α is the decay rate (set to 0.9 in [46]). Note that template scoring in Eq. (9.2) does not take into account the uncertainty associated with the type-annotation of the query. The intuition is that by considering the transitions between templates (based on subsequent queries from which they were generated) in a sufficiently large query log, noise will be eliminated and meaningful transition patterns will surface. This will be explained next.

Extending the Query-Flow Graph   In the extended query-flow graph, referred to as *query-template flow graph*, vertices represent not only queries but templates as well. We let the set $\mathcal{U}$ denote all templates that are generated by queries in the search log: $\mathcal{U} = \bigcup_{q \in Q} \mathcal{U}_q$. In addition to query-to-query transition edges (E_{qq}) of the original query-flow graph, we now have two additional types of edges: (1) query-to-template edges (E_{qu}) and (2) template-to-template edges (E_{uu}).

- There is a directed edge between query q and template u iff $u \in \mathcal{U}_q$. The corresponding edge weight $w(q, u)$ is set proportional to the query-template confidence score, which is given in Eq. (9.2).
- There is a directed edge between templates u_i and u_j iff (i) they have the same placeholder type and (ii) there is at least one *supporting edge* $(q_i, q_j) \in E_{qq}$ such that $u_i \in \mathcal{U}_{q_i}$, $u_j \in \mathcal{U}_{q_j}$, and the substituted query segment is the same in q_i and q_j. The set of all support edges is denoted as $E_s(u_i, u_j)$. Then, the edge weight between u_i and u_j is set proportional to the sum of edge weights of all supporting query pairs:

$$w(u_i, u_j) \propto \sum_{(q_i, q_j) \in E_s(u_i, u_j)} w(q_i, q_j) \, .$$

For example, for the template-to-template edge "$\langle city \rangle$ *hotels*" $\to$ "$\langle city \rangle$ *restaurants*," the set of support edges includes {"*Paris hotels*" $\to$ "*Paris restaurants*," *New York hotels*" $\to$ "*New York restaurants*," ... }.

Normalization is performed in both cases to ensure that the outgoing edge weights of graph vertices sum up to 1.

Generating Query Recommendations   Using the regular query-flow graph, candidate query recommendations q' for an input query q would be those for which there exists a directed edge $(q, q') \in E_{qq}$. With the extended query-template flow graph, candidate recommendations also include queries that have not been observed in the logs before but can be instantiated via a template. Specifically, there needs to be a mapping edge $(q,u) \in E_{qu}$ and a template-to-template transition edge $(u,u') \in E_{uu}$, such that by instantiating u' with the entity extracted from q, it yields q' as the result. Then, candidate recommendations are scored according to the following formula:

$$score(q';q) = w(q,q') + \sum_{\substack{u \in \mathcal{U}_q \\ (u,u') \in E_{uu} \\ ins(u',q,u)=q'}} w(q,u)\, w(u,u')\,,$$

where $ins(u',q,u)$ denotes the query that is the result of instantiating template u' based on query q and template u. Given that edge weights are normalized, the resulting score will be in the range $[0,1]$ and "can be interpreted as the probability of going from q to q' by one of the feasible paths in the query-template flow graph" [46].

9.1.2.4   Entity Relationships

Bordino et al. [12] extend the query-flow graph with entity relationship information, and Huang et al. [25] capitalize on this idea for generating query recommendations. Suppose that two queries q_i and q_j appear in the same session and they mention entities e_i and e_j, respectively. Then, in addition to the flow from q_i to q_j in the query-flow graph, we can also utilize the relationships between e_i and e_j in the knowledge base. More formally, the enhanced graph, referred to as *EQGraph* in [12], has query vertices V_Q and entity vertices $V_{\mathcal{E}}$, with query-to-query edges E_{qq}, entity-to-query edges E_{eq}, and entity-to-entity edges E_{ee}. See Fig. 9.4 for an illustration.

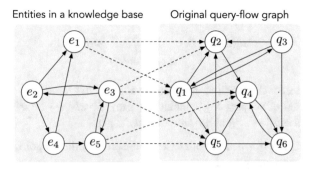

Fig. 9.4 Entity-query graph (EQGraph), extending the regular query-flow graph with entities from a knowledge base

Entity-to-Query Edges Each entity is connected to all the queries that mention it. The edge between entity e and query q is given a weight proportional to the relative frequency of that query in the log:

$$w(e,q) = \frac{c(q)}{\sum_{q':e \in \mathcal{E}_{q'}} c(q')},$$

where $c(q)$ is the number of times q appears in the query log, and $\mathcal{E}_{q'}$ denotes the set of entities that are extracted from q' via entity linking. Note that the outgoing edge weights sum up to one for each entity vertex.

Entity-to-Entity Edges Edge weights $w(e,e')$ represent the transition probability between a pair of entities e and e'. Bordino et al. [12] derive these weights based on the query log, by considering all query-to-query transitions $q \rightarrow q'$, where q mentions e and q' mentions e':

$$w(e,e') = 1 - \prod_{\substack{(q,q') \in E_{qq} \\ (e,q),(e',q') \in E_{eq}}} \left(1 - \frac{w(q,q')}{|\mathcal{E}_q| \times |\mathcal{E}_{q'}|}\right).$$

This formulation distributes the probability mass uniformly among the possible $(|\mathcal{E}_q| \times |\mathcal{E}_{q'}|)$ entity-to-entity transitions derived from $q \rightarrow q'$.

Generating Query Recommendations Huang et al. [25] generate query recommendations by computing personalized PageRank [26] on the EQGraph, starting from entities. The core of their approach lies in the idea that instead of considering the entities that are mentioned in the query ($\mathcal{E}_q$), they consider related entities from the knowledge graph. This set of related entities, denoted as $\mathcal{E}_R$, is derived based on the notion of *meta-paths* [45]. A meta-path M in the knowledge graph is a sequence of entity types $y_1, \ldots, y_n$ connected by predicates (i.e., relationship types) $p_1, \ldots, p_{n-1}$, such that $M = y_1 \xrightarrow{p_1} y_2 \ldots y_{n-1} \xrightarrow{p_{n-1}} y_n$. Each of these meta-paths represents a specific direct or composite ("multi-hop") relationship between two entities. Two entities may be connected by multiple meta-paths; a natural approach, followed in [25], is to select the shortest meta-path between them to represent their relationship. Let $\mathcal{M}$ be the set of meta-paths over the entity types in the KG, and $\mathcal{M}_y \subset \mathcal{M}$ be the set of outgoing meta-paths for type y. Related entities are collected by performing a path-constrained random walk [29] on knowledge graph predicates, with each meta-path $M \in \mathcal{M}_y$, for each of the types associated with the linked entities in the query ($y \in \mathcal{T}_e, e \in \mathcal{E}_q$). Each of these related entities $e \in \mathcal{E}_R$ accumulates weight, $w(e)$, based on the various meta-paths it can be reached on. See Algorithm 9.1 for the detailed procedure.

The most relevant queries, with respect to the related entities $\mathcal{E}_R$, are returned as recommendations. Specifically, for each of the related entities, $e \in \mathcal{E}_R$, personalized PageRank is performed on the EQGraph, starting from e with initial probability

---

Algorithm 9.1: Related entity finding for query recommendation [25]

Input: query-flow graph, knowledge graph, query q
Output: related entities $\mathcal{E}_R$ with weights w

1 $\mathcal{E}_q \leftarrow$ perform entity linking on q
2 $\mathcal{E}_R \leftarrow \emptyset$
3 **foreach** $e \in \mathcal{E}_q$ **do**
4   **foreach** $y \in \mathcal{T}_e$ **do**
5   **foreach** $M \in \mathcal{M}_y$ **do**
6              $\mathcal{E}' \leftarrow pathConstrainedRandomWalk(e, M)$
7   **foreach** $e' \in \mathcal{E}'$ **do**
8                  $\mathcal{E}_R \leftarrow \mathcal{E}_R \cup \{e'\}$
9                  $w(e') \leftarrow w(e') + \frac{1}{|\mathcal{E}_q| \times |\mathcal{E}'|}$
10   **end**
11   **end**
12   **end**
13 **end**
14 **return** $\mathcal{E}_R$, w

$w(e)$. The resulting probability distributions are aggregated, and the top-k queries with the largest aggregated score are offered as recommendations.

Note that this recommendation method considers only the entities mentioned in the query but not the other contextual terms in the query. It means that if two queries q and q' mention the same entities ($\mathcal{E}_q = \mathcal{E}_{q'}$) then generated recommendations will be exactly the same for the two.

9.1.3 Query Building Interfaces

In Chap. 4, we have seen that leveraging semantically enriched queries, referred to as *keyword++ queries*, yields improved retrieval performance. Such keyword++ queries may contain annotations of specific entities, target types, or relationships. One way to obtain those annotations is via automated techniques aimed at query understanding—which we have discussed in Chap. 7. Alternatively, it may be delegated to the user to provide semantic annotations for queries, and thereby more explicitly express the underlying information need. This, however, can be challenging for ordinary users, due to their unfamiliarity with the underlying knowledge base. Furthermore, even those that are acquainted with the knowledge base will find it problematic to navigate the large space of entities, types, and relationships without some support. In order to aid users in the process of formulating complex queries, specialized query building interfaces have been proposed [6, 24, 42]. A common feature of these systems is that they provide context-sensitive suggestions. The STICS system offers suggestions for entities and categories, as users type query terms; a screenshot is shown in Fig. 9.5. Schmidt et al. [42] present a corpus-

Fig. 9.5 Screenshot of the STICS system [24], http://stics.mpi-inf.mpg.de/

Fig. 9.6 Screenshot of the Broccoli system [6], http://broccoli.informatik.uni-freiburg.de/

adaptive extension, where the ranking of candidate suggestions also takes into account the underlying document collection. That is, they only suggest entities and categories "that lead to non-empty results for the document collection being searched" [42]. They further introduce a data structure for storing pre-computed relatedness scores for all co-occurring entities, in order to keep response times below 100 ms. Another example is the Broccoli system, which targets expert users and allows them to incrementally construct tree-like SPARQL queries, using similar techniques (i.e., corpus-based statistics) for generating suggestions [6]. Figure 9.6 presents a screenshot of the system.

9.2   Entity Cards

In recent years, there has been an increasing trend of surfacing structured results in web search in the form of various *knowledge panels*. Being served with rich and focused responses, users no longer need to engage with the organic (document-oriented) search results to find what they were looking for. This marks a paradigm shift in search engines evolving into *answer engines*. One group of knowledge panels, often referred to as *direct displays*, provide instant answers to a range of popular information needs, e.g., weather, flight information, definitions, or how-to questions. Some direct displays invite users to engage and interact with them (like currency conversion or finance answers), while others yield a clear inline answer (such as dictionary or reference answers) with no further interaction expected. Our focus in this section will be on another type of knowledge panel, called *entity card*, which summarizes information about a given entity of interest. Unlike direct displays, whose mere goal is to provide a succinct answer, entity cards intend to serve an additional purpose—to present the user with an overview of a particular entity for contextualization and further exploration.

> An entity card portrays a summary of a selected entity, commonly including the entity's name, type, short description, a selection of key attributes and relationships, and links to other types of relevant content.

Entity cards are an integral component of modern search engine result pages, on both desktop and mobile devices [14, 28]. Triggered by an entity-bearing query, a rich informational panel is displayed (typically on the top-right of the SERP on a desktop device), offering a summary of the query entity, as shown in Fig. 9.7a, b.

It has been long known that providing query-biased snippets for documents in the result list positively influences the user experience [49]. Entity cards may be viewed as the counterpart of document snippets for entities, and, as we shall show in this section, may be generated in a query-dependent fashion. It has been shown that entity cards attract users' attention, enhance their search experience, and increase their engagement with organic search results [14, 36]. Furthermore, when cards are relevant to their search task, users issue significantly fewer queries [14] and can accomplish their tasks faster [36].

We shall begin with an overview of what is contained in an entity card. Then, we will focus on the problem of selecting a few properties from an underlying knowledge base that best describe the selected entity, with respect to a given query, which will serve as the *factual summary* of that entity.

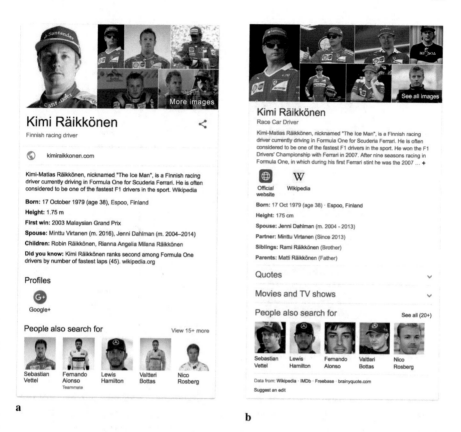

Fig. 9.7 (**a**) Entity card in Google. (**b**) Entity card in Bing

9.2.1   The Anatomy of an Entity Card

Entity cards are complex information objects that are dynamically generated in response to an entity-oriented query, after determining the intended entity for that query (cf. Sect. 7.3). Figure 9.8 shows a card layout that is commonly used in contemporary web search engines, comprising (1) images, (2) the name and type of the entity, (3) a short textual description, (4) entity facts, and (5) related entities. Additionally, depending on the type of the entity and the intent of the search query, other components may also be included, such as maps, quotes, tables, or forms.

Most elements of entity cards have their own set of associated research challenges. An entity may be associated with multiple types in the KB. For example, the types of ARNOLD SCHWARZENEGGER in Freebase include, among others, tv.tv_actor, sports.pro_athlete, and government.politician. The problem of selecting a single "main" type to be displayed on the card has been addressed using both context-independent [7] and context-dependent [50] methods. For emerging entities, that already have some facts stored about them in the KB, but

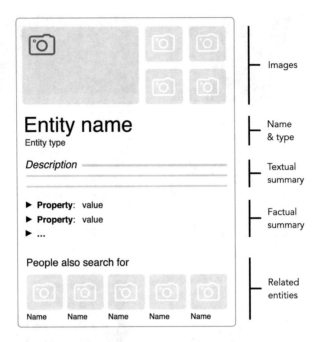

Fig. 9.8 Common entity card layout

lack a Wikipedia-style summary, natural language descriptions may be produced automatically [17, 41]. The factual summary, a truncated view of facts about the entity, is a central element of entity cards. We devote the remainder of this section to this very problem. Finally, related entity suggestions typically utilize search logs and entity co-occurrence information; we shall discuss specific methods in Sect. 9.3. Clicking on one of the related entities typically launches a new query with the related entity.

9.2.2 Factual Entity Summaries

The problem of generating informative entity summaries from RDF data has generated considerable interest over the recent years [16, 22, 23, 47, 48]. Since descriptions of entities in a knowledge base typically include hundreds of factual statements, "for human users it becomes difficult to comprehend the data unless a selection of the most relevant facts is provided" [47]. Below, we present the approach by Hasibi et al. [23] that is shown to be more effective than other relevance-oriented fact ranking methods, which employ variations of the PageRank algorithm [16, 47, 48]. Notably, it considers facts with both entity and literal object values.

Let $\mathcal{F}_e$ denote all the facts stored about a given entity e in the knowledge base $\mathcal{K}$. That is, $\mathcal{F}_e$ consists of all predicate-object pairs out of those SPO triples, where e appears as subject:

$$\mathcal{F}_e = \{(p,o) : (s,p,o) \in \mathcal{K}, s = e\} .$$

For notational convenience, we shall use the shorthand f to denote a single fact, which is essentially a property-value pair; further, we shall write p_f and o_f to denote the predicate and object elements of the fact, respectively. Since $\mathcal{F}_e$ is typically large (on the order of hundreds), the challenge is to select a small subset of them, to be displayed in the summary section of the entity card, that are deemed to have the highest utility for the user. Hasibi et al. [23] argue that factual entity summaries serve a dual purpose: "they offer a synopsis of the entity and, at the same time, can directly address users' information needs." Therefore, when selecting which facts to include in the summary, one should consider both their general importance, irrespective of any particular information need, as well as their relevance to the given user query. These two quantities are combined under the notion of *utility*. The utility of fact f is defined as a linear combination of its general importance and its relevance to the user query q:

$$utility(f,q) = \alpha \, importance(f) + (1 - \alpha) \, relevance(f,q) . \qquad (9.3)$$

For simplicity, importance and relevance are given equal weights in [23]. The generation of factual summaries is addressed in two stages. First, facts are ranked according to their utility. Then, the top-ranked facts are visually arranged in order to be displayed on the entity card.

9.2.2.1 Fact Ranking

The ranking of facts is approached as a learning-to-rank problem, using two main groups of features, aimed at capturing either the *importance* or the *relevance* of a fact. To learn the ranking function, target labels are collected via crowdsourcing for each dimension separately on a 3-point scale (0..2). Then, the two are combined with equal weights (cf. Eq. (9.3)), resulting in a 5-point scale (0..4). Below, we shall introduce some of the most effective features developed for relevance and importance, respectively. Table 9.2 provides an overview. For a complete list of features, we refer to [23].

Importance features are mostly based on statistics derived from the knowledge base. We introduce the concepts of *fact frequency* and *entity frequency*, which are loosely analogous to collection frequency and document frequency in document retrieval. Specifically, the *fact frequency of object* is the number of SPO triples in the KB with a given object value:

$$FF_o(o_f) = \left|\{(s,p,o) : (s,p,o) \in \mathcal{K}, o = o_f\}\right| .$$

Table 9.2 Features for fact ranking

Group	Feature	Description
Importance		
	$NEF_p(f)$	Normalized entity frequency of the fact's predicate
	$typeImp(f,e)$	Type-based importance of the fact's predicate
	$predSpec(f)$	Predicate specificity
	$isEntity(f)$	Whether the fact's object is an entity
Relevance		
	$lexSim_o(f,q)$	Lexical similarity between the fact's object and the query
	$semSimAvg_o(f,q)$	Semantic similarity between the fact's object and the query
	$iRank(f,q)$	Inverse rank of the fact's object
	$conLen(q)$	Context length

Entity frequency of predicate is the number of entities in the KB that have at least one fact associated with them with a given predicate:

$$EF_p(p_f) = \left| \{ e \in \mathcal{E} : \exists f' \in \mathcal{F}_e, p_{f'} = p_f \} \right| \ .$$

Another type-aware variant of the above statistic considers only those entities that, in addition to having the given predicate, also are of a given type y:

$$EF_p(p_f, y) = \left| \{ e \in \mathcal{E} : y \in \mathcal{T}_e, \exists f' \in \mathcal{F}_e, p_{f'} = p_f \} \right| \ .$$

Furthermore, we define the *type frequency of a predicate* to be:

$$TF_p(p_f) = \left| \{ y : \exists e \in \mathcal{E}, f' \in \mathcal{F}_e, p_{f'} = p_f, y \in \mathcal{T}_e \} \right| \ .$$

With the help of these statistics, we define the following features for measuring the importance of fact f for entity e:

- *Normalized entity frequency of predicate* is the relative frequency of the fact's predicate across all entities:

$$NEF_p(f) = \frac{EF_p(p_f)}{|\mathcal{E}|} \ ,$$

  where $|\mathcal{E}|$ is the total number of entities in the KB.
- *Type-based importance* also considers the frequency of the fact's predicate, but with respect to the types of the entity. Following [52], it is estimated using:

$$typeImp(f,e) = \sum_{y \in \mathcal{T}_e} EF_p(p_f, y) \log \frac{|\mathcal{T}|}{TF_p(p_f)} \ ,$$

  where $|\mathcal{T}|$ is the total number of types in the KB.

- *Predicate specificity* aims to identify facts with a common object value but a rare predicate:

$$predSpec(f) = FF_o(o_f) \log \frac{|\mathcal{E}|}{EF_p(p_f)} \; .$$

For example, the fact (*capital*, OTTAWA) would have relatively high predicate specificity, given that the object is frequent, while the predicate is relatively rare.
- *IsEntity* is a binary indicator that is true if the fact's object is an entity.

Relevance features capture the relevance of a fact with respect to the search query.

- *Lexical similarity* is measured by taking the maximum similarity between the terms of the fact's object and of the query using:

$$lexSim_o(f,q) = \max_{t \in o_f, t' \in q} \left(1 - dist(t,t')\right) \; ,$$

where *dist*() is a string distance function, taken to be the Jaro edit distance in [23].
- *Semantic similarity* aims to address the vocabulary mismatch problem, by computing similarity in a semantic embedding space. Specifically, we compute the average cosine similarity between terms of the fact's object and of the query:

$$semSimAvg_o(f,q) = \frac{\sum_{t \in o_f, t' \in q} \cos(\mathbf{t}, \mathbf{t'})}{\left|\{t : t \in o_f\}\right| \times \left|\{t' : t' \in q\}\right|} \; ,$$

where $\mathbf{t}$ denotes the embedding vector corresponding to term t. Hasibi et al. [23] use pre-trained Word2vec [33] vectors with 300 dimensions. The denominator is the multiplication of the number of unique terms in the fact's object and in the query, respectively.
- *Inverse rank* promotes facts with an entity that is highly relevant to the query as the object value. Entities in the KB are ranked with respect to the query. Then,

$$iRank(f,q) = \frac{1}{rank(o_f, \mathcal{E}_k(q))} \; ,$$

where $\mathcal{E}_k(q)$ is the set of top-k ranked entities returned in response to q, and *rank*() returns the position of an entity in the ranking (or ∞ if the entity cannot be found among the top-k results).
- *Context length* is the number of query terms that are not linked to any entity:

$$conLen(q) = \left|\{t : t \in q, t \notin linked(q)\}\right| \; ,$$

where *linked*(q) denotes the set of query terms that are linked to an entity.

Fig. 9.9 Structure of an
entity summary displayed on
an entity card. Image is based
on [23]

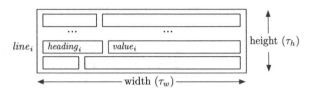

9.2.2.2   Summary Generation

The ranked list of facts we just obtained is yet to be arranged into a summary that
can be presented to the user. Visually, a summary consists of a number of lines, each
subdivided into heading and value parts. Additionally, it has a maximum display
size, defined in terms of the maximum number of rows (τ_h) and the maximum
number of characters within each row (τ_w); see Fig. 9.9. A straightforward approach
is just to fill this structure with the top-ranked facts, by using the predicate label
from the KB as the heading and the subject as the value part in each line. There are,
however, some additional considerations that, if addressed, can yield higher quality
summaries.

- There might be semantically identical predicates, even within a single KB, e.g.,
  `<foaf:homepage>` and `<dbp:website>` in DBpedia. Such duplicates need to be
  identified and resolved, such that only one of them is included in the summary.
- There may be multiple facts with the same predicate, e.g., parents of a person or
  founders of a company. While these constitute separate facts, the object values of
  these so-called multi-valued predicates can be concatenated together into a single
  list for display purposes.

Hasibi et al. [23] address these issues with a summary generation algorithm, shown
in Algorithm 9.2. Input to this algorithm is the list of top-k facts, generated by the
fact ranking step, denoted as $\hat{\mathcal{F}}_e$. The first line of the summary generation algorithm
creates a mapping from predicates in $\hat{\mathcal{F}}_e$ to their human-readable labels; these are
commonly provided as part of the KB schema. Predicates that are mapped to the
same label are then recognized as semantically identical. The mapping function
may implement additional heuristics, specific to the underlying knowledge base.
The summary is built in three stages. First (lines 2–8), up to τ_h unique line headings
are selected. Second (lines 9–14), the values for each line are assembled. This is the
part where multi-valued predicates are grouped. Third (lines 15–24), the heading
and value parts are combined for each line.

Algorithm 9.2: Summary generation [23]

Input: ranked list of facts $\hat{\mathcal{F}}_e$, max height τ_h, max width τ_w
Output: entity summary *lines*

1   $M \leftarrow$ predicate-name mapping from $\hat{\mathcal{F}}_e$
2   *headings* $\leftarrow$ []   /* Determine line headings */
3   **foreach** $f \in \hat{\mathcal{F}}_e$ **do**
4   *label* $\leftarrow M[p_f]$
5   **if** *(label $\notin$ headings) and ($|headings| \leq \tau_h$)* **then**
6   *headings*.append $\big((p_f, label)\big)$
7   **end**
8   **end**
9   *values* $\leftarrow$ []   /* Determine line values */
10   **foreach** $f \in \hat{\mathcal{F}}_e$ **do**
11   **if** $p_f \in$ *headings* **then**
12   *values*[p_f].append(o_f)
13   **end**
14   **end**
15   *lines* $\leftarrow$ []   /* Construct lines */
16   **foreach** $(p_f, label) \in$ *headings* **do**
17   *line* $\leftarrow label +$ ':'
18   **foreach** $v \in$ *values*[p_f] **do**
19   **if** len(*line*) + len(v) $\leq \tau_w$ **then**
20   *line* $\leftarrow line + v$ /* Add comma if needed */
21   **end**
22   **end**
23   *lines*.append(*line*)
24   **end**

9.3 Entity Recommendations

Earlier in this chapter, we have discussed tools that help users with expressing their information needs and with getting direct answers to those needs. There are also situations where users' information goals are less clearly defined, and they would just like to browse and explore, without looking for anything specific. Examples include learning about people in the news or exploring future travel destinations. Therefore, in addition to traditional search assistance tools, such as query suggestions and direct answers, exploration and discovery should also be regarded as central elements of next-generation search systems [55]. This section presents recommendation techniques that enable exploration, with the goal of increasing user engagement.

Specifically, our objective is to provide *related entity recommendations* (a.k.a. *related entity suggestions*) to users, based on their context. We shall consider multiple contexts that may serve as input data: (1) a particular entity (Sect. 9.3.1), (2) the user's current search session (Sect. 9.3.2), and (3) a given entity in a particular text context (Sect. 9.3.3).

The *entity recommendation* task is approached as a ranking problem: given some context (e.g., a particular entity or a search session) as input, return a ranked list of entities $\langle e_1, \ldots, e_k \rangle$ from an entity catalog $\mathcal{E}$ that are related to the user's context.

How does this problem relate to other tasks that have been discussed earlier in this book? A core component underlying all recommendation methods is a measure of *relatedness* between an input entity and a candidate entity. Pairwise entity relatedness has already been used in entity linking, for entity disambiguation (cf. Sect. 5.6.1.3), and those methods are applicable here as well. Another related task is that of finding similar entities (cf. Sect. 4.5), which also boils down to a pairwise entity measure. The similar entity finding task, however, has a different objective— it aims to complete an input set of entities, with similar entities. Consequently, it measures pairwise *entity similarity* as opposed to *entity relatedness*.

The degree of entity relatedness may be estimated using simple measures of statistical association, based on entity co-occurrence information (e.g., in search logs or in Wikipedia articles). Another family of methods makes use of entity relationship information stored in knowledge graphs and employs graph-theoretic measures (e.g., PageRank). Yet another group of approaches infers relatedness based on the content (i.e., attributes or descriptions) of entities.

In addition to receiving entity recommendations, users may also be interested in finding out *why* certain suggestions were shown to them. The problem of explaining recommendations boils down to the task of generating a human-readable description of the relationship between a pair of entities. This is discussed in Sect. 9.3.4.

9.3.1   Recommendations Given an Entity

We start by discussing the case of recommending entities related to a given input entity. A common application scenario is that of entity cards in web search, which are triggered by an entity-bearing query. These cards often include a "People also search for" section, displaying entities that are related to the query entity; see Fig. 9.8. This task may be formalized as the problem of estimating the probability of a candidate entity e', given an input entity e, $P(e'|e)$.

Blanco et al. [9] present the Spark system (with previous versions of the system described in [27, 58]), which had been powering related entity suggestions in Yahoo! Web Search. Spark extracts several signals (over 100 features) from a variety of proprietary and public data sources (including Yahoo!'s knowledge graph and web search logs, and social media platforms Flickr and Twitter) and combines them in a learning-to-rank framework. The training data consists of 47K entity pairs, labeled by professional editors on a five-point scale. Aggarwal et al. [2] show that comparable accuracy may be achieved by utilizing only publicly available data, in particular, Wikipedia, and using only 16 features.

Table 9.3 Features for related entity recommendation, given an input entity

Group	Feature	Description
Co-occurrence		
	$P(e, e')$	Joint probability $(c(e, e'; C)/|C|)$
	$P(e'|e)$	Conditional probability $(c(e, e'; C)/c(e; C))$
	$P(e|e')$	Reverse conditional probability $(c(e, e'; C)/c(e'; C))$
	$PMI(e, e')$	Pointwise mutual information
	$WLM(e, e')$	Wikipedia link-based measure (cf. Eq. (5.4))
Graph-theoretic		
	$PR(e)$	PageRank score of the entity (cf. Eq. (4.4))
Content-based features		
	$\cos(\mathbf{e}, \mathbf{e}')$	Cosine similarity between vector representations of entities
Popularity		
	$c(e; C)$	Frequency of the entity
	$P(e)$	Relative frequency of the entity $(c(e; C)/|C|)$

Unary features are computed for both e and e'. All statistics are computed over some data collection C, where $|C|$ denotes the total number of items (documents, queries, etc.); $c(e; C)$ is the frequency of entity e, i.e., the number of items in which e occurs; $c(e, e'; C)$ denote the number of items in which e and e' co-occur

We distinguish between four main groups of features: *co-occurrence features*, *graph-theoretic features*, *content-based features*, and *popularity features*. Popularity features and graph-theoretic features are unary, expressing the importance of an entity on its own. The remaining features are binary, capturing the strength of association between two entities. Table 9.3 presents a non-exhaustive selection of features.

Co-occurrence features Intuitively, entities that are observed to occur frequently together are likely to be related to each other. One question here is what data collection C to use for extracting co-occurrence information. Another question is what co-occurrence statistic to compute based on those observations. Prior work has considered a wide variety of data sources, including search logs [9], web pages [3], Wikipedia [2, 35, 44], Twitter [9], and Flickr [9]. Co-occurrence measures include joint, conditional, and reverse conditional probabilities, pointwise mutual information, KL divergence, entropy, and the Wikipedia link-based measure (WLM) [35].

Graph-theoretic features The most commonly used feature in this group is the PageRank score of an entity in the knowledge graph. PageRank may also be computed on a hyperlink graph obtained from the Web [9]. For details on PageRank and for additional centrality measures, we refer back to Sect. 4.6.

Content-based features This set of features aims to capture the similarity between a pair of entities based on their content. A standard approach is to represent entities either as term vectors or embedding vectors, and then compute the cosine similarity of those vectors. See Sect. 4.5.1 for alternative ways of measuring pairwise entity similarity.

Popularity features   Popularity is based on the frequency of an entity in a given data source, e.g., search queries and sessions, or number of views or clicks in web search. Additional popularity features were discussed in Sects. 4.6.1 and 5.6.1.1.

9.3.2   Personalized Recommendations

Rather than suggesting entities related to a given input entity, in this section we discuss methods that provide personalized entity recommendations based on the user's current search session. A number of approaches have been proposed for learning models for specific domains, such as movies or celebrities [8, 57]. Such *model-based methods* rely on manually designed domain-dependent features, related to specific properties of entities (e.g., the genre of a movie or how many pop singers were viewed by a specific user). There is an obvious connection to make here to traditional item-based recommender systems (e.g., the ones used in e-commerce systems), such as collaborative filtering [19]. One main difference is that in collaborative filtering approaches the user-item matrix is given. For entity recommender systems, user preferences of entities are more difficult to observe. Another difference is the sheer scale of data (i.e., millions of entities vs. thousands of products in an e-commerce scenario) coupled with extreme data sparsity. Fernández-Tobías and Blanco [20] perform personalized entity recommendations using a purely collaborative filtering approach. Inspired by nearest neighbor techniques, these *memory-based methods* exploit user interactions that are available in search logs. Since they do not depend on descriptions or properties of entities, memory-based methods generalize to arbitrary types of entities and scale better to large amounts of training data than model-based methods.

Next, we present three probabilistic methods for estimating $P(e'|s)$, the probability of recommending entity e' to a user based on her *current search session* s.[4] These methods are named after how item-to-item similarity aggregation is performed: entity-based, query-based, or session-based. In their paper, Fernández-Tobías and Blanco [20] define multiple alternatives for each component of these models. Here, we will discuss a single option for each, the one that performed best experimentally. According to the results reported in [20], the entity-based method performs best, followed by the query-based and then the session-based approaches.

We shall use the following notation below. Let Q be the set of unique queries in the search log and S be the set of all user sessions. For a given session $s \in S$, $Q_s \subset Q$ denotes the set of queries issued and $\mathcal{E}_s \subset \mathcal{E}$ denotes the set of entities clicked by the user within that search session.

---

[4]For notational consistency, we shall continue to denote the candidate entity recommendation, which is being scored, by e'.

9.3.2.1 Entity-Based Method

The intuition behind the first method is that an entity e' is more likely to be relevant to the user's current session s if it is similar to other entities that have previously been clicked by the user in the same session. Formally, this is expressed as:

$$P_{EB}(e'|s) = \sum_{e \in \mathcal{E}_s} P(e'|e) P(e|s) \,,$$

where $P(e'|e)$ captures the similarity between a pair of entities and $P(e|s)$ expresses the relevance of e given the search session s. Pairwise entity similarities are estimated in a collaborative fashion, by measuring the co-occurrence of entities within all user sessions using the Jaccard coefficient. To estimate the relevance of a clicked entity e within a session s, we aggregate the importance of e for each query q in that session, weighted by the query likelihood in that session:

$$P(e|s) = \sum_{q \in \mathcal{Q}_s} P(e|q,s) P(q|s) \,. \tag{9.4}$$

A given entity's relevance may be measured based on *dwell time*, i.e., how much time the user spent on examining that result, relative to all other entities that were returned for the same query:

$$P(e|q,s) = \frac{dwell(e,q,s)}{\sum_{e' \in \mathcal{E}_s} dwell(e',q,s)} \,, \tag{9.5}$$

where $dwell(e,q,s)$ is the time spent on examining entity e for query q in session s. In case the user did not click on e as a result to q, $P(e|q,s)$ is taken to be 0.

The probability $P(q|s)$ in Eq. (9.4) captures how important that query is within its session. It tends to reason that more recent queries should be considered more important, as they represent more accurately the user's current interests (which may have shifted over time). This notion of temporal decay is formally expressed as:

$$P(q|s) \propto e^{-(t_s - t_q)} \,, \tag{9.6}$$

where e is the mathematical constant (the base of the natural logarithm), t_q is the timestamp of query q, and t_s is the timestamp of the last query in the session.

9.3.2.2 Query-Based Method

The second approach works by first identifying queries from other sessions in the search log that are potentially relevant to the current session. Then, it retrieves entities from those sessions. Formally:

$$P_{QB}(e'|s) = \sum_{\substack{q \in \mathcal{Q} \\ q \notin \mathcal{Q}_s}} P(e'|q) P(q|s) \,, \tag{9.7}$$

where $P(q|s)$ is the probability that query q is relevant to the current session s and $P(e'|q)$ measures how relevant e is for query q (across all sessions). Note that, in contrast to the entity-based method, the queries q we aggregate over are not present in the current session. Rather, these are chosen from the queries submitted by other users, performing similar tasks. The relevance of an entity given a query is estimated by considering all sessions in the search log that contain that query:

$$P(e'|q) \propto \sum_{s' \in S} P(e'|q,s)P(q|s)P(s) \,, \tag{9.8}$$

where as before, $P(e'|q,s)$ is measured using dwell time (cf. Eq. (9.5)) and $P(q|s)$ is estimated based on temporal decay (cf. Eq. (9.6)). Note that $P(q|s)$ here expresses the probability of choosing the query q from session s containing that query ($q \in \mathcal{Q}_s$) and is not to be confused with $P(q|s)$ in Eq. (9.7), where it is used to capture the relevance of a query that is not observed in the given session ($q \notin \mathcal{Q}_s$). Finally, $P(s)$ is assumed to be uniform for simplicity.

The query relevance probability, $P(q|s)$ in Eq. (9.7), is defined to select queries from the search log ($q \in \mathcal{Q} \setminus \mathcal{Q}_s$) that are similar to the ones in the current session:

$$P(q|s) = \sum_{q' \in \mathcal{Q}_s} P(q|q')P(q'|s) \,,$$

where $P(q|q')$ expresses the similarity between a pair of queries and is computed based on the co-occurrence of q and q' within all sessions in the search log using the Jaccard coefficient. As before, $P(q'|s)$ uses the temporal decay estimator (cf. Eq. (9.6)).

9.3.2.3   Session-Based Method

The last approach works by finding sessions similar to the current session, then recommending entities from those sessions:

$$P_{SB}(e'|s) = \sum_{s' \in S} P(e'|s')P(s'|s) \,,$$

where $P(e'|s')$ is the importance of an entity given a session, computed as given by Eq. (9.4). The pairwise session similarity, $P(s'|s)$, is estimated based on entity embeddings. Specifically, Fernández-Tobías and Blanco [20] use Word2vec [33] (where sessions correspond to documents and entities within sessions correspond to words within documents) and extract embedding vectors of dimension 100. The

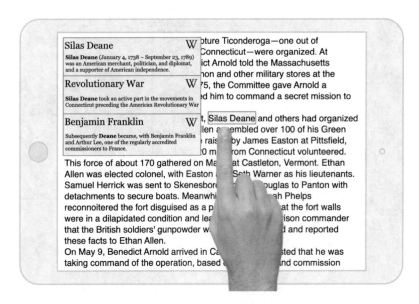

Fig. 9.10 An example of contextual entity recommendations. Image is based on [31]

similarity between two sessions is then computed based on the distance between the centroids of entities within them:

$$P(s'|s) \propto \left(\left\| \frac{1}{|\mathcal{E}_s|} \sum_{e \in \mathcal{E}_s} \mathbf{e} - \frac{1}{|\mathcal{E}_{s'}|} \sum_{e' \in \mathcal{E}_{s'}} \mathbf{e}' \right\| \right)^{-1},$$

where $\mathbf{e}$ is the embedding vector corresponding to entity e.

9.3.3 Contextual Recommendations

Web search is a prominent application area for entity recommendations but is certainly not the only one. Entity recommendation may also be offered directly within the application where content is consumed. As one such example, Lee et al. [31] present the scenario of a user reading a document on a tablet or e-reader device. At some point, the user might stumble upon an entity that she wishes to learn more about. Instead of leaving the application and switching to a web search engine to query for that entity, the user might just highlight and tap on an entity of interest. She will then be presented with a list of contextually relevant entities, as it is shown in Fig. 9.10.

According to the outlined scenario, the input to the *contextual entity recommendation* problem consists of an input entity e and some context c. Specifically, the

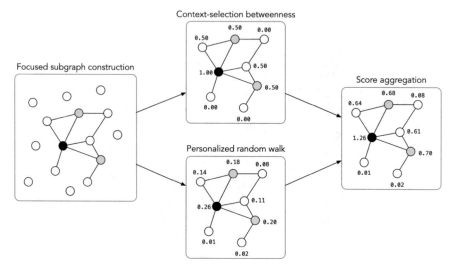

Fig. 9.11 Overview of the approach by Lee et al. [31]. The input entity (e) vertex is marked black, contextual entity vertices (E_c) are marked gray

context is a window of text around the selected entity mention (100 terms before and after in [31]). The approach proposed by Lee et al. [31] consists of three main steps, which are illustrated in Fig. 9.11.

1. A *focused subgraph* is extracted from the underlying knowledge graph G. The vertices of this focused subgraph are $V = \{e\} \cup \mathcal{E}_c \cup \mathcal{E}'_c$, where e is the input entity, $\mathcal{E}_c$ is the set of context entities, recognized in c by performing entity linking, and $\mathcal{E}'_c$ is the set of entities reachable from $\mathcal{E}_c$ via paths of length one in the knowledge graph. The edges between these vertices are induced from G.
2. Each candidate entity e' in the focused subgraph is scored using two different methods: context-selection betweenness and personalized random walk. *Context-selection betweenness* (CSB) captures the extent to which the candidate entity e' serves as a bridge between the input entity e and the context entities $\mathcal{E}_c$. Intuitively, a higher CSB score means that the candidate entity plays a more important role in connecting the input and context entities. Formally, CSB considers all shortest paths between the input and context entities, which go through the candidate entity:

$$CSB(e') = \frac{1}{Z} \sum_{e'':e' \in SP(e,e'')} \frac{w(e,e'')}{|SP(e,e'')| \times l(e,e'')},$$

where $SP(e,e'')$ is the set of all shortest paths between the e and e'', and $l(e,e'')$ is the length of that path. Each path between the input and a context entity is weighted by their semantic distance, based on the Wikipedia link-based measure

(WLM, cf. Eq. (5.4)):[5]

$$w(e, e'') = \max\left(WLM(e, e'') - \gamma, 0\right).$$

The threshold γ is used for emphasizing context entities that are semantically related to the input entity ($\gamma = 0.5$ in [31]). The normalization factor is set to:

$$Z = \sum_{e'' \in \mathcal{E}_c} \frac{w(e, e'')}{l(e, e'')}.$$

The other scoring method is *personalized random walk* (PRW, a.k.a. personalized PageRank [26]), which aims to measure the relevance of entities given the user's selection. To compute these scores, the random jump probabilities are initialized as follows. The input entity vertex is assigned probability $0 < x_e < 1$, the context entity vertices are assigned probability $x_c / |\mathcal{E}_c|$, where $0 \leq x_c \leq x_e$, and all other entity vertices are assigned probability 0. Lee et al. [31] use $x_e = 0.05$ and $x_c = 0$ in their experiments, and report that $x_c > 0$ does not lead to significant improvements.

3. The final score for each entity vertex is computed by taking a weighted combination of the context-selection betweenness and personalized random walk scores:

$$score(e'; e, \mathcal{E}_c) = \alpha \frac{|\mathcal{E}_c|}{|V|} CSB(e') + |V| \times PRW(e'),$$

where $|V|$ is the number of vertices in the focused subgraph and α is a scaling factor. The multipliers serve normalization purposes, making the two scores compatible.

9.3.4 Explaining Recommendations

Thus far in this section, we have presented both non-personalized and personalized methods for recommending related entities (e') given an input entity (e). In addition to the recommendations themselves, users might also be interested in finding out *why* certain entities were shown to them. This brings us to answering the question: How are the input entity and the recommended entity related? Such explanations are offered, e.g., on entity cards in modern web search engines by hovering the mouse over a recommended entity; see Fig. 9.12. Another typical application scenario for explaining entity relationships is timeline generation [3].

---

[5]Mind that we define WLM as a similarity measure, as opposed to a distance measure, hence the equation differs from the one in [31].

Fig. 9.12 Excerpt from a
Google entity card displayed
for BARACK OBAMA. When
hovering over a related entity,
an explanation of the
relationship is shown

One of the earliest works addressing the problem of explaining the relationship
between two entities is the *dbrec* music recommendation system [38]. It offers
explanations in the form of a list of shared property-value pairs between the
input and recommended entities, as shown in Fig. 9.13. This form of presentation,
however, was considered as "too geeky" by 6 out of the 10 test subjects participating
in the user evaluation [38]. Instead, human-readable descriptions that verbalize the
relationship are more natural to use. The use of natural language has also shown
to improve confidence in decision-making under uncertainty [21]. Three main lines
of approaches have been proposed in prior work for generating natural language
descriptions of entity relationships: (1) by manually defining templates [3], (2)
by retrieving sentences from a text corpus [54], and (3) by automatically creating
templates for a specific relationship and then filling the template slots for a new
relationship instance [53]. Below, we briefly elaborate on the latter two.

All existing approaches solve a simplified version of the task of explaining entity
relationships, by focusing on a specific relationship between a pair of entities. This
corresponds to generating a textual description for an SPO triple, where the subject
is e, the predicate is p, and the object is e'. We shall refer to the triple (e, p, e')
as *relationship instance*. When referring to predicate p, we shall use the terms
predicate and relationship interchangeably. As it is illustrated in Fig. 9.13, entities
may be connected via multiple relationships. Selecting p from the set of possible
predicates that connect two entities remains an open research challenge to date.

9.3.4.1   Explaining Relationships via Sentence Ranking

Voskarides et al. [54] approach the task as a sentence ranking problem: automatically extracting sentences from a text corpus and ranking them according to how
well they describe a given relationship instance (e, p, e').

Fig. 9.13 Explanation for recommending ELVIS PRESLEY for the input entity JONNY CASH from the dbrec music recommendation system. Figure taken from Passant [38] (C) Springer 2010, reprinted with permission

First, a set $\mathcal{X}$ of candidate sentences is extracted from a corpus of documents. In [54], this corpus is Wikipedia. Other document collections may also be used, as long as documents are pertinent to the entities of interest. A sentence is considered as a candidate if (1) it originates from the Wikipedia page of e or e' and contains a mention to the other entity or (2) it mentions both e and e'. In order to make sentences self-contained outside the context of the source document, pronouns "she" and "he" are replaced with the name of the respective entity. Further, sentences are annotated with entities by performing entity linking. As an illustration, consider the sentence "He gave critically acclaimed performances in the crime thriller Seven...," which, after these enrichment steps, becomes "BRAD PITT gave critically acclaimed performances in the crime thriller SEVEN..."

Next, candidate sentences $x \in \mathcal{X}$ are ranked using supervised learning. Four groups of features are employed:

- *Textual features* consider the importance of the sentence on the term level. These include sentence length, aggregated IDF scores, sentence density [30], and fractions of verbs, nouns, and adjectives.
- *Entity features* characterize the sentence based on the mentioned entities. These include, among others, whether e and e' are linked in x, and the distance between the positions of their mentions. Another group of features focuses on other entities mentioned in the sentence and whether those are related to e and e'.

Fig. 9.14 Dependency graph for the sentence "Brad Pitt appeared in the American epic adventure film Troy," using entities and predicates from DBpedia

- *Relationship features* indicate whether the relationship p occurs in x. Exact term-based matching has low coverage (e.g., "spouse" vs. "husband" or "married"), therefore both synonym-based matches (using Wordnet) and word embeddings (using Word2vec [33]) are considered.
- *Source features* describe the position of x and the number of occurrences of e and e' in the document from which x originates.

Voskarides et al. [54] train their models on a set of manually annotated sentences, using a five-level graded relevance scale. They show that learning relationship-specific models, as opposed to a single global model, can yield additional improvements.

9.3.4.2 Generating Descriptions of Relationships

The previous approach is limited by the underlying text corpus, which may not contain descriptions for certain relationship instances. Voskarides et al. [53] propose to overcome this by automatically generating descriptions. The idea is to learn how a given relationship p is typically expressed (in the document corpus), and create sentence templates for that relationship. Then, for a new relationship instance, the appropriate template can be instantiated.

As before, it is assumed that a given relationship between two entities can be expressed as a single sentence. This sentence should mention both e and e', and possibly other entities that may provide additional contextual information. The following example sentence is given as an illustration in [53] for the (BRAD PITT, stars in, TROY) relationship instance: "Brad Pitt appeared in the American epic adventure film Troy." It not only verbalizes the "stars in" predicate but also mentions other entities and attributes (the film's genre and origin) to offer additional context. To be able to provide such contextual information, each sentence is augmented with an entity dependency graph. In this graph, vertices represent entities and edges represent relationships (predicates). The graph is created by traversing paths in the knowledge base between each pair of entities that are mentioned in the sentence. See Fig. 9.14 for an illustration.[6]

---

[6]In their paper, Voskarides et al. [53] use Freebase as the underlying knowledge base and pay special attention to compound value type (CVT) entities. CVT entities are specific to Freebase, and are used for modeling attributes of relationships (e.g., date of a marriage). For ease of presentation, we will not deal with CVT entities in our discussion.

The template creation process then takes as input, for each predicate, a set of relationship instances, sentences describing those relationship instances, and entity dependency graphs corresponding to those sentences. The following sequence of steps are performed:

1. Entities sharing the same predicates are clustered together across the dependency graphs. This will group entities of the same type, such as persons and films.
2. A compression graph G_C is created from sentences where vertices are either words or entity clusters.
3. G_C is traversed for finding valid paths between all pairs of entity cluster vertices. A path is considered valid if (i) it contains a verb and (ii) it can be observed as a complete sentence at least once in the corpus.
4. A template is constructed from each path, which is supported by a minimum number of sentences in the corpus.

With a set of templates at hand, generating a description for a new relationship instance (e, p, e') goes as follows. First, the available templates for predicate p are ranked. Two template scoring functions are presented in [53], one based on cosine similarity of TF-IDF term vectors and another using feature-based supervised learning. The top-ranked template is then instantiated by filling its slots with entities from the knowledge base. If multiple instantiations exist, then one of those is chosen randomly. If the template cannot be instantiated, then we proceed to the next template in the ranking.

9.4 Summary

This chapter has introduced search assistance tools that help users (1) express their information needs, (2) get direct answers, and (3) explore related content with the help of entities. We have started with query assistance features, such as query auto-completion and query recommendation, which users would expect today as standard functionality from a modern search engine. An issue with traditional methods, which rely solely on query logs, is that of coverage. That is, they fail to provide meaningful suggestions for long-tail queries. We have discussed how knowledge bases may be utilized to alleviate this problem, yielding small but significant improvements over traditional methods. Next, we have looked at entity cards, a new type of search result presentation that has been adopted by major web search engines and intelligent personal assistants. Each card presents a concise summary of a specific entity, and can satisfy the user's information need directly, while also encouraging further engagement with search results and exploration of related content. We have addressed, in detail, the question of which facts to highlight about an entity in the limited screen space that is available on the card. Finally, we have presented techniques for promoting exploratory search by recommending related entities to users. We have further discussed how to generate a natural language explanation for the relationship between an input entity and a recommended related entity.

9.5   Further Reading

Web search result pages are becoming increasingly complex, featuring direct displays, entity cards, query suggestions, advertisements, etc., arranged in a nonlinear page layout. With these novel interfaces, determining the user's satisfaction with the search results is becoming more difficult. Research in this area includes the topics of evaluating whole page relevance [4], understanding how users examine and interact with nonlinear page layouts [13, 36], and detecting search satisfaction without clicks [28, 56].

Entity cards are the most widely used and universally applicable tools for summarizing entity information, but there are other possibilities that could serve users better in certain application scenarios. One such alternative that has garnered research interest is *entity timelines*, which organize information associated with an entity, arranged along a horizontal time axis. Timeline visualizations are often coupled with interactive features to enable further exploration. For example, Rybak et al. [40] visualize how a person's expertise changes over the course of time. Tuan et al. [51] and Althoff et al. [3] generate a timeline of events and relations for entities in a knowledge base.

In this chapter, we have focused on the algorithmic aspects of generating entity recommendations. For a study on how people interact with such recommendations, see, e.g., [34].

References

1. Agarwal, G., Kabra, G., Chang, K.C.C.: Towards rich query interpretation: Walking back and forth for mining query templates. In: Proceedings of the 19th international conference on World wide web, WWW '10, pp. 1–10. ACM (2010). doi: 10.1145/1772690.1772692
2. Aggarwal, N., Mika, P., Blanco, R., Buitelaar, P.: Leveraging Wikipedia knowledge for entity recommendations. In: Proceedings of the ISWC 2015 Posters & Demonstrations Track co-located with the 14th International Semantic Web Conference, ISWC '15. Springer (2015)
3. Althoff, T., Dong, X.L., Murphy, K., Alai, S., Dang, V., Zhang, W.: TimeMachine: Timeline generation for knowledge-base entities. In: Proceedings of the 21th ACM SIGKDD International Conference on Knowledge Discovery and Data Mining, KDD '15, pp. 19–28. ACM (2015). doi: 10.1145/2783258.2783325
4. Bailey, P., Craswell, N., White, R.W., Chen, L., Satyanarayana, A., Tahaghoghi, S.M.M.: Evaluating whole-page relevance. In: Proceedings of the 33rd International ACM SIGIR Conference on Research and Development in Information Retrieval, SIGIR '10, pp. 767–768. ACM (2010). doi: 10.1145/1835449.1835606
5. Bar-Yossef, Z., Kraus, N.: Context-sensitive query auto-completion. In: Proceedings of the 20th International Conference on World Wide Web, WWW '11, pp. 107–116. ACM (2011). doi: 10.1145/1963405.1963424
6. Bast, H., Bäurle, F., Buchhold, B., Haussmann, E.: Semantic full-text search with broccoli. In: Proceedings of the 37th International ACM SIGIR Conference on Research and Development in Information Retrieval, SIGIR '14, pp. 1265–1266. ACM (2014). doi: 10.1145/2600428.2611186

7. Bast, H., Buchhold, B., Haussmann, E.: Relevance scores for triples from type-like relations. In: Proceedings of the 38th International ACM SIGIR Conference on Research and Development in Information Retrieval, SIGIR '15, pp. 243–252. ACM (2015). doi: 10.1145/2766462.2767734

8. Bi, B., Ma, H., Hsu, B.J.P., Chu, W., Wang, K., Cho, J.: Learning to recommend related entities to search users. In: Proceedings of the Eighth ACM International Conference on Web Search and Data Mining, WSDM '15, pp. 139–148. ACM (2015). doi: 10.1145/2684822.2685304

9. Blanco, R., Cambazoglu, B.B., Mika, P., Torzec, N.: Entity recommendations in web search. In: Proceedings of the 12th International Semantic Web Conference, ISWC '13, pp. 33–48. Springer (2013). doi: 10.1007/978-3-642-41338-4_3

10. Boldi, P., Bonchi, F., Castillo, C., Donato, D., Gionis, A., Vigna, S.: The query-flow graph: Model and applications. In: Proceedings of the 17th ACM Conference on Information and Knowledge Management, CIKM '08, pp. 609–618. ACM (2008). doi: 10.1145/1458082.1458163

11. Bonchi, F., Perego, R., Silvestri, F., Vahabi, H., Venturini, R.: Efficient query recommendations in the long tail via center-piece subgraphs. In: Proceedings of the 35th International ACM SIGIR Conference on Research and Development in Information Retrieval, SIGIR '12, pp. 345–354. ACM (2012). doi: 10.1145/2348283.2348332

12. Bordino, I., De Francisci Morales, G., Weber, I., Bonchi, F.: From Machu_Picchu to "rafting the urubamba river": Anticipating information needs via the entity-query graph. In: Proceedings of the Sixth ACM International Conference on Web Search and Data Mining, WSDM '13, pp. 275–284. ACM (2013). doi: 10.1145/2433396.2433433

13. Bota, H.: Nonlinear composite search results. In: Proceedings of the 2016 ACM on Conference on Human Information Interaction and Retrieval, CHIIR '16, pp. 345–347. ACM (2016). doi: 10.1145/2854946.2854956

14. Bota, H., Zhou, K., Jose, J.M.: Playing your cards right: The effect of entity cards on search behaviour and workload. In: Proceedings of the 2016 ACM on Conference on Human Information Interaction and Retrieval, CHIIR '16, pp. 131–140. ACM (2016). doi: 10.1145/2854946.2854967

15. Cai, F., de Rijke, M.: A Survey of Query Auto Completion in Information Retrieval, vol. 10. Now Publishers Inc. (2016)

16. Cheng, G., Tran, T., Qu, Y.: RELIN: Relatedness and informativeness-based centrality for entity summarization. In: Proceedings of the 10th International Conference on The Semantic Web - Volume Part I, ISWC'11, pp. 114–129. Springer (2011). doi: 10.1007/978-3-642-25073-6_8

17. Dalvi, B., Minkov, E., Talukdar, P.P., Cohen, W.W.: Automatic gloss finding for a knowledge base using ontological constraints. In: Proceedings of the Eighth ACM International Conference on Web Search and Data Mining, WSDM '15, pp. 369–378. ACM (2015). doi: 10.1145/2684822.2685288

18. Dehghani, M., Rothe, S., Alfonseca, E., Fleury, P.: Learning to attend, copy, and generate for session-based query suggestion. In: Proceedings of the 2017 ACM on Conference on Information and Knowledge Management, CIKM '17, pp. 1747–1756. ACM (2017). doi: 10.1145/3132847.3133010

19. Desrosiers, C., Karypis, G.: A Comprehensive Survey of Neighborhood-based Recommendation Methods, pp. 107–144. Springer (2011)

20. Fernández-Tobías, I., Blanco, R.: Memory-based recommendations of entities for web search users. In: Proceedings of the 25th ACM International on Conference on Information and Knowledge Management, CIKM '16, pp. 35–44. ACM (2016). doi: 10.1145/2983323.2983823

21. Gkatzia, D., Lemon, O., Rieser, V.: Natural language generation enhances human decision-making with uncertain information. In: Proceedings of the 54th Annual Meeting of the Association for Computational Linguistics, ACL' 16. The Association for Computer Linguistics (2016)

22. Gunaratna, K., Thirunarayan, K., Sheth, A.: FACES: Diversity-aware entity summarization using incremental hierarchical conceptual clustering. In: Proceedings of the Twenty-Ninth AAAI Conference on Artificial Intelligence, AAAI'15, pp. 116–122. AAAI Press (2015)
23. Hasibi, F., Balog, K., Bratsberg, S.E.: Dynamic factual summaries for entity cards. In: Proceedings of the 40th International ACM SIGIR Conference on Research and Development in Information Retrieval, SIGIR '17. ACM (2017). doi: 10.1145/3077136.3080810
24. Hoffart, J., Milchevski, D., Weikum, G.: STICS: Searching with strings, things, and cats. In: Proceedings of the 37th International ACM SIGIR Conference on Research and Development in Information Retrieval, SIGIR '14, pp. 1247–1248. ACM (2014). doi: 10.1145/2600428.2611177
25. Huang, Z., Cautis, B., Cheng, R., Zheng, Y.: KB-enabled query recommendation for long-tail queries. In: Proceedings of the 25th ACM International on Conference on Information and Knowledge Management, CIKM '16, pp. 2107–2112. ACM (2016). doi: 10.1145/2983323.2983650
26. Jeh, G., Widom, J.: Scaling personalized web search. In: Proceedings of the 12th International Conference on World Wide Web, WWW '03, pp. 271–279. ACM (2003). doi: 10.1145/775152.775191
27. Kang, C., Vadrevu, S., Zhang, R., van Zwol, R., Pueyo, L.G., Torzec, N., He, J., Chang, Y.: Ranking related entities for web search queries. In: Proceedings of the 20th International Conference Companion on World Wide Web, WWW '11, pp. 67–68. ACM (2011). doi: 10.1145/1963192.1963227
28. Lagun, D., Hsieh, C.H., Webster, D., Navalpakkam, V.: Towards better measurement of attention and satisfaction in mobile search. In: Proceedings of the 37th International ACM SIGIR Conference on Research and Development in Information Retrieval, SIGIR '14, pp. 113–122. ACM (2014). doi: 10.1145/2600428.2609631
29. Lao, N., Cohen, W.W.: Relational retrieval using a combination of path-constrained random walks. Mach. Learn. $81(1)$, 53–67 (2010). doi: 10.1007/s10994-010-5205-8
30. Lee, H., Peirsman, Y., Chang, A., Chambers, N., Surdeanu, M., Jurafsky, D.: Stanford's multipass sieve coreference resolution system at the CoNLL-2011 Shared task. In: Proceedings of the Fifteenth Conference on Computational Natural Language Learning: Shared Task, CONLL Shared Task '11, pp. 28–34. Association for Computational Linguistics (2011)
31. Lee, J., Fuxman, A., Zhao, B., Lv, Y.: Leveraging knowledge bases for contextual entity exploration. In: Proceedings of the 21th ACM SIGKDD International Conference on Knowledge Discovery and Data Mining, KDD '15, pp. 1949–1958. ACM (2015). doi: 10.1145/2783258.2788564
32. Meij, E., Mika, P., Zaragoza, H.: An evaluation of entity and frequency based query completion methods. In: Proceedings of the 32nd International ACM SIGIR Conference on Research and Development in Information Retrieval, SIGIR '09, pp. 678–679. ACM (2009). doi: 10.1145/1571941.1572074
33. Mikolov, T., Sutskever, I., Chen, K., Corrado, G., Dean, J.: Distributed representations of words and phrases and their compositionality. In: Proceedings of the 26th International Conference on Neural Information Processing Systems, NIPS'13, pp. 3111–3119. Curran Associates Inc. (2013)
34. Miliaraki, I., Blanco, R., Lalmas, M.: From "Selena Gomez" to "Marlon Brando": Understanding explorative entity search. In: Proceedings of the 24th International Conference on World Wide Web, WWW '15, pp. 765–775. International World Wide Web Conferences Steering Committee (2015). doi: 10.1145/2736277.2741284
35. Milne, D., Witten, I.H.: An effective, low-cost measure of semantic relatedness obtained from Wikipedia links. In: Proceeding of AAAI Workshop on Wikipedia and Artificial Intelligence: An Evolving Synergy, pp. 25–30. AAAI Press (2008)
36. Navalpakkam, V., Jentzsch, L., Sayres, R., Ravi, S., Ahmed, A., Smola, A.: Measurement and modeling of eye-mouse behavior in the presence of nonlinear page layouts. In: Proceedings of the 22nd International Conference on World Wide Web, WWW '13, pp. 953–964. ACM (2013). doi: 10.1145/2488388.2488471

37. Ozertem, U., Chapelle, O., Donmez, P., Velipasaoglu, E.: Learning to suggest: A machine learning framework for ranking query suggestions. In: Proceedings of the 35th International ACM SIGIR Conference on Research and Development in Information Retrieval, SIGIR '12, pp. 25–34. ACM (2012). doi: 10.1145/2348283.2348290

38. Passant, A.: Dbrec: Music recommendations using DBpedia. In: Proceedings of the 9th International Semantic Web Conference on The Semantic Web - Volume Part II, ISWC'10, pp. 209–224. Springer (2010)

39. Reinanda, R., Meij, E., de Rijke, M.: Mining, ranking and recommending entity aspects. In: Proceedings of the 38th International ACM SIGIR Conference on Research and Development in Information Retrieval, SIGIR '15, pp. 263–272. ACM (2015). doi: 10.1145/2766462.2767724

40. Rybak, J., Balog, K., Nørvåg, K.: ExperTime: Tracking expertise over time. In: Proceedings of the 37th International ACM SIGIR Conference on Research and Development in Information Retrieval, SIGIR '14 (2014). doi: 10.1145/2600428.2611190

41. Saldanha, G., Biran, O., McKeown, K., Gliozzo, A.: An entity-focused approach to generating company descriptions. In: Proceedings of the 54th Annual Meeting of the Association for Computational Linguistics, ACL '16. The Association for Computer Linguistics (2016)

42. Schmidt, A., Hoffart, J., Milchevski, D., Weikum, G.: Context-sensitive auto-completion for searching with entities and categories. In: Proceedings of the 39th International ACM SIGIR Conference on Research and Development in Information Retrieval, SIGIR '16, pp. 1097–1100. ACM (2016). doi: 10.1145/2911451.2911461

43. Sordoni, A., Bengio, Y., Vahabi, H., Lioma, C., Grue Simonsen, J., Nie, J.Y.: A hierarchical recurrent encoder-decoder for generative context-aware query suggestion. In: Proceedings of the 24th ACM International on Conference on Information and Knowledge Management, CIKM '15, pp. 553–562. ACM (2015). doi: 10.1145/2806416.2806493

44. Strube, M., Ponzetto, S.P.: WikiRelate! - Computing semantic relatedness using Wikipedia. In: Proceedings of the 21st National Conference on Artificial Intelligence - Volume 2, AAAI'06, pp. 1419–1424. AAAI Press (2006)

45. Sun, Y., Han, J., Yan, X., Yu, P.S., Wu, T.: Pathsim: Meta path-based top-k similarity search in heterogeneous information networks. Proceedings of the VLDB Endowment 4(11), 992–1003 (2011)

46. Szpektor, I., Gionis, A., Maarek, Y.: Improving recommendation for long-tail queries via templates. In: Proceedings of the 20th International Conference on World Wide Web, WWW' 11, pp. 47–56. ACM (2011). doi: 10.1145/1963405.1963416

47. Thalhammer, A., Lasierra, N., Rettinger, A.: LinkSUM: Using link analysis to summarize entity data. In: Proc. of 16th International Web Engineering Conference, ICWE '16, pp. 244–261. Springer (2016). doi: 10.1007/978-3-319-38791-8_14

48. Thalhammer, A., Rettinger, A.: Browsing DBpedia entities with summaries. In: The Semantic Web: ESWC 2014 Satellite Events, pp. 511–515 (2014)

49. Tombros, A., Sanderson, M.: Advantages of query biased summaries in information retrieval. In: Proceedings of the 21st Annual International ACM SIGIR Conference on Research and Development in Information Retrieval, SIGIR '98, pp. 2–10. ACM (1998). doi: 10.1145/290941.290947

50. Tonon, A., Catasta, M., Prokofyev, R., Demartini, G., Aberer, K., Cudré-Mauroux, P.: Contextualized ranking of entity types based on knowledge graphs. Web Semant. 37–38, 170–183 (2016). doi: 10.1016/j.websem.2015.12.005

51. Tuan, T.A., Elbassuoni, S., Preda, N., Weikum, G.: CATE: Context-aware timeline for entity illustration. In: Proceedings of the 20th International Conference Companion on World Wide Web, WWW '11, pp. 269–272. ACM (2011). doi: 10.1145/1963192.1963306

52. Vadrevu, S., Tu, Y., Salvetti, F.: Ranking relevant attributes of entity in structured knowledge base (2016)

53. Voskarides, N., Meij, E., de Rijke, M.: Generating descriptions of entity relationships. In: Proceedings of the 39th European Conference on Information Retrieval, ECIR '17. Springer (2017). doi: 10.1007/978-3-319-56608-5_25

54. Voskarides, N., Meij, E., Tsagkias, M., de Rijke, M., Weerkamp, W.: Learning to explain entity relationships in knowledge graphs. In: Proceedings of the 53rd Annual Meeting of the Association for Computational Linguistics and the 7th International Joint Conference on Natural Language Processing (Volume 1: Long Papers), pp. 564–574. Association for Computational Linguistics (2015)
55. White, R.W.: Interactions with Search Systems. Cambridge University Press (2016)
56. Williams, K., Kiseleva, J., Crook, A.C., Zitouni, I., Awadallah, A.H., Khabsa, M.: Is this your final answer?: Evaluating the effect of answers on good abandonment in mobile search. In: Proceedings of the 39th International ACM SIGIR Conference on Research and Development in Information Retrieval, SIGIR '16, pp. 889–892. ACM (2016). doi: 10.1145/2911451.2914736
57. Yu, X., Ma, H., Hsu, B.J.P., Han, J.: On building entity recommender systems using user click log and Freebase knowledge. In: Proceedings of the 7th ACM International Conference on Web Search and Data Mining, WSDM '14, pp. 263–272. ACM (2014). doi: 10.1145/2556195.2556233
58. van Zwol, R., Pueyo, L.G., Muralidharan, M., Sigurbjornsson, B.: Ranking entity facets based on user click feedback. In: Proceedings of the 4th IEEE International Conference on Semantic Computing, pp. 192–199 (2010)

Chapter 10
Conclusions and Future Directions

Today, the importance of entities has been broadly recognized and entities have become first-class citizens in many information access systems, including web, mobile, and enterprise search; question answering; and personal digital assistants. Entities have also become a meeting point for several research communities, including that of information retrieval, natural language processing, databases, and the Semantic Web. Many of the methods and tools we have described in this book, such as ranking entities, recognizing and linking entity mentions in documents and queries, or displaying entity cards, are now integral components of modern search systems.

Is this the end of the road? Certainly not. It would be going too far to label those core tasks, like entity ranking and entity linking, as "solved." Obviously, there is still (plenty of) room for improvement. Also, it is not yet clear which techniques will be the "BM25's" of the entity world, as stable and reliable solutions. Only time will tell. Nevertheless, we have reached a point where these methods are "good enough" to be used as basic building blocks in more complex systems. Perhaps it is time to look beyond these core tasks. As we are approaching the end of this book, we shall attempt to look into the future and gauge what lies ahead. Many of the things we will discuss here have already begun to happen, while some other elements, or their exact form, are more of a speculation.

In Sect. 10.1, we shall summarize our progress so far. Where are we now and how did we get here? Then, in Sects. 10.2 and 10.3 we will attempt to look ahead and discuss some anticipated future developments. We will conclude with some final remarks in Sect. 10.4.

© The Author(s) 2018
K. Balog, *Entity-Oriented Search*, The Information Retrieval Series 39,
https://doi.org/10.1007/978-3-319-93935-3_10

10.1 Summary of Progress

Let us take a step back and distill the progress achieved over the past years, organized around three main thematic areas. We shall also briefly mention open issues; we will elaborate on some of these in more detail in Sect. 10.2.

10.1.1 Data

We start by discussing data, as developments in the data landscape have been instrumental to the progress made thus far. Specifically, the availability of large-scale knowledge bases has played a key role in transforming the search experience. Many information access applications utilize knowledge bases as a rich, structured repository of entities, and, to a lesser extent, for ontological background knowledge. Knowledge about particular entities may be used to complement the traditional (document-oriented) search results, allow for direct answers and various knowledge panels, and facilitate content exploration and discovery. Knowledge bases also enable machine understanding of natural language text, by using entities as a pivot to connect unstructured and structured data sources (Chap. 5). In turn, massive volumes of unstructured documents may be utilized to populate KBs with additional entities and their properties (Chap. 6).

Open Issues Knowledge bases are inherently incomplete and keeping them up-to-date requires a continuous effort. Automatic knowledge acquisition is an active area of research. Open challenges include the discovery of long tail and emerging entities, and the quality of data (correctness and trustworthiness of facts); see Sect. 10.3.3 for further data-related issues.

10.1.2 Retrieval Methods

A significant portion of the book has been devoted to entity retrieval methods. Early approaches build on document retrieval techniques and focus on how to adopt those for various types of data, from unstructured to structured (Chap. 3). More recent approaches utilize the rich structure associated with entities in knowledge bases (Chap. 4). Many—in fact, most—of the other tasks we have addressed in this book were also cast as ranking problems, for instance, disambiguating entities that may refer to a particular mention in text (Sect. 5.6), filtering documents that contain vital information about an entity (Sect. 6.2), identifying target types of a query (Sect. 7.2), finding interpretations of a query (Sect. 7.3.4), or determining which facts to display on an entity card (Sect. 9.2.2). For all these tasks, the current state of the art involves a discriminative learning approach, i.e., learning-to-rank, employing a rich set of carefully designed features.

Open Issues It appears to be a "safe" recipe to tackle any ranking problem by hand-crafting a large set of features, then throwing machine learning at it. Indeed, the importance of feature engineering is not to be underestimated. Nevertheless, one might argue that this general approach can even get rather mechanical, and scientifically less interesting, after a while. Neural methods, especially deep learning, hold the promise of learning directly from raw data, without such labor-intensive feature engineering. Extending traditional IR models to incorporate word embeddings has already proven effective for various entity-related search tasks, see, e.g., [5, 7, 13, 16, 17, 19]. Developing end-to-end architectures, which more fully embrace neural modeling, is an exciting and active research direction [12, 21]. Yet, it remains to be seen if deep learning can categorically outclass other approaches, and whether it will surpass all other forms of machine learning and take over the entire field of IR (as it did with computer vision, speech recognition, and machine translation). Even if it does, one might say that all this means is that feature engineering will get replaced by network engineering. Another issue here will surely be the availability of training data. In that regard, industry has a distinct advantage over academia, as target relevance labels may be derived on a much larger scale from usage data.

While the core entity-oriented retrieval tasks described above both merit and have the potential for further improvement, another open issue is how to combine these into more complex useful applications. After all, our eventual goal should be aiding users in achieving their goals, i.e., completing their tasks, which goes far beyond the ranking of items; see Sect. 10.3.2.

10.1.3 Understanding and Interacting with Users

Users increasingly expect search engines to understand them and respond to their information needs more directly than just serving documents matching the query terms. Today, the search box functions more like a "request box," and queries are answered by rich search result pages, including direct answers and interactive widgets (maps, currency conversion, etc.). We have looked at how to utilize entities and types to understand information needs (Chap. 7) and to provide an enhanced search experience (Chap. 9).

Open Issues Search has become a consumer experience. Major search engines are continuously introducing new types of "functional" results (interactive widgets), enabling users to do more and more, without leaving the SERP. Result presentation and interacting with entities still offer plenty of opportunities for research and innovation. One recent line of work focuses on actionable knowledge bases, i.e., identifying potential actions that can be performed on a given entity [4].

Another open issue in this area is that search (or, more broadly, information access) is moving from desktop to mobile and from text to voice. Personal digital assistants are increasingly being used to respond to natural language questions. We can say that search is becoming a conversation between humans and machines; see Sect. 10.3.1.

10.2   A Peek into the Future

In this section, we present a fictional conversation that takes place sometime in the not-too-distant future, between a user, say a male university professor, and an intelligent personal assistant, simply referred to as "AI." This conversation could in fact happen on any device, but for the sake of illustrating certain points, we shall assume that the device is a mobile phone and that the user interacts with the device via spoken natural language. The conversation, which will later be referred to as *scenario*, is accompanied by some narrative.

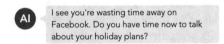

AI   I see you're wasting time away on Facebook. Do you have time now to talk about your holiday plans?

The first thing to notice is that it is the AI that initiates the conversation. Based on the user's current activity and past behavioral patterns, it decides that now would be a good time to address a future information need.

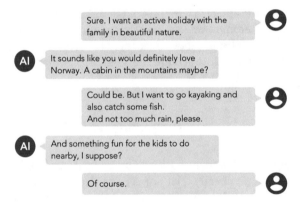

Sure. I want an active holiday with the family in beautiful nature.

AI   It sounds like you would definitely love Norway. A cabin in the mountains maybe?

Could be. But I want to go kayaking and also catch some fish.
And not too much rain, please.

AI   And something fun for the kids to do nearby, I suppose?

Of course.

The AI refines the requirements iteratively by asking a series of questions. Observe that it has knowledge of the user's background (family situation). After having the initial requirements clarified, it comes up with a specific suggestion:

AI   How does **Oltedal** sound?
People visiting there have been quite successful with catching lake trout, based on what I found on Instagram.

There is also a **theme park** and **horseback riding**, both within 50 kms.

Links are boldfaced and underlined in the response text; the user could follow these for more information. The photos are also clickable.

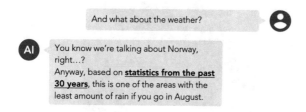

Two things are worth pointing out in the AI's response. One is that it has a (certain) sense of humor. Humor is an essential human communication behavior. The other is that it is able to give a compact answer to the question, and backs it up with a link to the source (evidence) that the answer was based on.

The AI presents a list of (personalized) accommodation options. After examining the results, the user selects one of the items and asks for further information.

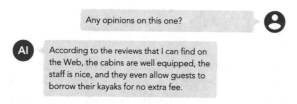

With this reply, the AI demonstrates some impressive summarization skills. It focuses on aspects that are likely of interest to the user.

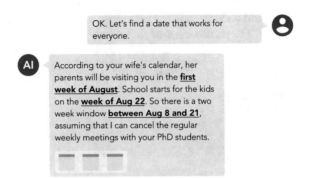

At this point, we are moving away from what was a kind of exploratory search scenario to a different type of information access problem, where the AI helps to automatically manage scheduling.

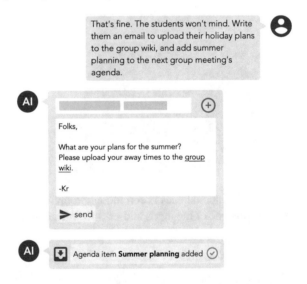

First, the AI prepares a draft of the requested email message in the user's style. The user can refine the text before sending. Then, the AI adds the agenda item to the team wiki and displays a success notification received from the wiki software.

The AI attempts to check availability via an online booking system. Realizing that it is currently not accessible, it decides to resort to more conventional means of communication and calls the place over the phone. Next, it composes an email message, combining the user's language model with a flavor of "kind." After reviewing—and perhaps correcting the AI's over-the-top romantic vibe here-and-there—the user decides to send off the message.

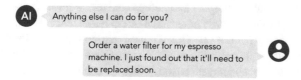

With this task completed, the AI asks if it could be of any more assistance. Then, it receives a request of a different nature.

10.3   Future Research Directions

Below, we discuss a number of directions and areas for future research. Along the way, we will occasionally make references to certain elements of the scenario presented in the previous section.

10.3.1   Understanding and Interacting with Users

Search Is a Conversation For many years, keyword queries have been the *lingua franca* of information access. This, however, is changing. With the emergence of the mobile-over-desktop culture of information consumption, and the advancement in voice recognition technologies, voice search is gaining ground. In 2015, Google reported that the volume of mobile search has surpassed that of desktop search in several countries.[1] As of 2016, around 20% of queries on Google mobile devices are voice input in the USA [14]. Voice queries are not only longer on average than text queries but also use richer language [9]. But there is more. Voice search facilitates the possibility of a speech dialogue with the user. Such natural language interfaces are already a reality, as manifested in personal digital assistants, such as the Google Assistant, Apple's Siri, Microsoft's Cortana, or Amazon's Alexa.

Conversational search offers many possibilities, such as the ability to ask the user for clarification, if needed. It also presents many challenges, as the system no longer returns massive search engine result pages, somewhere on which the user hopefully finds what she was looking for. The response needs to be more "intelligent," i.e., comprehensive and spot-on. In this regard, voice-based result presentation that enables a completely hand-free interaction with the user still has a long way to go [9]. There is also a need for novel evaluation measures that can capture user satisfaction in a conversational setting. A good conversation entails more than just the fulfillment of an information need; among others, it should flow and be engaging, be just about the right length, and, occasionally, even humorous.

Anticipating Information Needs The traditional way of information access is *reactive*: The system responds to a user-issued request. A *proactive* system, on the other hand, "would anticipate and address the user's information need, without requiring the user to issue (type or speak) a query" [3]. Hence, this paradigm is also known as *zero-query search* [1]. Our scenario started out with the AI proactively bringing up a future information need. We observed proactive recommendations in later parts of the conversation too, when considering additional criteria in exploratory search (activities for kids) and when figuring out how to schedule the vacation (cancelling meetings). Some of today's personal digital assistants are already capable of pre-fetching information cards based on users' behavioral patterns or upcoming events (e.g., Google Now and Microsoft Cortana). Recent research has focused on a number of specific problems in this space, including modeling user interests [20], predicting when users will perform a repetitive task again in the future [15], identifying what information needs people have in a given context [3], and determining the right context for pushing proactive recommendations [6]. With intelligent devices capable of sensing the user's environment (location, and even pulse rate or blood pressure using wearable devices), there are increasingly more contextual signals that may be utilized. Notably, current work

---

[1] https://adwords.googleblog.com/2015/05/building-for-next-moment.html.

is limited to near-term information needs. The area of anticipating more long-term information needs (such as reminding a user months in advance about planning a vacation or finding a school for a child that is going to go to school next year) has not been explored yet.

Verification and Explainability As we move away from ranked lists of items to direct answers and summaries, it becomes crucial to allow for the verification of the system's responses. What is the right form of explanation? In many cases, providing access to the raw data is sufficient. We have seen several examples of this in our scenario, when the AI provided links to pages about weather statistics, reviews, and calendars. In other cases, it may not be possible to refer to a single source; then, the user should be granted access to some intermediate data representation. It is an open issue how to make those intermediate representations suitable for human consumption. These questions also relate to the broader problem area of providing explanations of algorithmic decisions that significantly affect an individual (particularly legally or financially), which is to be a human right according to the European Union General Data Protection Regulation ("right to explanation"), to take effect in 2018 [8].

Personalization In our scenario, we could observe a high degree of personalization, including the interaction with the user, the generation of responses, and the language usage when executing tasks on the user's behalf. Personal digital assistants are expected to deliver such a personalized user experience. To be able to do that, they will need to get to "know" the user, her habits, preferences, and the things she cares about. With human assistants, there is often a more-or-less clear separation between work and private matters. This is not the case with digital assistants; most users would likely use the same personal AI for any and all kinds of business they encounter. This brings up many issues around trust, privacy, and data protection. Digital assistants must be aware of the user's momentarily situation and context too.

10.3.2 Complex Information Needs and Task Completion

Major web search engines have made a great progress with answering one-shot queries with rich search result pages, thereby putting the bar rather high regarding the search experience. Users now expect intelligent personal assistants to respond with direct answers as opposed to a ranked list of results. Thus, it may be fitting to refer to these systems no longer as search engines but as *answering engines*. Intelligent agents are further capable of assisting users in "getting things done," such as making calendar appointments or setting reminders. However, neither web search engines nor digital assistants have the capability yet to handle truly complex tasks, such as the holiday planning in our scenario. These complex information needs require a better understanding and modeling of the user's high-level goals. It requires no less than a paradigm shift, from answering engines to *task-completion engines* [2]. Entities will continue to play a key role here, for modeling users, tasks, and context.

10.3.3   Data and Knowledge

On-the-Fly Information Extraction Despite all automatic knowledge acquisition efforts, there will always be long-tail entities that are not contained in any knowledge base. Moreover, even if the entities in question are present in a knowledge base, it is not possible to capture all information associated with them, due to the finite vocabulary of knowledge base predicates. Consequently, we will continue to come across information needs to which the answer is "out there" in some digital form, but not yet contained in a knowledge base. For example, in our scenario, this could be the case with some accommodations at obscure locations. These situations may be handled by on-the-fly information extraction techniques.

Personal Knowledge Base In our scenario, the user has made numerous references to entities he was in some way related to: "my kids," "my wife," "my group," "my espresso machine," etc. These entities constitute the users' *personal knowledge base*, i.e., the universe of things he cares about. Throughout interactions with the user, the entities of this universe may be mapped onto the same data representation model that knowledge bases use. It is also possible to make "same-as" links to other knowledge repositories that contain the same entity (e.g., the espresso machine). Some entities, however, will reside only in the user's personal KB. What is powerful about this idea is that the same methods and techniques we have discussed for general-purpose KBs are readily applicable to a personal KB. The problem thus boils down to the automatic population and maintenance of the personal KB.

Commonsense Knowledge Knowledge bases have largely focused on accumulating factual knowledge about specific entities. An intelligent system, such as a personal digital assistant, however, needs a much broader understanding of the world. Simple statements like "things fall down, not up" and "open the door before entering" are obvious to humans but not to machines. To endow computers with common sense is one of the long-standing goals of AI research. Some projects, such as Cyc [10] or ConceptNet [11], have begun to amass large collections of such commonsense knowledge. However, "there is still a long way to go for computers to learn what every child knows" [18].

10.4   Concluding Remarks

Reaching the end of this book, it may be appropriate to have a moment of reflection. Information technology has changed and will continue to change our lives. We are increasingly more surrounded by intelligent autonomous systems (which we like to call AI): personal assistants, self-driving cars, smart homes, etc. There are some thought-provoking open questions here related to responsibility: If a fatal accident happens involving an autonomous vehicle or a disastrous decision is made based on false information served by a search engine (which perhaps retrieved it from some underlying knowledge base), who is responsible for that? Surely, the company behind the given product should take some responsibility. But then, would it ultimately come down to the individual software engineer who wrote the corresponding piece of code (or to the knowledge engineer who was responsible for that entry ending up in the KB)? Or would the blame be put on the end user, who did not study or consider carefully enough the terms of usage? These are important and challenging regulatory issues on which conversations have already started.

We are now in the third AI spring, which draws mixed reactions from people: great excitement, overblown expectations because of the hype, and fear. Technological singularity, i.e., the emergence of an (evil) artificial superintelligence that would cause the human race to go extinct—in the author's opinion—is merely a dystopia that Hollywood loves to portray in speculative fiction. Technology itself is not good or evil—it depends on how we use it. It appears though that as time goes on, increasingly more technology will be "forced" on us. Yet, we have the free will and responsibility decide what technology we want to use or adopt. Importantly, technology should enable and not distract us on that awesome journey, with its ups and downs, that is called Life. Along the way, we should take the time to contemplate on the deeper questions of existence, being, and id*entity*—searching for the answers to those questions is what it means to be a human. No computer system, however intelligent, will ever be able to do that for us.

References

1. Allan, J., Croft, B., Moffat, A., Sanderson, M.: Frontiers, challenges, and opportunities for information retrieval: Report from SWIRL 2012 the Second Strategic Workshop on information retrieval in lorne. SIGIR Forum **46**(1), 2–32 (2012)
2. Balog, K.: Task-completion engines: A vision with a plan. In: Proceedings of the First International Workshop on Supporting Complex Search Tasks, SCST '15 (2015)
3. Benetka, J.R., Balog, K., Nørvåg, K.: Anticipating information needs based on check-in activity. In: Proceedings of the 10th ACM International Conference on Web Search and Data Mining, WSDM '17, pp. 41–50. ACM (2017). doi: 10.1145/3018661.3018679
4. Blanco, R., Joho, H., Jatowt, A., Yu, H., Yamamoto, S.: NTCIR Actionable Knowledge Graph task (2017)
5. Blanco, R., Ottaviano, G., Meij, E.: Fast and space-efficient entity linking for queries. In: Proceedings of the Eighth ACM International Conference on Web Search and Data Mining - WSDM '15, pp. 179–188. ACM (2015). doi: 10.1145/2684822.2685317
6. Braunhofer, M., Ricci, F., Lamche, B., Wörndl, W.: A context-aware model for proactive recommender systems in the tourism domain. In: Proceedings of the 17th International Conference on Human-Computer Interaction with Mobile Devices and Services Adjunct, MobileHCI '15, pp. 1070–1075. ACM (2015). doi: 10.1145/2786567.2794332
7. Garigliotti, D., Hasibi, F., Balog, K.: Target type identification for entity-bearing queries. In: Proceedings of the 40th International ACM SIGIR Conference on Research and Development in Information Retrieval, SIGIR '17. ACM (2017). doi: 10.1145/3077136.3080659
8. Goodman, B., Flaxman, S.: European Union regulations on algorithmic decision-making and a "right to explanation". ArXiv e-prints (2016)
9. Guy, I.: Searching by talking: Analysis of voice queries on mobile web search. In: Proceedings of the 39th International ACM SIGIR Conference on Research and Development in Information Retrieval, SIGIR '16, pp. 35–44. ACM (2016). doi: 10.1145/2911451.2911525

10. Lenat, D.B.: CYC: A large-scale investment in knowledge infrastructure. Commun. ACM **38**(11), 33–38 (1995). doi: 10.1145/219717.219745

11. Liu, H., Singh, P.: ConceptNet - A practical commonsense reasoning tool-kit. BT Technology Journal **22**(4), 211–226 (2004). doi: 10.1023/B:BTTJ.0000047600.45421.6d

12. Mitra, B., Craswell, N.: Neural models for information retrieval. ArXiv e-prints (2017)

13. Pappu, A., Blanco, R., Mehdad, Y., Stent, A., Thadani, K.: Lightweight multilingual entity extraction and linking. In: Proceedings of the Tenth ACM International Conference on Web Search and Data Mining, WSDM '17, pp. 365–374. ACM (2017). doi: 10.1145/3018661.3018724

14. Pichai, S.: Google I/O 2016 keynote (2016)

15. Song, Y., Guo, Q.: Query-less: Predicting task repetition for nextgen proactive search and recommendation engines. In: Proceedings of the 25th International Conference on World Wide Web, WWW '16, pp. 543–553 (2016). doi: 10.1145/2872427.2883020

16. Van Gysel, C., de Rijke, M., Kanoulas, E.: Learning latent vector spaces for product search. In: Proceedings of the 25th ACM International on Conference on Information and Knowledge Management, CIKM '16, pp. 165–174. ACM (2016a). doi: 10.1145/2983323.2983702

17. Van Gysel, C., de Rijke, M., Worring, M.: Unsupervised, efficient and semantic expertise retrieval. In: Proceedings of the 25th International Conference on World Wide Web, WWW '16, pp. 1069–1079 (2016b). doi: 10.1145/2872427.2882974

18. Weikum, G., Hoffart, J., Suchanek, F.: Ten years of knowledge harvesting: Lessons and challenges. IEEE Data Eng. Bull. **39**(3), 41–50 (2016)

19. Xiong, C., Power, R., Callan, J.: Explicit semantic ranking for academic search via knowledge graph embedding. In: Proceedings of the 26th International Conference on World Wide Web, WWW '17, pp. 1271–1279. International World Wide Web Conferences Steering Committee (2017). doi: 10.1145/3038912.3052558

20. Yang, L., Guo, Q., Song, Y., Meng, S., Shokouhi, M., McDonald, K., Croft, W.B.: Modeling user interests for zero-query ranking. In: Proceedings of the 38th European Conference on IR Research, ECIR '16, pp. 171–184 (2016). doi: 10.1007/978-3-319-30671-1_13

21. Zhang, Y., Mustafizur Rahman, M., Braylan, A., Dang, B., Chang, H.L., Kim, H., McNamara, Q., Angert, A., Banner, E., Khetan, V., McDonnell, T., Thanh Nguyen, A., Xu, D., Wallace, B., Lease, M.: Neural information retrieval: A literature review. ArXiv e-prints (2016)

Index

© The Author(s) 2018
K. Balog, *Entity-Oriented Search*, The Information Retrieval Series 39,
https://doi.org/10.1007/978-3-319-93935-3

Printed in the United States
By Bookmasters